Thanasis Daradoumis, Santi Caballé, Angel A. Juan, and Fatos Xhafa (Eds.)

Technology-Enhanced Systems and Tools for Collaborative Learning Scaffolding

Studies in Computational Intelligence, Volume 350

Editor-in-Chief
Prof. Janusz Kacprzyk
Systems Research Institute
Polish Academy of Sciences
ul. Newelska 6
01-447 Warsaw
Poland
E-mail: kacprzyk@ibspan.waw.pl

Further volumes of this series can be found on our homepage: springer.com

Vol. 328. Raoul Huys and Viktor K. Jirsa (Eds.)
Nonlinear Dynamics in Human Behavior, 2010
ISBN 978-3-642-16261-9

Vol. 329. Santi Caballé, Fatos Xhafa, and Ajith Abraham (Eds.)
Intelligent Networking, Collaborative Systems and Applications, 2010
ISBN 978-3-642-16792-8

Vol. 330. Steffen Rendle
Context-Aware Ranking with Factorization Models, 2010
ISBN 978-3-642-16897-0

Vol. 331. Athena Vakali and Lakhmi C. Jain (Eds.)
New Directions in Web Data Management 1, 2011
ISBN 978-3-642-17550-3

Vol. 332. Jianguo Zhang, Ling Shao, Lei Zhang, and Graeme A. Jones (Eds.)
Intelligent Video Event Analysis and Understanding, 2011
ISBN 978-3-642-17553-4

Vol. 333. Fedja Hadzic, Henry Tan, and Tharam S. Dillon
Mining of Data with Complex Structures, 2011
ISBN 978-3-642-17556-5

Vol. 334. lvaro Herrero and Emilio Corchado (Eds.)
Mobile Hybrid Intrusion Detection, 2011
ISBN 978-3-642-18298-3

Vol. 335. Radomir S. Stankovic and Radomir S. Stankovic
From Boolean Logic to Switching Circuits and Automata, 2011
ISBN 978-3-642-11681-0

Vol. 336. Paolo Remagnino, Dorothy N. Monekosso, and Lakhmi C. Jain (Eds.)
Innovations in Defence Support Systems – 3, 2011
ISBN 978-3-642-18277-8

Vol. 337. Sheryl Brahnam and Lakhmi C. Jain (Eds.)
Advanced Computational Intelligence Paradigms in Healthcare 6, 2011
ISBN 978-3-642-17823-8

Vol. 338. Lakhmi C. Jain, Eugene V. Aidman, and Canicious Abeynayake (Eds.)
Innovations in Defence Support Systems – 2, 2011
ISBN 978-3-642-17763-7

Vol. 339. Halina Kwasnicka, Lakhmi C. Jain (Eds.)
Innovations in Intelligent Image Analysis, 2010
ISBN 978-3-642-17933-4

Vol. 340. Heinrich Hussmann, Gerrit Meixner, and Detlef Zuehlke (Eds.)
Model-Driven Development of Advanced User Interfaces, 2011
ISBN 978-3-642-14561-2

Vol. 341. Stphane Doncieux, Nicolas Bredeche, and Jean-Baptiste Mouret(Eds.)
New Horizons in Evolutionary Robotics, 2011
ISBN 978-3-642-18271-6

Vol. 342. Federico Montesino Pouzols, Diego R. Lopez, and Angel Barriga Barros
Mining and Control of Network Traffic by Computational Intelligence, 2011
ISBN 978-3-642-18083-5

Vol. 343. Kurosh Madani, Antnio Dourado Correia, Agostinho Rosa, and Joaquim Filipe (Eds.)
Computational Intelligence, 2011
ISBN 978-3-642-20205-6

Vol. 344. Atilla Elçi, Mamadou Tadiou Kon, and Mehmet A. Orgun (Eds.)
Semantic Agent Systems, 2011
ISBN 978-3-642-18307-2

Vol. 345. Shi Yu, Lon-Charles Tranchevent, Bart De Moor, and Yves Moreau
Kernel-based Data Fusion for Machine Learning, 2011
ISBN 978-3-642-19405-4

Vol. 346. Weisi Lin, Dacheng Tao, Janusz Kacprzyk, Zhu Li, Ebroul Izquierdo, and Haohong Wang (Eds.)
Multimedia Analysis, Processing and Communications, 2011
ISBN 978-3-642-19550-1

Vol. 347. Sven Helmer, Alexandra Poulovassilis, and Fatos Xhafa
Reasoning in Event-Based Distributed Systems, 2011
ISBN 978-3-642-19723-9

Vol. 348. Beniamino Murgante, Giuseppe Borruso and Alessandra Lapucci (Eds.)
Geocomputation, Sustainability and Environmental Planning, 2011
ISBN 978-3-642-19732-1

Vol. 349. Vitor R. Carvalho
Modeling Intention in Email, 2011
ISBN 978-3-642-19955-4

Vol. 350. Thanasis Daradoumis, Santi Caballé, Angel A. Juan, and Fatos Xhafa (Eds.)
Technology-Enhanced Systems and Tools for Collaborative Learning Scaffolding, 2011
ISBN 978-3-642-19813-7

Thanasis Daradoumis, Santi Caballé, Angel A. Juan,
and Fatos Xhafa (Eds.)

Technology-Enhanced Systems and Tools for Collaborative Learning Scaffolding

Springer

Thanasis Daradoumis,
University of the Aegean,
Greece / Open University of Catalonia,
Spain

Angel A. Juan,
Open University of Catalonia,
Spain

Santi Caballé,
Open University of Catalonia,
Spain

Fatos Xhafa,
Technical University of Catalonia,
Spain

ISBN 978-3-642-26792-5 ISBN 978-3-642-19814-4 (eBook)

DOI 10.1007/978-3-642-19814-4

Studies in Computational Intelligence ISSN 1860-949X

Softcover reprint of the hardcover 1st edition 2011

Typeset & Cover Design: Scientific Publishing Services Pvt. Ltd., Chennai, India.

Printed on acid-free paper

9 8 7 6 5 4 3 2 1

springer.com

Preface

Technology-Enhanced Systems and Tools for Collaborative Learning Scaffolding

Computer-Supported Collaborative Learning (CSCL) is one of the most influencing research paradigms dedicated to improve teaching and learning with the help of modern information and communication technology. On the one hand, collaborating in small groups may constitute a powerful means for promoting and enhancing learning, work and social interaction. Recent studies of e-learning have pointed out that involving learners in collaborative learning activities could positively contribute to extending and deepening their learning experiences, test out new ideas, improve learning outcomes and increase learner satisfaction, at the same time decreasing the isolation that can occur in an e-learning setting. On the other hand, during task realization, peers learning and working via CSCL technology and methods need guidance and support in order to collaborate effectively and achieve their tasks and learning goals successfully.

Many researchers have acknowledged the need of adequate systems, methods and tools to help the members of learning or working groups with their mindful and appropriate learning and work. Such frameworks are essential in all types of education and working settings. Essentially, in web-based education and blended education, the existence of this kind of tools is crucial for the teachers' and students' more effective involvement.

This book reports important research work and experiences that investigate on the improvement of on-line collaborative processes through the development of collaborative e-learning systems and applications which are empowered with intelligent methods and techniques. As a result, collaborative systems and applications are to be more powerful and flexible and also more adaptable to collaboration demands and thus provide better support, feedback and monitoring to a variety of online learning and working activities – both at individual and group levels. Moreover, the book appeals for providing software developers and researchers in the field of CSCL with fresh and innovative ideas that allow them to extend current capabilities and functionalities of e-learning platforms. The goal here is to make an efficient use of these technology-enhanced platforms in a distributed environment where adaptive learning designers, content creators, interaction analysts, service providers, and users – either instructors, learners, or academic coordinators– share similar learning and work experiences.

As a matter of fact, despite the considerable progress in this field, there are still plenty of issues to investigate on how to employ the emergent computational technologies to fully support online collaborative learning and working activity. Four such issues concern systems which employ methods that foster collaborative learning, on the one hand, through distributed e-learning repositories, content creation and customization, or social networks and, on the other hand, through collaborative ontologies building, educational multiplayer video games, and adaptive collaboration support; the use of models for analyzing students' interactions and structuring sequences of activities; and finally, the development of models that support more powerful e-collaboration settings through the use of personalization and adaptation techniques.

As a result, this book presents up-to-date research approaches for developing technology-enhanced systems and tools to support functional online collaborative learning and work settings. The book covers the needs and interests of a wide range of readers, giving them the opportunity to deepen further on the above four issues and also to extend their knowledge to areas other than the ones they are used to work with. Moreover, the merge of all these synergies represents an attractive challenge that will yield systems capable of providing more effective answers on how to improve and enhance on-line collaborative learning and working experiences.

Among the many features highlighted in the book, which is the result of important research on technology-enhanced systems for Collaborative Learning Scaffolding, we could distinguish the following:

The study of frameworks and infrastructures that foster collaborative learning through the application of different methods. Such methods include, on the one hand, the design and use of a distributed e-learning repository, content creation and customization, or social networks. These methods allow communities of learners to use collaborative web-technologies for lifelong learning, whereby tutors are able to create virtual social networks for sharing knowledge and views among them which are related to the teaching activity. Social networks are an important space for sharing learning resources and an opportunity for the collective construction of knowledge. In addition, specific tools are developed that facilitate the effective creation, dissemination and customization of learning materials.

On the other hand, other methods include collaborative ontologies building, educational multiplayer video games, and adaptive collaboration support. The collaborative ontology is a research domain linked to concepts of "extended cognitive context" and knowledge building in co-participation. The use of ontology engineering methodologies and ontology authoring tools demonstrates that a collaborative approach to ontology authoring, development and harmonization fortifies the process of ontology engineering, and is linked to a strong awareness of how the dynamic and participatory review may impact on the good maintenance of a domain. Besides it, the design of collaborative activities in educational video games requires the development of new relationships and premises in group tasks. More specifically, a set of design guidelines has to be defined to favour collaborative processes between group members and to retain the many advantages that this type of learning offers. Finally, personalization features can be very useful to adapt the collaborative learning experience to the student needs, since they can serve in supporting management, tracking and evaluation tasks in collaborative settings.

Interaction analysis approaches. There is a strong need for methodologies and tools that analyze students' interactions and structuring sequences of activities with the aim to increase students' collaborative behaviors, performance and group organization. To this end, the book explores learning environments that make it possible to examine both new ways in which students and instructors collaborate and to provide new evidence that addresses one of the fundamental problems faced by students – procrastination. Another approach provides logfile-based interaction analysis techniques that can be used to support CSCL activities by communicating the state of evolving group knowledge. Yet, other flexible solutions can be provided by applying Collaborative Learning Flow Patterns (CLFPs) to structure sequences of activities in real contexts, which may support teachers in organizing groups of students taking advantage of the intrinsic constraints defined by a CLFP.

Implementing adaptation methods in computer-supported collaborative learning (CSCL) activities. Adaption methods for learning aim to intelligently regulate the settings of the learning activity in order to tailor the experience to learner's needs and preferences. Adaptive Collaboration Support is an emergent research area which attempts to exploit adaptive techniques in order to support user activities that take place in e-collaboration settings. When participants engage in an activity-oriented process, this process may be supported or optimized in a number of ways. This book explores ways of adaptive interventions and adaptive storytelling in activities that take place in the context of educational games, adaptive collaborative design-pattern in the context of an open-source learning design-based environment, recommendation of learning materials to students that collaborate in an e-learning environment, and supporting teachers in order to design effective collaborative learning tasks. All in all, the impact of employing adaptive forms of support in CSCL settings has proved to provide encouraging results and a continuous effort is still active toward this direction.

Introduction

This book consists of 13 chapters organized into four major areas: (i) *Frameworks and infrastructures that foster collaborative learning through a distributed e-learning repository, content creation and customization, or social networks,* (ii) *Methodologies and tools for analyzing students' interactions and structuring sequences of activities with the aim to increase students' collaborative behaviours, performance and group organization,* (iii) *Models that enhance collaborative e-learning experiences through collaborative ontologies building, educational multiplayer video games, and adaptive collaboration support,* and (iv) *Systems and tools that use personalization and adaptation techniques to support the development of more powerful e-collaboration settings.*

Frameworks and infrastructures that foster collaborative learning through a distributed e-learning repository, content creation and customization, or social networks

The chapters in this area are organized as follows:

The first chapter by Bessis et al. presents the combined results and over-arching conclusions of a two-phase project that assesses a pilot distributed e-learning repository for computer-based assessment resources. On the one hand, the online repository not only offers easy access to online quality materials, but also communities of learners may overlap and interact through a distributed repository, using such collaborative web-technologies for lifelong learning, particularly as their size and content expand into large scale digital libraries. On the other hand, the online repository enables tutors to share knowledge about the courses and could well serve to create a virtual network among tutors for sharing knowledge and views related to the teaching activity. The results of this study show the benefits of using the e-learning distributed repository to support learning processes of a community of learners.

Fernàndez and Gil-Rodríguez, in chapter 2, present an approach that takes advantage of the potential capabilities of social networks and uses them as collaborative learning platforms for sharing learning resources and achieving a collective construction of knowledge. This work focuses on investigating whether social networks do provide a real learning opportunity, in what way and what are its characteristics. More specifically, this chapter offers an initial approach to the changes in the methods and practices of academic staff and students when learning through these social networks, for instance Facebook This learning experience is based on an open learning methodology and on the use of open education resources. Several results and recommendations were drawn which can be of practical use for other possible learning projects using Facebook.

Chapter 3, by Grasman et al., proposes a learning framework - the "E-Warehouse" framework - which is an innovative concept that provides an integrated learning environment for collaborative enterprise learning. It is based on an enabling computing foundation, learning pillars, and an integrating roof, which allows for cross-functional applications. Curriculum materials are developed to incorporate theory, homework, case studies, and other learning material by utilizing reference textbooks, journal articles, and other pertinent publications. These resources are converted into instructional material using multimedia tools in order to facilitate its effective dissemination and customization. Two software packages have been developed associated with E-Warehouse. The first allows the system administrators to create content modules, and the second package allows a user to customize the content. Additionally, a functional quiz application has been developed, which allows a user to select from a repository of questions, add their own questions, and grade the quizzes.

Methodologies and tools for analyzing students' interactions and structuring sequences of activities with the aim to increase students' collaborative behaviours, performance and group organization

The rise of a new class of collaboration tools for analyzing students' interactions encourages new ways of examining and evaluating the collaborative process and makes possible new modalities of knowledge extraction, feedback provision and monitoring facilities. The chapters in this area address several important issues and shed new light to this challenging research topic.

Atkisson and Brent, in chapter 4, propose an approach that collects interaction data from a CSCL environment which then analyses to examine student behaviours and performance. The analysis reveals both new ways in which students and instructors collaborate and provides new evidence that addresses one of the fundamental problems faced by students – procrastination. A new element is added, which is a deadline that is assigned to students. The work examines student behaviour as time to deadline approaches and interprets that behaviour using a rational framework based on Temporal Motivation Theory. Both qualitative and quantitative data are presented to highlight changes in student behaviour and performance as time to deadline approaches.

In chapter 5, Kahrimanis et al. provide an extensive and comprehensive overview of logfile-based interaction analysis techniques that can be used for the support of CSCL activities. Logfiles capture information about the content and the process of collaboration. This information can then be analyzed by automated or semi-automated analysis tools. The objective of this analysis is often to support participants, in several ways: explicitly, by providing feedback to them in order to regulate their practices, or by making adaptive changes to some aspects of the collaborative setting; or implicitly, by making available to them representations of their activities. This chapter presents the most common approaches used in interaction analysis, while it particularly emphasizes recent innovative efforts to reap the advantages of machine learning techniques in order to overcome common shortcomings of previous approaches.

Chapter 6, by Pérez-Sanagustín et al., suggests the application of Collaborative Learning Flow Patterns (CLFPs) to structure sequences of activities in real contexts, thus organizing groups of students according to the constraints imposed by the pattern. Sometimes an adjustment and adaptation of the group structures to a new context is needed. If the collaborative pattern is complex, this group redefinition might be difficult and time consuming to be carried out in real time. In this context, technology can help on notifying the teacher which incompatibilities exist between the actual context and the constraints imposed by the pattern. This chapter presents a flexible solution for supporting teachers in the group organization profiting from the intrinsic constraints defined by a CLFP codified in IMS Learning Design. A prototype of a web-based tool for the TAPPS and Jigsaw CLFPs and the preliminary results of a controlled user study are also presented.

Models that enhance collaborative e-learning experiences through collaborative ontologies building, educational multiplayer video games, and adaptive collaboration support

This area is introduced by chapter 7 (Mangione et al.) which presents a pedagogical approach for collaborative ontologies building. The collaborative ontology is a research domain linked to concepts of "extended cognitive context" and knowledge building in co-participation. Collaborative ontology authoring not only fortifies the process of ontology engineering, but also indicates that the collaborative ontology development and harmonization is not well supported by any of the existing ontology authoring tools or environments. These tools do not use a relevant pedagogical collaborative frame, as a collaborative writing approach for shaping the design features of cooperative building. Also the process of ontology building does not take into account what we can call “rich tagging”, that is the extraction of ontologies maturing through text produced and shared at a networking layer. This chapter presents a CSCL driven “ontology design model”. In this model, the ontology building process is maintained and validated by the encounter of 1) top-down level, where the collaborative writing scripts directs the development of authoring tools for the collaborative ontologies design and 2) bottom-up level, where the collective learning spaces such as forums and wikis, revisited by a semantic structure, are functional to the ontology extraction and validation in the learning experience.

Chapter 8, by Padilla et al., presents a model of a Video Game Supported Collaborative Learning (VGSCL) system which enhances collaborative e-learning experiences through the analysis of the quality of collaboration and re-adaptation of the proposed game. Assessing collaboration that occurs during an educational video game allows the system to take new parameters (not previously detected) into account and re-adapt the game to make both the play and learning experiences more enjoyable and effective. These new parameters refer to interesting data about players, groups, and the game process itself. These data can be used to model players’ activities during the game taking three perspectives into account: 1) educational and recreational goals, 2) educational and recreational tasks, and 3) interaction between players/students.

Bayon et al., in chapter 9, go a step further by proposing a model that is enhanced with personalization features which enable the system to adapt the learning experience to the student needs and achieve a more effective collaboration. In particular, they propose a framework that provides adaptive collaboration support for a CSCL environment framed in an open and standards-based learning management system. The proposal combines adaptation rules defined in IMS Learning Design specification and dynamic support through recommendations via an accessible and adaptive guidance system. A partial prototype of this approach has been implemented and a formative evaluation was carried out to guide the on-going work. The implementation offers CSCL courses following a methodology called Collaborative Logical Framework and has been run in a real world scenario.

Systems and tools that use personalization and adaptation techniques to support the development of more powerful e-collaboration settings

The last area of this book is comprised by four chapters that raise important issues related to the design and application of personalization and adaptation techniques that can be used to support the development of more intelligent collaborative environments. Thus, in the first case, Kickmeier-Rust et al., in chapter 10, advocate that conventional methods of educational adaptation, are often not suitable in the context of games, as they may force an interruption of the game experience and thus, destroy immersion and engagement of the player. For this reason, they enhance the educational potential of computer games through a strong personalization and adaptation to the individual needs and preferences. More specifically, the approach presented in this chapter allows embedding instruction into the game experience and narrative, through non-invasive assessment of knowledge and motivation, as well as the delivery of various types of adaptive interventions, and adaptive storytelling.

The next chapter 11, by Kordaki and Siempos, presents an innovative description of the Jigsaw collaboration method, in the form of an online, adaptive collaborative design-pattern that has been constructed taking into account adaptation techniques, within the context of open-source learning design-based environments, such as LAMS. This method is described with special reference to the learning of essential issues in Computer Science and especially in the area of programming languages. The innovative description of the Jigsaw collaborative method within LAMS is based on the fact that: (a) the tasks assigned to the expert groups consist of investigation of real world scenarios and not merely the study of learning material as is usually proposed, (b) adaptive techniques are integrated with the method and (c) for the design of the collaborative learning activity, an intuitive learning design tool like LAMS is used.

Chapter 12, by Lichtnow et al., presents an approach for recommendation of learning materials to students in an e-learning environment. The aim here is to increase the current system's personalization capabilities for students in different scenarios making use of recommendation techniques. The recommendation is produced considering learning materials' properties, student's profile and the context of use. In addition, the process of recommendation is improved through students' collaboration. In the context of this work, a learning material is a link to a Web page or a paper available on the Web and previously stored in a private repository. The process of collaboration occurs during student's evaluations of the recommendations. These student's evaluations are used by the system to produce new recommendations for other students. The main features of the recommendations aspects are described and some examples are also used to discuss and illustrate how to provide this personalization.

The last chapter 13, by Magnisalis and Demetriadis, discusses a pattern-based approach in which it is shown how educators' ideas can provide the basis for adaptation patterns which, in turn, can be expressed in IMS-LD modeling language. In particular, this chapter presents representative and selective design case studies exemplifying the implementation of the core specification of an Adaptation

Pattern (Input, Rules, Model and Output) on the basis of using tools compliant to IMS-LD. The authors analyze what is necessary for implementing an adaptation pattern and discuss the benefits of the pattern-based approach. Finally, they highlight what issues would be important toward integrating the adaptation pattern capabilities in LD compliant tools for collaborative learning design.

Final Words

Technology-Enhanced Systems and Tools for Collaborative Learning Scaffolding is a major research theme in CSCL and CSCW research community. It comprises a variety of research topics that span from the study of frameworks and infrastructures that foster collaborative learning and work through the application of different methods (distributed e-learning repositories, content creation and customization, social networks, collaborative ontologies building, and educational games) to the use of personalization and adaptation techniques to support the development of more powerful e-collaboration settings, including methodologies and tools for analyzing students' interactions with the aim to increase students' collaborative behaviors, performance and group organization. Researchers will find in this book the latest trends in these research topics. Academics will find practical insights on how to use conceptual and experimental approaches in their daily tasks. Developers from CSCL community can be inspired and put in practice the proposed models and evaluate them for the specific purposes of their own work and context.

Finally, we would like to thank the authors of the chapters and also the referees for their invaluable collaboration and prompt responses to our enquiries, which enabled completion of this book on time. We gratefully acknowledge the support and encouragement received from the editors of Springer, Thomas Ditzinger and Heather King.

We hope the readers of this book will find it a valuable resource in their research, development and educational activities in online environments.

Barcelona, Spain
January, 2011

The Editors of the Book

Thanasis Daradoumis, University of the Aegean, Greece /
Open University of Catalonia, Spain

Santi Caballe, Open University of Catalonia, Spain

Angel A. Juan, Open University of Catalonia, Spain

Fatos Xhafa, Technical University of Catalonia, Spain

Acknowledgements

"This book has been partially supported by the Spanish MICINN project "INCoS2009" (Ref: TIN2009-07489-E), as well as by the European Commission under the Collaborative Project ALICE: Adaptive Learning via Intuitive/Interactive, Collaborative and Emotional System, VII Framework Programme, Theme ICT-2009.4.2: Technology-Enhanced Learning, Grant Agreement n. 257639."

Contents

User-Centred Evaluation and Organisational Acceptability of a Distributed Repository to Support Communities of Learners

Nik Bessis[1], Peter Norrington[1], Fatos Xhafa[2], and Giorgio Venturi[1]

[1] Institute for Research in Applicable Computing,
University of Bedfordshire, Luton, Bedfordshire, LU1 3JU, UK
{nik.bessis,peter.norrington}@beds.ac.uk,
giorgio.venturi@closertag.com

[2] Departament de Llenguatges i Sistemes Informàtics,
Universitat Politècnica de Catalunya, Spain
fatos@lsi.upc.edu

Abstract. This chapter presents the combined results and over-arching conclusions of a two-phase project to assess a pilot distributed e-learning repository for computer-based assessment resources. This investigation takes place within the larger concerns of communities of practice, such as learners interacting with computer-based assessments or the tutors delivering courses or developing themselves as professionals. On the one hand, the online repository not only offers easy ac-cess to online quality materials, but communities of learners may overlap and in-teract through a distributed repository, using such collaborative web-technologies for lifelong learning, particularly as their size and content expand into large scale digital libraries. On the other hand, the online repository enables tutors to share knowledge about the courses and could well serve to create a virtual network among tutors for sharing knowledge and views related to the teaching activity. The online repository presented in this chapter has been evaluated in with real us-ers in two phases. The first phase investigated direct users perceptions of the re-pository and its usability and pedagogical effectiveness with a toolkit of six ex-perimental methods for triangulation. The second phase consisted of interviews with senior staff involved in educational management roles to assess the wider or-ganisational perspective on the acceptability of such a repository. The results of this study showed the benefits of using the e-learning distributed repository to support learning processes of a community of learners and the importance of eva-luating the online repository with real learners in order to achieve its organiza tional acceptability.

Keywords: e-learning distributed repository, organisational acceptability, user-centred evaluation, computer-based assessment, community of learners, web technologies, user adaptability, lifelong learning.

T. Daradoumis et al. (Eds.): Technology-Enhanced Systems and Tools, SCI 350, pp. 1 – 25.
springerlink.com

1 Introduction

Online repository systems are becoming a common approach in many universi-ties to storing teaching materials and offering them to students and tutors. Such repositories are usually linked to virtual learning environments and have shown great usefulness for enhancing both learning and teaching activities. However, in many experiences reported in the literature, the projects related to online e-learning repositories are not evaluated with real users nor do they achieve organ-isational acceptability of a wide range of stakeholders (teachers, students, library staff, central IT staff, etc). The lack of evaluation with real users as well as of or-ganizational acceptability can inhibit the development and uptake of online dis-tributed repositories as means of achieving long life learning goals.

This chapter presents the experience of a two-phase project, firstly investigating the practical user-centred evaluation of a distributed repository for learning resources, and secondly the organisational acceptability of such a repository, within the context of computer-based assessment (CBA), a commonly used method by tutors that makes use of computers to administer tests and exams, as well as make reports on them.

The chapter presents both a range of evaluation tools which may be of use to other researchers for triangulation of evaluations of repositories and results from users and managers of the benefits and drawbacks of repositories which researchers may wish to confirm or reject as applying to other systems, or investigate to discover whether repositories do in fact offer such benefits or drawbacks.

The chapter is organized as follows. We give some context and related work in Section 2. In Section 3 we introduce the EERN and DELTA project phases. The methods for the two phases are presented in Section 4. The results of the study are discussed in Section 5 and evaluations and conclusions are given in Section 6. We conclude the chapter in Section 7 with indications for future work.

2 Background

2.1 Context

Before moving towards the particulars of this project, we place this work within the wider international environment, where repositories for learning objects, such as this one, form a significant part of the lifelong learning environment and the development and maintenance of communities of practice to support this. As staff, students and organisations move within an increasingly international arena – physically, virtually, culturally – knowledge management in the educational context must consider the impact on communities of practice of computer mediation (Hildreth, Kimble et al 1998), as they are in distributed international commercial organisations (Hildreth, Kimble et al 2000).

One of the key features of the repository is its adaptability at two levels: individual (learner) level and institutional level. The former (addressed during the first phase) raises the need to adapt the learning repository interface to different learner and ability profiles, especially considering the global scale of online learning environments and the varied nature of communities of practice. The different profiles could vary significantly in terms of expertise (from novices to experienced users), backgrounds etc. In the later (addressed during the second phase), adaptability is needed to cope with the cross-administrative, cross-institution domain requirements for repository acceptability.

The lifelong learning environment stretches over many communities of practice (CoPs) (including students, tutors, examiners etc.) and a resource pool containing digital libraries of many kinds (CBA question repositories are but one). The many CoPs together may be called a community of curators, responsible in different ways for processes such as creation, testing, use, description etc. of the contents of the libraries, summarised in Fig 1. The membership of the communities changes in complex ways, as do the needs of the communities and their members.

Regional and national interest in developing distributed content repositories for teaching and learning materials, e.g. Bull and Danson (2004), has grown as the benefits of these have become apparent. Developing and maintaining repositories at a local, institutional level has obvious benefits insofar as these provide a local store of resources and practice for its members to draw on. The benefits that distributed repositories offer include practitioner access to a wider range of teaching and learning materials for individual use and development, standardisation and the raising of standards, and the development of CoPs. Practitioners as professionals are themselves lifelong learners who benefit from the opportunities such repositories offer, who may have careers spanning many decades which require continuous development. Depending on local organisational circumstances, critical mass for a community may not be available (and in difficult economic times may not be possible), so the inevitably distributed nature of organisations comes to the fore – in this project several institutions were involved in the first phase.

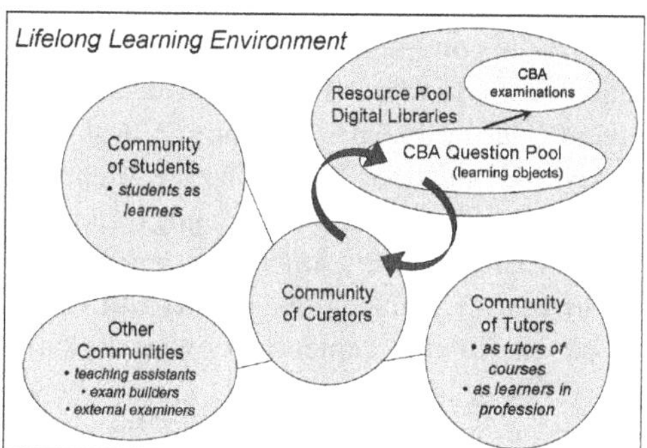

Fig. 1 Lifelong learning environment, communities of practice and digital libraries

This project's motivations are described in JISC (2005), briefly presented here:

1. Acceptance of vicarious learning and tertiary courseware into mainstream teaching and learning, to impact on assessment methods, increase reflectivity on the part of learners, and learner empowerment.

2. Greater sharing of good quality case material of effective practice, of particular importance for both initial teacher training and continuous professional development.
3. Standard ontology or metadata for tagging and accessing distributed and shared resources. Important for user groups to have an agreed, valid and easy-to-use metadata scheme.
4. An architecture that is interoperable with other systems; is extensible and is acceptable to relevant user communities (including developers).
5. Insights into the acceptability of sharing resources and of building sustainable communities of practice around resource sharing.

The first phase, the East of England Resource Network (EERN), is based on the JISC-funded DELTA system (Distributed e-learning Tool and Resource Architecture), utilising the power of the semantic web in cataloguing educational resources located in distributed repositories (JISC 2007). Each teaching and learning resource stored in DELTA may be classified with respect to a resource type (Cliff 2002), learning outcomes ("Generic Learning Activity") and stage of learning cycle ("Practice Activity"). The taxonomy of Generic Learning Activities is a derivation of Bloom's original taxonomy of learning outcomes – i.e. knowledge, comprehension, application, analysis, synthesis and evaluation (Bloom 1956). The taxonomy of Practice Activities is taken from the conceptualisation of learning as a three-stage cycle (Mayes and Fowler 1999): (a) conceptualisation, (b) construction and (c) dialogue. These three stages are described in DELTA as (a) "exploring and presenting subject matter", (b) "constructing and supporting learning" and (c) "reflecting on and discussing learning". DELTA aims to facilitate teacher and learner-controlled material review and rating(s) by the community. By making these reviews and ratings available to the community their value should increase within a 'grow in context' sharing paradigm. The options beyond the pedagogic taxonomy allow resources to be discussed by the community of practice.

The evaluation of the EERN phase started when DELTA had already completed one iteration of its development. During the first iteration, a usability walkthrough was conducted (DELTA Expert Evaluation 2005). Three evaluation qualities were specifically identified: usability, effectiveness of pedagogy and acceptability of DELTA to its end-users. The EERN phase members at the University of Bedfordshire (then the University of Luton) evaluated the quality of EERN with respect to usability, pedagogical needs and user acceptance (Fowler 2006a), and reported results in Venturi and Bessis (2006a).

Setting EERN within the wider context, it was funded as part of the JISC e-Learning Capital programme for the UK Higher Education and Further Education, one of a number of parallel projects, each with a different focus. For example, Jorum (see Rice at al 2007) had a similar remit across the sector – to collect and share learning and teaching materials – however, EERN was specifically intended to cover CBA applications of a distributed repository.

The second phase complemented and extended the first, particularly regarding evaluation, reported in Bessis and Norrington (2008). The University of

Bedfordshire Educational Resource Network (UBERN) aimed to transfer knowledge from the EERN phase to the University by supporting the deployment, piloting and evaluation of the acceptability of the Network. As a notionally complete repository of all local teaching and learning materials would not have been feasible within the phase scope, the conditions for the organisational acceptability of a distributed learning resources repository catalogue for CBAs were investigated, particularly as: the University has CBA significant experience; it meets the University's strategic goals (UoL 2006), it supports the implementation of professional standards, e.g. "The UK Professional Standards Framework for Teaching and Supporting Learning in Higher Education" (HEA 2006), and is also an area of teaching resources which has received little general attention.

The significant outstanding issue from the EERN phase, addressed in the UBERN phase, was organisational acceptability (i.e. at strategic, managerial level), concerned with embedding the new approach into practice at the individual and institutional level, including both 'bottom-up' (e.g. inclusion in courses) and 'top down' activities (e.g. inclusion in institutional strategies) (Fowler 2006a:15).

2.2 Related Work

Examples of other investigations into e-learning repositories, which further indicate the international dimension to this research, include:

- EducaNext, an online repository, the continuing service implementation of the Universal Exchange for Pan-European Higher Education (UNIVERSAL) project, see Maillet (2003)
- Context eLearning with Broadband Technologies (CELEBRATE), a large-scale 30-month demonstration project covering 500 schools, co-ordinated by European Schoolnet and supported by the European Commission's Information Society Technologies Programme (IST), see McCormick et al (2004)
- PROLEARN, a 48-month demonstration project covering 20 partners in technology enhanced professional learning (using social software), co-ordinated by Learning Lab Lower Saxony, Hannover and supported by the European Commission's Information Society Technologies Programme (IST), see Wolpers and Grohmann (2005).

This work is intended to contribute to the literature in this area, although direct comparisons with the following are not straightforward, as the projects were conducted in different domains in different ways. Vrasidas and Retalis (2004) provide a review of issues arising from UNIVERSAL, but without the same level of detail in methodology or qualitative results as here. Massart (2006), reporting on CELEBRATE, indicates two major areas of concern (notwithstanding positive acceptability) were technical issues and difficulty achieving a critical mass of resources in the repository. Simon, Oberhumer et al (2007) report that EducaNext stalled due to lack of objects created, and that in general an appropriate business model for a repository needs to be defined before a technological implementation.

3 Introduction to EERN and DELTA

3.1 Methodological Background

The background to the combined EERN/UBERN project is not based on formal theories. Rather it is an exploratory project, placed within a much larger project environment led by major organisations in the UK's Higher Education sector. Nevertheless, the underpinning to the project landscape was provided by the E-Learning Framework (ELF) (see Wilson et al 2004) based on: "based on a service-oriented factoring of a set of distributed core services required to support e-Learning applications, portals and other user agents."

A full description of that landscape would be beyond the scope of this chapter, including the relationships between the funders, HEFCE (the UK's Higher Education Funding Council), and JISC, the technical development body, with its co-operative partner representing Higher Education practitioners, the Higher Education Academy (HEA). Furthermore, the Digital Repositories programme 2005-7 was a part of the much larger Distributed e-Learning Programme.

Two important points should be borne in mind: firstly, for both phases, they were not "testing before adoption"; rather "testing as proof-of-concept", as "incubator projects"; and secondly, for the second phase, this was for development of CBA examinations, not lectures, these having different delivery modes, which may not involve the presence of course lecturers.

3.2 EERN/DELTA Goals

The overall aim of EERN was to pilot an enhanced version of DELTA within the region that will benefit practitioners and learners alike and will contribute to the building of a community of practice within the region. To achieve this aim the pilot had to demonstrate the following quality characteristics (JISC 2007):

Validity: EERN must be able to meet a wide range of valid user requirements (organisational as well as pedagogical)
Reliability: the EERN system should be robust and simple to maintain
Usability: the system should be easy to use and with the required level of performance
Acceptability: the deployment and use of EERN needs to be acceptable to a wide range of stakeholders (including teachers, learners, trainers, managers etc.).

In addition to these pilot specific objectives there was also a regional agenda to address. Of particular relevance were the East of England Development Agency's (EEDA) priorities, which included (Fowler 2006a):

1. Ensuring strong links between regional universities, research institutes, and the private sector in part by establishing close ties between the SME (Small Medium Enterprise) community and the research base, i.e. the Higher Education sector, to ensure maximum exploitation of intellectual property.

2. Maintaining and building upon the quality of research establishments in the region in part by promoting the strengths and distinctive roles of the region's universities and research establishments to the UK business community.

The DELTA project aim is to develop a new range of tools using open standards and specifications that will allow practitioners and learners to share e-learning resources, such tools being (DELTA, undated):

Distributed:	with localized resources, under the control of the resource provider.
Extensible:	providing a generic framework, allowing new types of resources to be defined.
Open:	based on open interfaces which allow a range of users and applications to access the system.

3.3 *DELTA/EERN Architecture*

The DELTA system allows resources to be submitted, searched and retrieved, with mark-up using standardized metadata, from across the EERN network (Fig 2). The resources are distributed across a range of repositories held in any number of institutions, not centrally on the DELTA system. DELTA is defined as (DELTA, 2005): "A service-orientated architecture is provided through WSDL (Web Services Description Language), UDDI (Universal Description, Discovery and Integration) and SOAP (Simple Object Access Protocol). Providing that a well-defined set of interfaces is maintained across the framework, components from disparate organisations may be easily incorporated." Thus DELTA may be viewed as a potential interface, not necessarily a final interface.

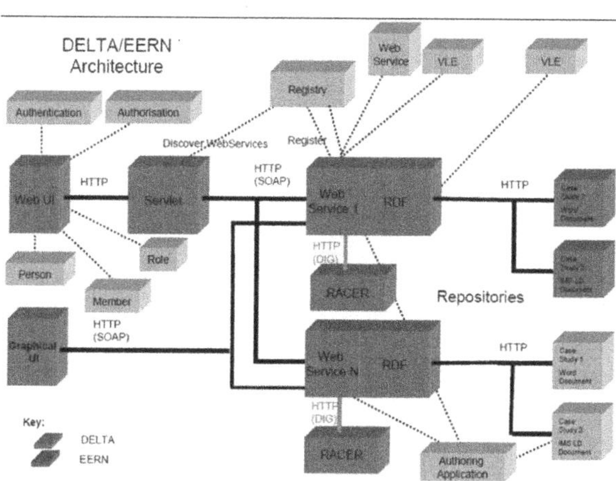

Fig. 2 DELTA/EERN architecture (from EERN 2005: 2)

3.4 DELTA's User Interface

Following, we present a brief description based on the DELTA User Guide 1.3, Scott and Fowler (2006), with illustrative images from the project documentation: EERN (2005), Scott and Fowler (2006) and Fowler (2006b).

The user interface goal of DELTA was primarily usability, resting on the EERN system goals of validity, reliability and acceptability. The DELTA interface aims at a straightforward, uncluttered initial approach, which leads to more sophisticated ways of discovering resources, according to the user requirements. DELTA's advantage over basic resource repository is to offer standardized marking up or tagging in terms of resource pedagogy based on an underlying pedagogic ontology (Fig 3), not just standard metadata such as title, subject, educational level, etc.. Since DELTA is aimed at practitioners ranging from the novice to the

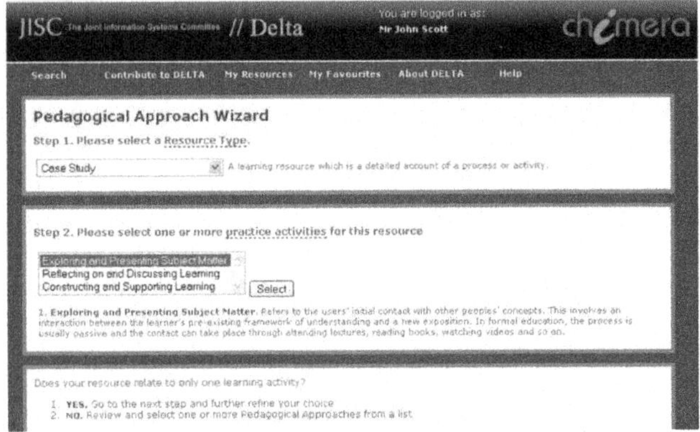

Fig. 3 Pedagogical approach wizard (adapted from Fowler 2006b: 16)

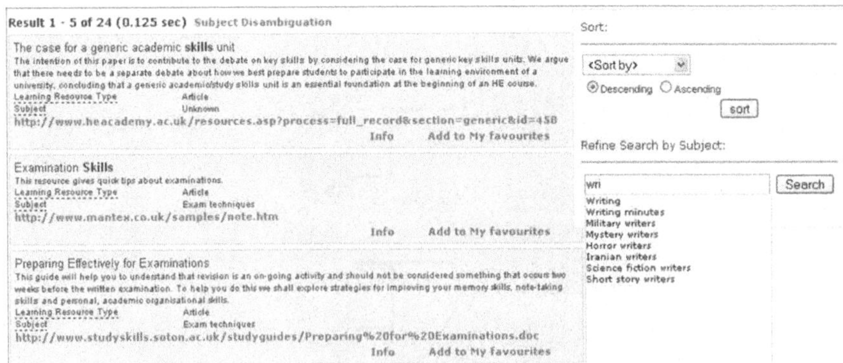

Fig. 4 Search results with options to refine search by subject (adapted from Scott and Fowler 2006: 19)4 Methods used in Phase 1 and Phase 2

experienced, this categorization is intended to assist practitioners think about the core pedagogic principles that affect their choice of learning resources for any given learning activity when selecting a resource (Fig 4); to assist them to think about the pedagogic principles that underpin resources they are sharing (Fig 3), and to encourage continuous professional development in developing multiple resources within a community offering reviews and ratings of resources.

4 Methods Used in Phase 1 and Phase 2

4.1 Phase 1: EERN Evaluation and Analysis

The EERN evaluation involved four main sub-phases:

1. Elicitation of system goals by claims analysis on user scenarios. After expert and stakeholder review, twenty-four initial claims were synthesised to nine.
2. Definition of an evaluation plan including users, tasks and a set of qualitative and quantitative evaluation methods.
3. Data collection in the actual context of use.
4. Analysis of the results, cross-examining the results obtained by each technique.

We note that in this work we describe the quantitative methods used in Venturi and Bessis (2006a); these methods form our toolkit, which we believe others may find interesting and useful. However, as the results and statistical analysis were of situational, usability interest to inform the direction of the project's development, we refer the interested reader to that paper, rather than repeating the results or more than an overview of the statistical analysis here.

Scenarios and Claims Analysis

Scenarios are stories about users and their activities that can "focus designers on the needs and concerns of people in the real world" (Carroll 2000). They provide high-level, abstract representations of user-system interactions. Claims are hypotheses about the "effect of the features on the user activities and their outcomes" (Rosson and Carroll 2002), and are elicited from each scenario.

Each claim is a natural language statement made up of three components:

a) the feature(s) under scrutiny (for example, "simple and advanced search functionalities")
b) the quality(ies) assessed (for example, "easy to use")
c) the user expertise involved (for example, "novices").

Claims analysis identified the system goals for evaluation. Although such an analysis is usually made to evaluate different design options against claims, we used the method to turn the scenarios' rich description into items to be evaluated.

For example, one scenario dealt with teaching practitioners searching and retrieving learning resources from DELTA to support learners with the development of study and research skills. New DELTA users may need to refine the search results by subject ("study or research skills") or pedagogical approach(es), in this case "face-to-face" or "individual" learning. Two claims were elicited from this scenario, and are shown in Table 1 with the Evaluation Goals (EG) they were translated to. The full set of EGs are presented and discussed in Section 5.1.

Not all claims have the same level of generality. For example, the first claim is more specific compared to the second, as it relates both to the easiness of reading search results and filtering them. However, both claims relate to usability, or rather, "Quality of Use" (Bevan 1998). Other types of issues were discovered which were not strictly related to the concept of quality of use, such as copyright. Hence, the issue of how to evaluate these aspects required addressing. EERN evaluation has a complex socio-technical nature involving both a technical (DELTA) and a social system (HE and FE organisations).

Table 1 Scenario claims translated into evaluation goals

Claim components: f – *feature*; q – *quality*; u – *user*

	Claim		Evaluation Goal
1	Simple and advanced search functionalities will be easy to use for people new to DELTA (novices)	EG1	Are (f) "simple and advanced search functionalities" (q) "easy to use" for (u) "people new to DELTA (novices)"?
2	Search results will be easy to interpret and filter by advanced search for novices	EG2	Are (f1) "search results" and (f2) "filtering by advanced and pedagogical search" (q) "easy to under-stand" for (u) "novices"?

The nine central claims were then translated into Evaluation Goals. Finally, these were associated to three quality dimensions (QD):

- Quality of use (U)
- Learning and Pedagogical effectiveness (P)
- Acceptability (A).

It was felt to be crucial to identify whether users were likely to accept DELTA, actively contribute their own experiences, or whether they would be relatively passive users. The evaluation was thus defined to include qualitative methods (such as one-to-one interviews) in order to better understand the motivations behind users' behaviour. On the other hand, the opportunity to employ quantitative techniques like usage logs and usability tests presented itself. Given the range of techniques available here, the interesting possibility of triangulating qualitative and quantitative results arose. Triangulation is a process of cross-checking findings derived from both quantitative and qualitative approaches (Deacon, Bryman et al 1998). Thus, an "evaluation toolkit" was formed, comprising a selection of one quantitative, three qualitative and two hybrid research methods:

- Web questionnaire (quantitative and qualitative)
- Usability tests (quantitative and qualitative)
- Logs analysis (quantitative)

- Interviews (qualitative)
- User diaries (qualitative)
- Pedagogy workshop (qualitative).

A total of 29 end users participated in the evaluation. Approximately half of them (15) belonged to FE, 8 to HE, while the remaining 6 belonged to both sectors. Fifteen of the subjects were direct users (lecturers, tutors, module/unit developers, library/learning resource officers) whilst the remaining fourteen were indirect users (IT managers and support officers, academic and course managers). Not all of the users were involved in all evaluation methods. 14 out of 15 the direct users participated in the usability tests; 3 direct users participated in the pedagogy workshop; 4 indirect users were interviewed.

Evaluation methods

Web questionnaire

A questionnaire was handed to the participants of EERN before their training with DELTA. The first section of the questionnaire sampled participant demographics, professional role and experience in e-learning repositories. Profiled participants were to be chosen for the other evaluation methods. The second section probed participant perception of positive and negative aspects of e-learning repositories and personal expectations towards them. A qualitative, thematic analysis was applied to this data and the issues found categorised.

Usability tests

Fourteen direct users were involved in the usability tests. Scenarios were used to identify tasks, which were piloted with students prior to their deployment with actual (direct) users. Users carried out six types of task:

- Logging in
- Searching by full text
- Searching by field
- Searching by pedagogical approach
- Reviewing a resource
- Creating a resource.

The "think aloud" test protocol for usability tests (Someren, Barnard et al 1994) was carried out for each task. Relevant metrics were recorded for quality of use (completion rate, errors, task time). After all tasks were completed, users were prompted about the way they perceived the usability of the system (e.g. "how did you find using …?"). This provided qualitative feedback. After that, they filled in a psychometric questionnaire (Brooke 1996) to generate an assessment of the perceived usability. The results were analysed, identifying errors from the notes taken during the observation and ranking impact errors. Error impact was calculated by multiplying their frequency with estimated severity. Severity was ranked in three categories: "showstopper" (leading the user to the failure of task), "major" (giving the user serious problems in completing the task) and "minor".

Usage logs analysis
Logs of user activities were recorded over a five month period, consisting of:

- Date and time of request;
- Specific actions taken, namely: User sessions; Default searches; Advanced searches; Search by Subjects; and Resources created.

Interviews
Two support staff and two managers were interviewed in their own work contexts. Themes to discuss were identified in advance, probing for an understanding of the indirect user's own perception of DELTA. The focus was on organisational issues (e.g. curriculum, ownership of learning materials). The interview transcripts were given a qualitative, thematic analysis.

User diaries: The Direct DELTA users were prompted during DELTA training to write an e-diary including their concerns or impressions. Users were explicitly encouraged to report both positive and negative issues. Five out of the sixteen templates sent out were returned (31%). Qualitative, thematic analysis was made of the diaries' content.

Pedagogy workshop: The aim of this workshop was to reflect on and discuss the way DELTA classifies learning resources, with a specific attention to the underlying pedagogical ontology. The workshop included an individual and a group session on the classification of resources. In the individual session users classified a learning resource (HEA 2003) being watched by an observer, who occasionally prompted the user to reflect or give an explanation of his/her behaviour. In the group session, two subject matter experts carried out a similar task discussing step-by-step their choices with their practitioners. Qualitative, thematic analysis was made of the event transcription.

The event was conducted not as an evaluation activity, but as an occasion for professional development, following the "co-operative inquiry" approach:

> Everyone is involved in the design and management of the inquiry; everyone gets into the experience and action that is being explored; everyone is involved in making sense and drawing conclusions; thus everyone involved can take initiative and exert influence on the process. (Heron and Reason 2001)

Although only three participated in the event, outcomes were very positive; they were especially enthusiastic about their direct involvement in the project.

4.2 Phase 2: UBERN Evaluation and Analysis

The UBERN evaluation picked up on the otherwise unexplored issue of organisational acceptability, rather than individual acceptability.

For evaluation purposes we were based on Computer Based Assessment techniques developed in the e-learning research community (Scalise and Gifford 2006). A questionnaire was prepared for tutors and lecturers authoring and assessing CBAs to investigate user acceptability issues of a CBA repository.

The "Computer Based Assessment (CBA) Practitioner Questionnaire" was developed, based on the "Web survey questionnaire" used in the EERN extension phase (Venturi and Bessis 2006b Appendix XI: 95 ff.), but adapted to the particular needs of the CBA evaluation. The questionnaire was emailed to relevant tutors and lecturers; anyone taking part in the interviews for organisational acceptability (see below) was excluded as it was felt that their inclusion could, at this point, bias results towards practitioners with explicitly higher level knowledge of CBAs. Due to the short time-scale imposed by the change of plans and coincidence with significant institutional academic processes, only four questionnaires were returned. However, these were from CBA authors who rated themselves as Experienced and Very experienced, and contained some very useful responses. Responses of particular interest are quoted to support the organisational acceptability interviews.

The "Computer Based Assessment Indirect User Interview" was the focus of the UBERN phase, identifying institutional members having different management functions and responsibilities for CBA strategy and implementation and conducting semi-structured interviews with them to investigate organisational acceptability issues.

The institutional members were identified by functional areas, based on considerations of and extensions to indirect user stakeholders in the EERN phase, see Venturi, Bessis et al (2006) and also Venturi and Bessis (2006a). Seven areas and representative staff were identified, of whom six were available for interview. These six interviews are close to the optimum obtainable for discussion of strategic educational issues within this institution.

The Indirect User Interview was developed as a series of questions to support a semi-structured interview. Although the questions were presented in sequence, participants were free to move forwards or backwards as ideas came to them. Where answers appeared to the interviewer to fall under a question other than the one being directly answered, the participant was asked for clarification and answers were recorded under the question the participant found most relevant. Responses were recorded by hand, as taping the interviews and reviewing and transcribing them was not possible.

To support the interviewer, a list of 32 prompts was derived from the "Thematic analysis of beliefs and expectations of the respondents" (Venturi and Bessis 2006b: 88-94). The prompts were not shown to the participants and were only for use if the participants "ran dry". In fact, the prompts were not referred to at all as the participants all talked fluently with very little intervention by the interviewer.

5 Results and Discussion

5.1 Phase 1: EERN

DELTA fulfilled four of the nine goals (Table 2: EG1, 2, 5, 7) and fell short on four others (EG3, 4, 8, 9). The evaluation team did not reach an agreement on goal (EG6) as it was felt too early to make a valid conclusion about the acceptability of the system. The results were communicated to the DELTA design team for improvement in the successive development iteration. Issues were mapped to the three evaluation goals categories (quality of use, pedagogical effectiveness and acceptability) and coupled with design recommendations where appropriate.

Table 2 DELTA Evaluation goals (EGs) and quality dimensions (QDs) obtained from claims analysis

Claim components: f – *feature*; q – *quality*; u – *user*
QDs: U – *Quality of use*; P – *Pedagogical effectiveness*; A – *Acceptability*

EG		QD
EG1	Are (f) simple and advanced search functionalities (q) easy to use for (u) practitioners new to DELTA (novices)?	U
EG2	Are (f1) search results and (f2) filtering by advanced and pedagogical search (q) easy to understand for (u) novices?	U
EG3	Is (f) pedagogical classification (q) easy to use for (u) practitioners with limited knowledge of pedagogy?	U
EG4	Are (f) DELTA ontologies (q) understandable and valuable for the (u) practitioners, in order to understand the way they teach and think about ways to improve it?	P
EG5	Is (f) help on copyright policies (q1) easy to access from the home page and (q2) understand for (u) novices?	U
EG6	Are the activities of (f1) retrieving and (f2) sharing learning resources with DELTA (q) acceptable for (u) practitioners?	A
EG7	Is the (f1) registration and (f2) log-in process (q) easy for (u) novices?	U
EG8	Is the (f) registration process (q) clear for (u) novices in explaining how to set up DELTA?	U/A
EG9	Is the (f) pedagogical summary (concluding the creation process) (q) clear for (u) practitioners with limited knowledge of pedagogy in explaining the pedagogical choices available?	U/P

Quality of use
The simple and advanced search, and filtering tasks were easy to carry out (EG1, 2). The results from the usability tests showed that users performed significantly better the second time they performed a search with respect to task time (N=14, $p<0.05$). The same did not apply to pedagogical search, which took significantly longer than other types of advanced search even after one attempt. Classification of learning resources was a demanding task (EG3), with 6 people failing or asking for assistance and 3 people refusing to do the task at all (N=14). Overall, 44% could not classify learning resources without assistance during the usability tests. Copyright policies were included in the help section, but few people (two) consulted them (EG5) during the usability tests. DELTA lacked contextual help: the term "copyright" was ambiguous because it could refer both to the holder and the type of copyright. Results from the usability tests showed that logging-in (EG7) was straightforward (everyone completed the task). However, two user diaries indicated that the registration process was neither quick nor explained clearly (EG8)

that, before using DELTA, configuring the institution's proxy server was necessary. When users tried to access it, they received a misleading error message.

Qualitative analysis of the observations of usability tests revealed a total of 29 usability issues. Half of them were minor (15). However, 3 of them were ranked as "showstoppers":

1. Lack of contextual help explaining the vocabulary employed by DELTA
2. Generic and misleading error messages
3. Difficulties in recovering from errors; sometimes the back button of the browser didn't work.

Pedagogical effectiveness

Results from the pedagogy workshop showed that users of DELTA found it very difficult to comprehend the vocabulary of the ontologies (EG4), especially "Learning Resource Type", "Generic Learning Activity" and "Practice Activity". Help was provided on paper and on-line. Actually, many practitioners preferred to go ahead without looking for an explanation of the key terms. The information contained in the pedagogical summary was confusing and ambiguous (EG9). It did not show the difference between the choices made by the user and the choices made by the system. Practitioners wanted to associate the resource to more than one resource type and generic learning activity, but this was not possible due to the constraints embedded in the ontology. One lecturer stated that this bias could lead to people not finding the resources that they are looking for. For some direct users (e.g. library/learning resource officers) the advanced features of DELTA in terms of pedagogical classification could actually be a deterrent, as their knowledge of pedagogy might be less than or different from teaching staff.

Acceptability

With respect to end-user acceptability, user diaries revealed that some users were frustrated because after being registered they couldn't access DELTA. The system failed to provide clear instructions on how to set up the proxy server (EG8). Retrieving and sharing resources with DELTA was not evaluable for acceptability (EG6). During the five months of the trial we collected conflicting evidence. Some people were positively impressed by the quality of resources; others found them useful but not for their everyday teaching activities.

Log analysis showed that the search and retrieval of resources with DELTA was discontinuous, due to the division of the teaching calendar in two semesters.

EERN expects learning resources to be tagged by their authors. However, the ten resources tagged by practitioners should be considered few, compared to those tagged by DELTA staff (about 190). At the end of the project DELTA was still dependant on work undertaken by those staff. A learning officer also pointed out that DELTA did not store learning resources by itself, which could lead to maintainability and sustainability concerns. Quantity and quality of contents are perceived by practitioners as a benefit delivered by repositories. A thematic analysis from the qualitative data of the web questionnaire identified and ranked the three most important benefits from users' perspective. Results show that users do value efficiency, high quality content and a higher awareness of relevant pedagogy.

Some organisational and cultural constraints should be removed in order to get full acceptance. Users were burdened by heavy schedules and no time slot was specifically allocated to them for learning how to use DELTA. Practitioners have not all adopted a mindset of resource sharing. An academic manager stated during her interview that even intra-departmental sharing was proving hard to encourage.

The evaluation of EERN was a unique opportunity to reflect on the fitness for purpose of the six research methods for studying the usage of a distributed system in its actual context. The full benefit and nature of the triangulation of the methods was only apparent at the end of the evaluation.

1. Web questionnaires were essential for profiling participants in usability tests and the pedagogy workshop, offering insights into user acceptability issues.
2. Usability tests were extremely effective in discovering usability issues and in assessing their impact. The issues were confirmed by the pedagogy workshop and user diaries.
3. Usage logs were not very effective for triangulation purposes. They provided an overview of usage trends, but only for a limited period of time in this study.
4. The pedagogy workshop was useful for gaining a better insight into issues found by usability tests and for giving design advice on how to enhance users' understanding of pedagogy. Data clearly showed severe issues arising in the mechanics of classification process itself; practitioners couldn't complete the process without assistance from an expert.
5. User diaries were returned at the end of the evaluation process, while we were carrying out the interviews. User diaries were very effective in cross-examining results coming from other methods. These diaries confirmed evidence revealed by the usability tests and also the pedagogy workshop. Moreover, user diaries crucially revealed the difficulties met by practitioners in setting up DELTA.
6. Interviews with managers were useful for gaining an insight into organisational (the division of teaching activities into two semesters) and cultural issues (sharing not part of the mindset of people). Interviews explained – to some extent – the behaviours recorded by usage logs.

The three quality dimensions (quality of use, pedagogical effectiveness and acceptability) exhibit interdependency within this socio-technical system. Quality of use is the fundamental quality: if users do not find the system usable the other two dimensions are compromised. If the system is usable, pedagogical effectiveness is considered: if the system supports dialogue and reflectivity upon current teaching practices, then it will be employed by the user organisation. Acceptability is a higher-level issue; it is influenced by the two former qualities and by other organisational and cultural aspects.

The outcomes of the first phase largely focused on the processes of immediate engagement with the EERN system – e.g. logging in, accessing materials, adding materials, tagging materials – and their usability and pedagogical effectiveness as perceived by practitioners. What is lacking is a deeper understanding of the acceptability of these activities (EG6).

In fact, the acceptability of sharing resources is not simply determined by individual practitioners treated as isolated from their environment, but is influenced by wider environmental factors determining policy and practice. For example, new staff may come with ideas from previous environments, but will be expected to adapt to their new institution; yet all staff, new or not, in an institution contribute to an ongoing debate about the appropriateness of policies and practices within the institution and the institution's wider environment.

Thus, the Phase 2 investigation of the acceptability of such resource-sharing repositories at the level of managers has been important to understand the shared and changing culture into which such repositories might be introduced.

5.2 Phase 2: UBERN

The responses to the UBERN Indirect User Interview developed further understanding of these aspects, by enquiring about the range of benefits for practitioners or their students through the use of CBAs; the range of benefits and disbenefits for practitioners through the use of CBA repositories; and the overall benefits or otherwise of e-learning repositories. Managerial participants clearly identified a range of benefits gained through the use of CBAs for practitioners or their students:

- Consistent marking
- Time-saving
- Richer, personalised diagnostic assessment
- Tailored to users' needs
- Helpful for formative assessment and feedback, less certain for summative as "less thinking involved", useful tool but not "be all and end all".
- Reproducibility
- Alternative assessment mechanism to heavily text-based examinations
- Consistency of feedback
- Rapid testing of large numbers of students
- Relative ease to set-up
- CBAs may be more difficult at Master's level, but not impossible
- Makes exam fairer to students who do not have English as mother language
- Students like multiple choice more than essay writing.

Thus, seeking to build a level of community of practice above individual usage may indeed bring advantages. However, awareness of existing repositories was low. Four of the participants indicated that they were not aware of any repositories to assist with building CBAs, although they supposed there were some. Others questioned the effectiveness or continued development of those that exist.

Participants in managerial roles clearly identified a range of benefits for practitioners gained through the use of CBA repositories:

- Time-saving ("library of resources to draw on")
- Mechanism to easily acquire questions fit-for purpose ("analysis should support 'good questions', though there are issues around renewal and grade differentiation", "Test banks provide an armoury of suitable questions that can be modified mixed and matched")

- Web-based general availability
- Tagging questions as "not used before"
- Discovering best practice ("tap into experience")
- Promote CBA questions and answers easily
- Access to wider ideas of assessment methods (appropriate to different teaching levels and adaptation for domains).
- Brings some standardisation ("benchmarking would be useful")
- Can build in validation mechanism ("requires submission to a shared panel as a community process across institutions of validation members")
- Provide feedback to staff to improve style by offering options in system.

These are congruent with the benefits that end users identified in Venturi and Bessis (2006a; 2006b) regarding teaching and learning repositories in general. This indicates that there is awareness of the benefits at both user and managerial level which can be utilised in both top-down and bottom-up approaches to developing CBA repositories. Participants in managerial roles clearly identified a range of negative issues regarding CBA repositories:

- Questions may be overused ("but students need access to past exams, which is good pedagogic practice") (although we note that this is not current practice for CBAs in this institution)
- Questions may not be fresh
- Questions may not be secure (through leakage or in group work situations)
- Questions may not fit lecturers' style
- Intellectual property issues in protecting lecturers' and the University's work
- Workload putting resources in place (regarding academics' workload or "finding other ways around problems", "loss of diversity when others stop contributing")
- Inadequate breadth or depth
- Fit with paper-based resources (conversion from existing format and synchronisation)
- Locking into particular assessment methods
- Availability of service (client- or server-side, proxying, mirroring)
- Quality assurance ("across HE sector for subject domains", "verified questions" and standardisation)
- Impact of question type on incorporation into a CBA (regarding (a) differently scaled multiple choice questions and their effect on grading percentages, (b) discrimination of students' abilities by the questions)
- Staff development as a line manager issue ("possibly linking access to contribution")
- Often too tightly coupled to a book or course that is being examined.

Additionally identified were the use of a Strictly Proper Scoring Rule to control for guessing or the use of Hunt's scoring system; the need for questions to be rated for their facility index (FI) and discrimination index (DI); and a prediction of the Kuder-Richardson formula reliability of any final set of items selected after the item analysis procedure is complete.

There is much congruence with the negative issues that end users identified in Venturi and Bessis (2006a; 2006b) regarding teaching and learning repositories in general, though some additional ones are identified here (equally applying to repositories other than for CBAs). This indicates that there is awareness of the negative issues at both user and managerial level, which need to be addressed in both top-down and bottom up-approaches to developing CBA repositories.

All participants were enthusiastic about an e-learning repository and the possible gains. There was a general cautionary note about workload implications and balancing these against existing responsibilities. The following impacts on roles were identified:

- Offer an improvement
- Raise awareness and encourage staff
- Provide resources for use ("directions to 'how did they do that' ")
- Develop skills (staff and students in information retrieval, paradigm testing, professional networking, knowledge of field, although possibility of overload)
- Helping with cataloguing ("in an advisory capacity")
- Development and support of protocols and implementation (move to "business critical" application has important consequences)
- Maintaining academic quality (workload depending on degree to which quality assurance process is built in, provide consistent feedback).

Participants recognised that the question of where leadership for such a project should lie was a politically sensitive, though necessary question. It is inappropriate to link respondents to their views here, as future discussions should take place in an appropriate forum. The CBA resource network would be a significant undertaking, participants agreeing that it is an issue for academics to take a role in. The responses indicate that different aspects of leadership might be located in different areas, be it in a particular institution or in external bodies such as JISC.

Overall, Phase 2 met the project's needs, indicating clear, shared understanding at practitioner and managerial level of the issues, both positive and negative for such a repository. It also indicated that at managerial level there are additional issues which must be considered to make such a repository a workable system.

5.3 Towards a Fully Decentralized Learning Repository

Distributed learning repositories are meant to serve increasing number of users and can be a source for lifelong learning. Therefore, scalability can become an issue as the number of repository users could increase over time given that more universities are implementing online learning and teaching programmes or supporting face-to-face learning and teaching with online tools. Also, it is reasonable to expect that in the mid- or long-term, the number of documents in the repository would increase due to new material being added and improvement and versioning of existing materials. Thus, the repository should be scalable in terms of both user numbers and document volume. It should be noted that scalability has a direct impact on *response time* of users' requests on the repository and would, therefore, impact a repository's quality goals.

In our focus, the repository's scalability can be approached by using a farm of servers. Indeed, by taking advantage of the web-services based architecture of the repository, the repository can be distributed in different servers, with each server hosting a part of the whole document set. Then, user requests are dispatched to the server having the documents. In this way, by loading different servers, the scalability in both terms of users and documents can be achieved. In fact, the server farm approach could also be enhanced with replication techniques to overcome a single point of failure and would thus ensure availability of the repository.

6 Summative Evaluation and Conclusions

An evaluation of quality of use, pedagogical effectiveness and acceptability of DELTA has been conducted. Scenarios and claims analysis were used to define a set of evaluation goals. Qualitative and quantitative research methods were employed in the evaluation process and compared by triangulation. Teaching practitioners (lecturers, tutors and learning material developers) and other indirect users were involved in the process. DELTA fulfilled four out of nine goals that were defined and fell short on four others. Installation issues were highlighted, which can undermine the practical aspects of implementing DELTA, or indeed any system. Searching and filtering tasks were fairly easy even for novices. Classification of a learning resource was rather more demanding. The main usability issues were lack of contextual help, generic error messages and failures in recovering from errors. Practitioners found it difficult to understand the key terms of the pedagogical ontology. The needs of different practitioner learners and the points of failure for some of them indicate that adaptability of the system at the individual learner level is important for fundamental engagement with such systems. Insufficient evidence was found to make a conclusion on final acceptability, the ninth evaluation goal. However, the results show that there are several constraints related to organisational and cultural factors.

The EERN phase provided an interesting case to reflect on the fitness for purpose of the evaluation toolkit, which may be of interest to other researchers. The effectiveness of user diaries and pedagogical workshops was rewarding. User diaries pointed to usability and acceptability issues that were simply not discovered using other methods. The pedagogical workshop proved useful in explaining the pedagogical issues and providing recommendations about the ontology and the classification process.

The particular quantitative results and analysis presented in Venturi and Bessis (2006a) applied within the context of the usability testing of DELTA to discover pointers for redesigning that system. We believe that the results are indicative of issues that other systems may face, and as qualitative observations may thus be of interest. In a wider context, it would be useful to conduct comparisons of such repository systems, whether in terms of their conceptual structure, technical implementation or usability.

Thereafter, the UBERN phase addressed issues of organisational acceptability, successfully identifying several interrelated areas towards understanding the organisational acceptability of a CBA repository.

The CBAs practitioner questionnaire drew useful responses, despite few returns, which would be valuable rerun with more returns. The interviews with the managerial participants were successful in eliciting significant points of interest. Conducting these at other institutions may provide a wider range of concerns, whilst supporting for those expressed so far.

Participants in managerial roles clearly identified a range of positive and negative points regarding the use of CBAs and CBA repositories. These are congruent with the issues that end users perceive of teaching and learning repositories in general, seen in the EERN phase. This is a positive point as it shows that there are good prospects for organisational acceptance and implementation of a CBA repository at both user and managerial levels, subject to an appropriate structure.

Managerial participants also clearly identified a range of negative issues, again showing congruency with previously identified issues. This is a significant point as it shows that there are important issues to clarify and solve for organisational acceptance and implementation of a CBA repository at both user and managerial levels (leading to defining features of an appropriate structure). An example of this is the assessment of workload implications.

Sound and relevant pedagogic underpinning is required for resources, with suitable adaptation to specific areas of interest such as CBAs, most notably how the usefulness for a particular resource for assessment can be appropriately defined and captured in metadata.

As such a distributed repository by nature crosses administrative and institutional boundaries, it is clear that such systems must be adaptable over these domains. Such adaptability is indicated by the congruence of responses across managerial functions, and their indication of issues which have multiple owners.

Finally, although the distributed repository has been developed and evaluated in the context of the East of England Resource Network, we believe that the approach and lessons learned are valid to other online learning contexts as the repository has been developed and evaluated from a general purpose perspective.

7 Future Work

Further work would be required to substantiate claims of pedagogic effectiveness during periods of real-world use in setting assessments of various kinds, where for example these may be summative or formative, be of foundation, undergraduate or postgraduate level, or vary in the nature of the questioning technique.

Leadership issues around repository projects require discussion. It may be imagined that notwithstanding small-scale projects such as this, unless such issues are resolved there will be no actual leadership of any real projects. Nevertheless, the UBERN phase identified enthusiasm across all managerial roles interviewed for an e-learning repository and all foresaw gains. As ever, with a new project, there was a general cautionary note about workload implications and balancing these against existing responsibilities, but this cannot be evaluated at this point.

Overarching both phases a significant point arises, suggesting that a repository for e-learning objects may seem a simple idea. However, in practice, it becomes apparent that there is a wide range of stakeholders whose needs and perspectives must be taken into account for the repository to work. The repository must work on several different levels (describable in various ways), e.g.: a technical system; a human-computer interface; a system embodying concepts and practices; and a system integrated into wider organisational policies and practices.

These levels, and indeed the stakeholders, are not necessarily in conflict; in fact, as one may suppose that the aim of practitioners and managers alike is to provide a sound pedagogic environment for students, all are working in the same direction. However, to make exploration of these complex issues practicable, small scale, multi-phase projects such as this offer a constructive way to develop complex systems, and highlight issues relating to cross-domain adaptability before they become too difficult to re-engineer when embedded in a large scale system.

There may however be some interesting differences between stakeholder groups, or indeed between individuals, which would benefit from further exploration, again an issue of adaptability for different kinds of user. The tagging of repository items has, in the design discussed here, a pre-defined pool of standardized tags to promote consistent tagging across all users, assuming that they understand the tags in the same way. However, there are other approaches to tagging which might benefit users. There are many established Web 2.0 contexts in which users are able to apply their own tagging, either in the form of private tagging or tags shared across groups or communities of users, namely folksonomies. Such user-defined tagging may allow users greater flexibility in how they locate repository items, for example whether by using tags that relate to content, although the tag concept may not appear explicitly in the text, or by functional tagging, such as the example from one of the indirect users: tagging questions as "not used before", which would necessarily be time and location dependent.

Of course, putting items into the repository and tagging is not enough to enable active access to a repository's potential value; it is necessary to know what is in the repository. Traditional search facilities, whilst important, require a user to revisit a repository; what is missing here are options for notification services to draw a user back when something of possible value to the user has been added or amended; indeed deletions may also be of interest. Users may wish to set queries against a repository, whether running on a, say, scheduled, pull basis to sweep a repository for relevant changes, or on an ad hoc push basis at the time something is changed. Moreover, there may be users who wish for notifications to a variety of devices, such as mobile phones, which will necessitate consideration of the presentation of and level of detail in notifications. An alternative or indeed additional approach to keeping users informed of other useful material would be the inclusion of a recommender system in some form, enabling immediate and context sensitive notification of items of potential interest to the user.

The qualitative results from the organisational acceptability phase are presented as data towards further research into intra- and inter-organisational acceptability management of systems (for example, as a case study, as data for a meta-study, or as observations to inform a different research framework). They contribute towards writing statements that can contribute to making a business case (see Simon, Oberhumer et al 2007) for a repository at a local level and a distributed repository across institutions. We believe it is important to consider the variety of statements, as they may not apply equally across all parts of a single institution, never mind across different institutions, as these organisational domains may likely have different agendas and resources, notwithstanding any general goodwill towards such projects.

Returning finally to the international dimension, there are clearly interesting and challenging issues for such repositories, beyond those such as technical interoperability and scalability, for example the extent to which digital libraries can cross linguistic and cultural boundaries.

Acknowledgements

We thank JISC for funding the EERN phase and Bridges-CETL (Centre for Excellence in Teaching and Learning) at the University of Bedfordshire for funding the UBERN phase. Our thanks also to the many people involved with software development, technical support, and survey participation.

References

Bessis, N., Norrington, P.: Assessing the Acceptability of a Distributed Learning Resources Repository Catalogue for Computer Based Assessments. In: Proc. 7th IASTED Int. Conf., pp. 160–165 (2008)

Bevan, N.: Measuring Usability as Quality of Use. Softw. Qual. J. 4(2), 115–130 (1998)

Bloom, B.S. (ed.): Taxonomy of educational objectives: The classification of educational goals. Longman, New York (1956)

Brooke, J.: Software Usability Scale (SUS): a "quick and dirty" usability scale. In: Jordan, P.W., Thomas, B., Weerdmeester, B.A., McClelland, A.L. (eds.) Usability Evaluation in Industry, Taylor & Francis, London (1996)

Bull, J., Danson, M.: A briefing on computer-assisted assessment. LTSN Generic Centre – Assessment Series No 14. Learning and Teaching Support Network, York (2004)

Scalise, K., Gifford, B.: Computer-Based Assessment in E-Learning: A Framework for Constructing "Intermediate Constraint" Questions and Tasks for Technology Platforms. J. Technol., Learn., and Assess. 4(6) (2006)

Carroll, J.M.: Making Use: Scenario Based Design of Human Computer Interaction. MIT Press, Cambridge (2000)

Cliff, P.: RDN Resource Types (2002), `http://www.rdn.ac.uk/publications/cat-guide/types/` (accessed September 20, 2010)

Deacon, D., Bryman, A., Fenton, N.: Collision or collusion? A discussion of the unplanned triangulation of quantitative and qualitative research methods. Int. J. Soc. Res. Methodol. 1, 47–63 (1998)

DELTA Expert Evaluation. Project Deliverable. University of Essex (2005), `http://www.essex.ac.uk/chimera/delta/DELTAwalk-thruReport.pdf` Accessed September 20, 2010)

DELTA Homepage (2005), http://www.essex.ac.uk/chimera/DELTA/index.html (accessed September 20, 2010)

EERN. JISC Project & Quality Plan – EERN, Issue 1.0. JISC (2005), http://www.jisc.ac.uk/media/documents/programmes/distributedelearning/project_quality_plan_eern_1.2.pdf (accessed September 20, 2010)

Fowler, C.: East of England Educational Resource Network - EERN Final Report. 31 University of Essex (2006a), http://www.essex.ac.uk/chimera/projects/EERN/Evaluationreport-final.pdf (accessed September 20, 2010)

Fowler, C.: East of England Educational Resource Network - EERN Final Report. University of Essex (2006b), http://www.jisc.ac.uk/media/documents/programmes/distributedelearning/eernfinalreportdec06.doc (accessed September 20, 2010)

HEA. A guide on the teaching method of Problem Based Learning. Higher Education Academy, York (2003), http://www.materials.ac.uk/guides/pbl.asp (accessed September 20, 2010)

HEA. The UK Professional Standards Framework for Teaching and Supporting Learning in Higher Education. Higher Education Academy, York (2006), http://www.heacademy.ac.uk/assets/York/documents/ourwork/professional/Professional_Standards_Framework.pdf (accessed September 20, 2006)

Heron, J., Reason, P.: The Practice of Co-operative Inquiry: Research with rather than on people. In: Reason, P., Bradbury, H. (eds.) Handbook of Action Research: Participative Inquiry and Practice, pp. 179–188. Sage, London (2001)

Hildreth, P., Kimble, K., Wright, P.: Computer mediated communications and international communities of practice. In: Proc. of Ethicomp 1998, The Netherlands, pp. 275–286 (1998)

Hildreth, P., Kimble, K., Wright, P.: Communities of practice in the distributed international environment. J. Knowl. Manag. 4(1), 27–38 (2000)

JISC. JISC Project & Quality Plan - EERN. Issue 1.0. London, Bristol, JISC (2005), http://www.jisc.ac.uk/media/documents/programmes/distributedelearning/project_quality_plan_eern_1.2.pdf (accessed September 20, 2010)

JISC. EERN: The East of England's Educational Resource Network. London, Bristol, JISC (2007), http://www.jisc.ac.uk/whatwedo/programmes/programme_edistributed/eern.aspx (accessed September 20, 2010)

Maillet, K.: EducaNext A Service for Knowledge Sharing. The Universal Consortium (2003), http://internetng.dit.upm.es/ponencias-jing/2003/Educanext-Katherine%20Maillet.pdf (accessed September 20, 2010)

Massart, D.: CELEBRATE's Lessons, in 'What Went Wrong?' In: Duval, E. (ed.) Works. at the European Conf. on Technol-Enhanced Learn (EC-TEL), Crete, Greece (2006), http://fire.eun.org/wwwCelebrate.pdf (accessed September 20, 2010)

Mayes, J.T., Fowler, C.J.H.: Learning technology and usability: a framework for understanding courseware. Interact with Comput. 11, 485–497 (1999)

McCormick, R., Scrimshaw, P., Li, N., Clifford, C.: CELEBRATE Evaluation Report Version 2. Brussels, European Schoolnet, http://celebrate.eun.org/eun.org2/eun/Include_to_content/celebrate/file/Deliverable7_2EvaluationReport02Dec04.pdf (accessed September 20, 2004)

Rice, R., Burnhill, P., Rees, C., Robertson, A.: Repository Junction and Beyond at the EDINA (UK) National Data Centre. In: Kovács, L., Fuhr, N., Meghini, C. (eds.) ECDL 2007. LNCS, vol. 4675, pp. 543–545. Springer, Heidelberg (2007)

Rosson, M.B., Carroll, J.M.: Usability Engineering: Scenario Based Development of Human Computer Interaction. Academic Press, San Francisco (2002)

Scott, J., Fowler, C.: DELTA User Guide, Issue 1.3. University of Essex, 28 (2006), `http://www.essex.ac.uk/chimera/EERN/User%20Guide-Issue%201.3T.pdf` (accessed September 20, 2010)

Simon, B., Oberhumer, P., Kristöfl, R.: The Everlasting Dawn of Educational Brokers - A Search for Key Design. In: Proc. 1st Int. Works. on Learn. Object Discovery & Exchange (LODE), pp. 16–23 (2007), `http://ceur-ws.org/Vol-311/paper03.pdf` (accessed September 20, 2010)

Someren, W.M., Barnard, F.Y., Sandberg, A.C.J.: The think aloud method. A practical guide to modeling cognitive process. Harcourt Brace & Company, London (1994)

UoL. Education Strategy 2005-08. Luton, University of Luton (March 2006)

Venturi, G., Bessis, N.: User-centred Evaluation of an E-learning Repository. In: Proc. 4th Nordic Conf. Hum-Comput. Interact (NordiCHI), pp. 203–211 (2006a)

Venturi, G., Bessis, N.: Work package 5: Evaluation report. University of Luton, Luton (2006b)

Venturi, G., Bessis, N., Mayes, T.: Evaluation Plan of the EERN Pilots, Version 1.2. JISC (2006)

Vrasidas, C., Retalis, S.: Evaluation of the UNIVERSAL Exchange for Pan-European Higher Education. In: Proc 4th Panhellenic Congr. with international participation "ICT in Education" (Athens), pp. 627–634 (2004), `http://www.etpe.gr/files/proceedings/filessyn/A627-634.pdf`(accessed September 20, 2010)

Wilson, S., Blinco, K., Rehak, D.: An e-Learning Framework: A Summary. Presented at Altilab 2004-ELF, Redmond (July 15, 2004)

Wolpers, M., Grohmann, G.: PROLEARN: Technology Enhanced Learning and Knowledge Distribution for the Corporate World. Int. J. Knowl. Learn. 1(1/2), 44–61 (2005)

Facebook as a Collaborative Platform in Higher Education: The Case Study of the Universitat Oberta de Catalunya

Carles Fernàndez and Eva P. Gil-Rodríguez

Learning Technologies Office, Universitat Oberta de Catalunya, Av. Tibidabo 47
08035 Barcelona, Spain
{cfernandezba,egilrod}@uoc.edu

Abstract. The emergence and huge success of 2.0 applications on the internet and, more specifically, of social networks, has led to new questions being asked as to the potential of the latter as learning platforms. Social networks are an important space for sharing learning resources and an opportunity for the collective construction of knowledge. Several authors consider that the learning acquired from participating in social networks does not follow the parameters associated with onsite education, or with what we can now begin to call "traditional e-learning". Therefore, the need presents for considering the characteristics of this 2.0 learning: Do social networks provide a real learning opportunity? In what way and what are its characteristics? The purpose of this chapter is to offer an initial approach to the changes in the methods and practices of academic staff and students when learning using these social networks, specifically Facebook.

Keywords: 2.0 learning, social networks, open social learning, edupunk, connectivism.

1 Introduction

The purpose of this chapter is, as we mentioned above, to offer an initial approach to the changes in the methods and practices of academic staff and students when learning using social networks as Facebook, through the analysis and discussion of a specific learning experience which the Open University of Catalonia has carried out on this specific social network.

We shall begin, therefore, by describing the first educational experience carried out by the Open University of Catalonia (UOC) on the Facebook social network. It is worth mentioning that there are some, although few, previous experiences of the application of Facebook in the university education context (which are described in detail in the state of the art). In comparison to these previous experiences, ours

T. Daradoumis et al. (Eds.): Technology-Enhanced Systems and Tools, SCI 350, pp. 27 – 46.
springerlink.com

was based on creating a free, fully on-line course on the subject “2.0 Travel”, based on an open learning methodology (described in point 5 of the chapter) and on the use of open education resources, more specifically with a Creative Commons (CC) licence. The actual characteristics of the course required collaboration between researchers and technical personnel of several of the institute's teams, including the content management department and the technological area. The *Facebuoc* project, which is the name that was finally given to this experience, forms part of the actual research and innovation dynamics of the university using new technologies for learning and is a pioneer learning initiative within the university setting, in that the Facebook platform has been used as a virtual basic learning environment for one of the courses offered by the Open University.

2 State of the Art

It is true that new ideas and experiences have started to emerge with regard to how to use social networks in general and Facebook in particular to the benefit of learning and with educational objectives. Despite the fact that it is still difficult to find scientific research regarding the effectiveness of Facebook as a learning technology, several web resources show how, with a level of variable formality, faculties are threshing out specific uses for the tool and applying them to the classroom. We can find several representative examples of these uses on the web [1] [2] [3] [4].

A mention should be made, apart for being pioneer and for its great depth, of the work by Alejandro Piscitelli [4] at the University of Buenos Aires, which we will explore more, later. Other universities, particularly in the USA and Canada, have started to integrate Facebook as a platform for educational purposes of a different type. The University of Michigan, for example, uses Facebook to publish news and connect its students, Stanford uses the social network to share its research data and even for its lecturers' tutorials, or as a repository for its learning resources, Florida University makes similar use of it, etc.

In many cases, these initiatives have come about voluntarily and in a fairly non-institutionalised way, mainly thanks to the efforts of lecturers and study groups, particularly at primary and secondary levels of education. This has conditioned the fact that in-depth research on the educational impact of the social network has not been able to be carried out. In our case, we have carried out an experience that not only intended to be useful, practical and interesting for students, but also aimed to analyse the impact of Facebook on open learning methodologies.

2.1 Open Social Learning and 2.0 Social Networks

We do not intend to go over the origins and generalisation of the social network's concept and practice, but to focus on the use of social networks in education. As an extremely practical example, Juan José de Haro [5] makes a brief synthesis of the main uses of these networks in the learning context, highlighting aspects such as the centralisation of learning activities, the sensation of learning community, increase in the simplicity of communication, ease of team work, etc.

The breakthrough of the term web 2.0 is closely related to the significant increase in the use of social networks. Initially coined by Tim O'Reilly in 2004, it has most probably been overused with the aim of providing certain activities, applications, training courses, etc with special and superior characteristics. Although its definition suggests many generalisations, the majority of authors consider it to include a series of characteristics that can be explained in detail. According to Anderson [6] web 2.0 is related to six ideas: Individual production and User Generated Content, Harness the power of the crowd, Data on an epic scale, Architecture of participation, Network Effects and Openness.

Although many tools prior to 2004 already included some of these characteristics, in general we can state that there is a paradigm shift 'web 1.0 to web 2.0', which can be very clearly appreciated over the last few years: social networks have become the general and daily use applications for the majority of users, and in general have focused on their social or social interconnection potential and the transfer of power to the actual users in the generation of content and services.

However, and although the majority of these web 2.0 tools weren't created for learning purposes, their potential and obvious use in this field has meant that at both an individual and an institutional level a large part of the learning community has considered using them. The key question is, to what point have we known how to apply the merits of the 2.0 paradigm, such as the importance of authorship of the actual students, aspects of collective intelligence, a more horizontal knowledge management, etc. Antonio Bartolomé [7] cites specific elements of this paradigm that should be integrated in a new e-learning model, such as the use of the whole network as a potential learning platform, collective intelligence, tags, multiple devices and the syndication of content. The reflection as to whether we are already in an "e-learning 2.0" creation context is complex, although broadly speaking we all sense that our education systems which are more static and slow to integrate new technologies have not facilitated the emergence of more up-to-date learning models.

Either way, it is clear that since the proliferation of 2.0 tools there has been a significant increase in the number of learning actions aimed at the social construction of knowledge through the participation, interaction, collaboration and use of collective intelligence of internet users. Of the possibilities that these environments convey is the theoretical concept of "open social learning" (OSL) [8] [9] [10]. Its progressive exaltation has caused a great stir in the foundations of learning theories, creating a debate as to the suitability of open social applications for teaching and learning.

Connectivism as a learning theory explains that "learning is a process that occurs within nebulous environments of shifting core elements, not entirely under the control of the individual" [11]. Under this assertion, the key for learning is to establish connections between learning resource nodes that the student can access. These connections are more relevant than the learning that the student has accumulated until now. Informal learning, undervalued in other theoretical trends, becomes an essential aspect of the learning experience, as the nodes for establishing connections are present in multiple contexts, from more or less formal communities, to actual networks or daily events that surround the student. This is an alternative theory, complementary to constructivism, which aims to adopt aspects of the brain's neurological function and even of the theory of chaos to learning, which has a new status in a highly complex knowledge society, full of crypticism. In the words of Siemens himself, "our learning and information acquisition is a mashup. We take pieces, add pieces, dialogue, reframe, rethink, connect, and ultimately, we end up with some type of pattern that symbolizes what's happening "out there" and what it means to us" [11].

The specific impact on the learning activity leads to reflections such as, up to what point does the validation of too many substitute that of the expert, or if the actions of the group are more effective than a more active teaching role, then certification should be awarded based on more collective or horizontal criteria, for example. A real challenge, particularly for higher education institutions that traditionally cling to a more hierarchical, structured and parcelled form of knowledge. Again according to Siemens, Open Social Learning therefore relates the effectiveness of learning with "an instructor to focus less on lecturing and content presentation and more on assisting learners in creating personal learning or knowledge networks" [11].

Of course, criticism has arisen of various aspects of the Open Social Learning model from the connectivism movement. Jon Dron [12] suggests that the structure generated by social tools may not be educationally appropriate, as could occur in the case of gigantic networks that complicate the emergence of smaller spaces where knowledge could create difference of opinion (and therefore, wealth). Not to mention that these networks "are susceptible to intentional attack, whereby a malevolent or mischievous individual or group can bend the system to its purposes". Kischner and other researchers, on the other hand, point out that a group of students can only administer a competent guiding role to the learning activity when they have a sufficiently high, previous level of learning [13].

2.2 Uses of Facebook in Higher Education: Background

With regard to properly documented learning experiences with Facebook, to date we have only been able to access a few. Following is a brief review of the most relevant projects with regard to educational uses of this platform.

The Facebook Project, headed by professor Alejandro Piscitelli [4] is one of the pioneering and most famous examples of use of this social network for educational purposes. For this idea, hundreds of Communication Sciences Degree students from the University of Buenos Aires (Argentina) used Facebook as a platform for collaboratively creating audiovisual products.

Another, recently published experience refers to Computer Science studies at the universities of Newcastle and Durham [14]. Collaborative groups were created to develop software and it was observed how the students naturally opted for the social network Facebook instead of other proposed, collaborative technologies such as email, wikis, Skype, forums, etc.

The development of specific applications that use Facebook is one of the platform's keys to success and one of its possibilities. In this regard there are innumerable possibilities for use, given that many of these applications were not initially created for educational purposes, but creativity places few limits on educational adaptation. José Mª Villatoro [15] has made a fairly exhaustive list of some of the more interesting applications, among which are those for sharing books, work and research articles, taking notes, creating study groups, creating calendars, recording videos for students and creating courses based on predefined templates.

The distinction between the use of Facebook as an educational tool and the use of Facebook as a LMS, as in the core element for the teaching-learning activity, is also important. Although for the most part Facebook is being chosen as an additional tool within the educational process, there is the possibility of transforming it into a LMS, integrating administrative and content creation possibilities. For example, Udutech [16] is an application that adds these functions to the original Facebook platform.

To explore the specific uses, applications, benefits and advice for teaching staff in more detail, we recommend visiting a complete resource created by Onlinecollege.org [2] and Collegedegree.com [3]. We will not enter into a more detailed assessment of more social aspects which, although related and having a clear learning impact, go beyond the objectives of this chapter. For this, we recommend the University of Michigan's paper on the social capital of Facebook [17].

3 Objectives

In the case of the Open University of Catalonia, the idea arose from the continual motivation and need of the university's own teaching and teaching support community. The recognition of Facebook, having brought together more than 300 million users on its platform, made us consider the possibility of using this social network to develop more open, learning actions than those generally offered by the university's programmes. On the other hand, the observation of how millions of users used the social networks on a day to day basis to satisfy their own learning or knowledge needs in an informal way, offered us the ideal breeding ground for bringing together many users with very high learning interests and motivation for learning. In short, the idea was to propose a technological means for learning and a methodology that were more in line with what our potential students use day to day.

All this with the objective of proving that learning can be enormously promoted if we design educational methodologies that are closer to open social interaction, with a direction and power that falls more to the actual users and using tools that they are accustomed to using.

On analysis of the experience and considering the object of the study, qualitative and quantitative information has been observed and gathered, using an ethnographic methodology, from contributions carried out during the development of the learning action. This information has served as a basis for approaching new teaching and learning modes using these networks, obtaining results relating to the roles assumed by the lecturer and students, as well as the interaction characteristics between the participants.

4 The Learning Methodology

To better understand our experience it is important to understand the actual context of the Open University of Catalonia. UOC is one of the few universities to offer 100% on-line learning programmes, therefore the course offered would follow this principle. In addition, we did not consider a specific subject of one of the UOC's programmes for which certain activities could be carried out on Facebook, but a short course which would be entirely carried out on Facebook. Therefore, for this case study a learning experience was designed which took into account the open and informal characteristics of the Facebook social network: the course would be short and for free, would include subjects related to 2.0 internet user experiences, have educational content with Creative Commons B licence and be open to all Facebook users (regardless of whether or not they were students of the Open University of Catalonia).

A 5-week learning experience was designed, which began on 5 October 2009, on the subject: "2.0 Travel: on-line tools and resources", using two "standard" Facebook groups as learning spaces, one for the course in Spanish and one for the course in Catalan[1]. The objective of the 2.0 Travel course was for each student to become a reporter of their own journeys using the existing web 2.0 tools. To publicise the experience, as well as traditional university marketing mechanisms, viral marketing on the social networks was also used, which included quiz type applications and informal invitations to users linked to a profile, created to manage courses. 89 pre-enrolments were received, all of which were accepted, and therefore, a total of 89 participants were registered on the course, 52 on the Spanish edition and 37 on the Catalan edition, figures which were considered more than reasonable for a pilot experience.

The teaching methodology was completely open, developed based on opening forums and making them dynamic, opened either by the lecturer or by the students, on different subjects relating to the content of the course, arising during discussion. Making connections and creating mash-ups as products (basically as blogs) are the basis of the learning methodology. In this context, informal learning obviously becomes the main space where learning is produced.

Horizontality in the interaction and participation is without a doubt one of the main keywords of the methodology, and also the most challenging. Changing the traditional verticality which fills our education systems is not simply a matter of

---

[1] The courses can be accessed via the following website links: Spanish course http://www.facebook.com/inbox/?drop&ref=mb#/group.php?gid=190864356112 Catalan course http://www.facebook.com/profile.php?ref=name&id=1626712779#/group.php?gid=145066908847

methodological design, but is vital and requires a set of decisions and actions (particularly teaching-related) which promote these less hierarchical relations. This mainly implies a change in the role of the teaching staff, who under our terminology become facilitators (or even non-lecturers as we sometimes say) for the community. Under this methodology, the lecturer proposes, but does not impose; suggests, but does not give definitive answers; assesses but does not explain everything that must be done; traces possible paths, but does not mark a single route to achieve objectives. The ultimate objective is that the community self-manages its own learning process and understands that the lecturer is another resource, for a specific speciality, to enrich this process.

Therefore, and in order to understand the practical implications of our design, we included a facilitator, previously trained in these methodological aspects, who proposed forums, different forms of working, who actively participated without becoming the main character, who offered students the possibility of opening and managing discussion forums themselves, who provided a multitude of web resources for cooperating in the learning activities, who interwove the experiences of the students to promote a more cooperative work, who encouraged a higher participation limiting their own, who proposed practical and open activities based on which the students could use their previous experiences, etc. As suggested by Kischner [13], we do not offer students the full administrative and guiding role of the experience. We have to assume that this is the first time for most of these online learners to receive a training course on such a new Facebook-based methodology.

5 Assessment Methodologies

In order to obtain an approximation as close as possible to the results of our Facebook experience, we chose to use our own qualitative and quantitative methodologies. With regard to the qualitative methods, we based them on the use of an ethnographic study to thoroughly explore the most relational aspects of the research, in our case interaction, participation and cooperative work. This more qualitative part was complemented by the opinion of experts on social networks and on-line education. With regard to the research's quantitative dimension, we based this on the analysis of basic data for the course (objective participation) and on the analysis of an extensive survey which was sent to all the students who participated.

Therefore, the participation, interaction and cooperation analysis has been made from two points of view: that of the student, taken from a questionnaire which included items regarding various indicators and, for the purpose of exploring the relational network in more detail, we used the ethnographic method.

We have placed emphasis in the analysis of interaction, cooperation, technology and in general the behaviour of users within the system. In contrast, we have not focused that much on final learning outcomes. Obviously, learning is essential and the experience would not make sense without it, but we haven't gone in depth on the analysis of learning as much as we have on the overall methodology.

5.1 Analysis of the Experience at a Quantitative Level

With regard to the questionnaire sent to the students, which combined Likert scales with open questions, following are the indicators which were included:

Table 1 Indicators.

Indicator	
Methodology	Communication with fellow students via FB is very easy and efficient
	Communication with the expert/non-lecturer is very easy and efficient
	The methodology used has helped me to learn
	The dynamic established has helped me to participate
	The dynamic established has helped me to follow the subjects dealt with
	The methodology has enabled me to adapt to the non-course according to my needs
	The pace of the presentation of different subjects on the forums was appropriate
	The non-lecturer motivated me to participate and follow the activities
	The participation of other users motivated me to participate and follow the activities
	The group activities were suitable for the subject work
	The freedom that the proposed methodology provided to the non-course favoured my participation
Facilitator role	How do you value the role developed by the expert/non-lecturer? (in this case, on a scale of 1-10)
	The contributions of the expert/non-lecturer were adequate for following the non-course
	The contributions from other users have helped you follow the non-course
	When you requested help you received a response from the non-lecturer
	The non-lecturer role is necessary for the correct development of the non-course
Learning activities	The learning activities were suitable for the non-course objectives
	I have enjoyed doing the proposed activities during the non-course

Table 1 *(continued)*

The tool / social network	The social aspect of FB favours participation on the non-course The interface is suitable for carrying out the non-course Access to the course from my profile is easy The characteristics of Facebook favour my motivation to learn
Your participation and that of others	Assess your participation in general on the non-course (Very high, low,... NEVER). If you respond NEVER please give reasons for not participating. Did you participate in the group activities? (Very high, low,... NEVER). If you respond NEVER, please give reasons for not participating. When you requested help you received a response from other non-course users I have helped other users with my participation when they have requested it I have learnt by reading the messages on the non-course's different spaces, both from the non-lecturer and the other users I have learnt from participating on the different forums and activities of the non-course
Course philosophy	The open character of the non-course is a key factor for my participation My expectations at the beginning of the non-course have been met
Overall satisfaction	What is your overall evaluation of the UOC's Facebook course? On a scale of 1 to 10 What have you most liked about the course? Close with certain items: subject matter, environment, non-lecturer, non-methodology, materials, fellow students,... and leave an open question What have you least liked about the course? Close with certain items: subject matter, environment, non-lecturer, non-methodology, materials, fellow students,... and leave an open question Make an overall assessment of your experience (open question)

A total of 22 students responded to the questionnaire. The evaluation of each one of the items would excessively draw out this chapter, therefore we have included a series of items we consider key and a synthesis-type assessment of each one of the indicators.

5.2 Qualitative Approach: A Virtual Ethnography in the Facebook Context

To analyse the interaction developed on the forums of the different courses, virtual ethnography was used as study methodology. Virtual ethnography is based on "a researcher submerging themselves in the world being studied for a specific time and taking into account the relations, activities and significances forged between those participating in social processes", in the internet context [18]. Despite there being various studies which use virtual ethnography as methodology for extracting data, no research has been found that tries to analyse the social learning phenomenon within a virtual community. In general, research that uses virtual ethnography focuses more on analysing aspects relating to social interactions and the construction of a community. Also, this research has focused more on virtual communities which are not included within the scope of education.

The only research found within a learning scope [19] (Bielman, V., Putney, L., & Strudder, N., 2003) describes how the participants' interaction in an on-line class build the social culture of a distance learning class. Another study [20], analysed the interaction between lecturers and learners on a Masters Degree in Education delivered on-line. Interactivity, adaptability, discursivity and reflectivity categories, created by Laurillard [21], were used for virtual ethnography. Among the conclusions reached were the advantages provided by asynchronous communication for reflecting on answers, the creation of personal support communities beyond the purposes of the course. Disadvantages included the actual technology and aspects such as access, equito and support issues.

Thus, the majority of research that uses virtual ethnography focuses on describing an aspect of a specific virtual community. For example, in Chan, AHN (2009) [22] the function of on-line and off-line identity was examined in a group of working mothers in Hong Kong who were users of a virtual community. In Goodsell, TL (2008) [23], forms in which the communication and support within the on-line and off-line community were developed during a period of time were identified. In Skageby, J. (2009) [24], how Facebook users used social metadata such as gift-giving, "I like this" clicks, etc to enrich interpersonal communication was studied. In Skageby, J. (2008) [25] a photo-sharing network is described. In Lemai, Nguyen; Luba, Torlina; Konrad, Peszynski; Brian, Corbitt, (2006) [26] the social interactions, shared values, belonging, commitment, loyalty and in particular the importance of power relations are explored in two Vietnamese communities.

Another field that has also been extensively studied using virtual ethnography are multi-player, on-line games. The social and social interaction aspects of this type of games are also mainly focused on in these papers (Nicolas, Ducheneaut; Robert, JM.; Eric, Nickell, 2007 [27]; Moore, RJ (2009) [28].

Given the small amount of existing research in the learning field which uses ethnography, application of the inductive perspective was chosen, as according to the Grounded Theory [29], in which the hypotheses result from the data and not the other way around. The first step consists in preparing observation categories which will enable all of the events that occurred during the course to be gathered in as much detail as possible. The creation of these categories was based on the experience of the research team as open observer and non-participant in course forums through a brainstorming session. Ultimately, the results were summed up in three main categories: type of content given on the forums, reason for intervening and social or peer-to-peer learning. The corresponding results for these three categories are summarised in the following section.

6 Results

First of all, we present a series of basic data relating to participation:

Table 2 Participation of students in the different course activities.

Spanish edition	Catalan edition
52 users joined	37 users joined
14 participated	13 participated
12 created a blog	8 created a blog
8 blogs	4 blogs
9 discussion forums	7 discussion forums
155 posts in total	87 posts in total

Broadly speaking we can see that participation in the Spanish version of the course was slightly higher than in the Catalan edition. There is no clear conclusion on the matter, although the fact that UOC is a well-known university in South America and that the majority of the Spanish edition participants were from this continent provides a reasonable explanation.

What is immediately notable in Table 2 is that only 1 in 3 or 4 students that enrolled on the course participated actively. A student is considered to have participated actively when s/he has created a blog to report his/her travel experience and has also participated various times in the discussion forums. Therefore, one of the key questions we asked of our Facebook project is, why did only 25-35% of enrolled users commit to the course? In order to investigate this point, we created a special questionnaire which we sent to users who didn't actively participate, in order to discover the reasons for their early abandonment.

In summary, the non-participation analysis instrument gave the following five reasons for abandoning the course:

Table 3 Reasons for abandoning the course.

Reasons	Users (%)
"In the end, I didn't have enough time"	Almost 80%
"I was a bit lost within the Facebook structure"	25%
"I didn't want such a 'self-learning' methodology, I felt it was too open"	20%
"I didn't know when the course started – no notifications"	15%
"Personal issues"	30%

Our general non-participation analysis led us to reflect on the commitment of the level of implication of students with an initiative that did not pose formal obligations and which was also free. In fact, the level of commitment of any Facebook user is highly debatable. For example, Facebook as a social network enables the creation of groups with social change ideology at various levels, and it is simple to register on these groups, but the action commitment of so many thousands of users that these groups normally corner is relatively low. Our Facebook course opened an inscription process that did not demand anything of the potential students except to complete a simple survey. The fact that finishing the course or completing all of the activities was not compulsory and that UOC did not offer an official diploma for doing the course, facilitated the abandonment of all those users who did not have a strong commitment to the experience.

On the other hand, other specific reasons that the students mentioned for abandoning the course, must also be considered. Some stated that they did not understand the structure of the course, despite the Facebook groups working on a very simple structure, although somewhat different to other on-line courses. Even more interesting is the fact that some students expected a more guiding methodology, and decided to abandon the activities when they realised that the course facilitator did not play the role of classic lecturer, proposing all of the course activities, correcting them and ultimately driving the development of the course in a very specific way. What becomes obvious therefore is that proposing a learning activity on a network characterised for its informality, horizontality and power of the actual communities, does not guarantee that the user-student will assume this philosophy in his/her participation on an on-line course on Facebook. Finally, we must also point out that the lack of automatic notifications of the Facebook groups themselves led to some students forgetting the course start date and discovering that it had already begun when they showed an interest in it.

With regard to the remaining participation indexes, in total the students created 12 travel blogs (some individual and other collective) and opened 16 discussion forums on various subjects related to the course.

With regard to methodological aspects, the freedom perceived by the students was one of the aspects that we were most interested in with the activities on Facebook. A methodology based on the participation of the student community, self-management of the actual learning and horizontality aims to promote a level of motivation that leads the students to participate more freely and spontaneously. In this regard, we

Table 4 Results Items

Items	Edition	1*	2	3	4	5**
The degree of freedom increased my participation (%)	Catalan	0	0	17	50	33
	Spanish	8	0	0	26	66
I had fun participating in the activities (%)	Catalan	8	0	16	8	68
	Spanish	0	0	8	0	92
The role of the teacher is necessary in this course (%)	Catalan	0	0	0	15	85
	Spanish	0	0	20	0	80
The social dimension of Facebook enhanced my participation (%)	Catalan	0	12	0	24	64
	Spanish	8	0	16	0	76

*1 = I completely disagree.

** 5 = I completely agree.

posed the following item on the course evaluation questionnaire for the students, with a very positive result for both the Spanish and the Catalan editions.

Enjoyment, understood to be a key aspect for encouraging the intrinsic motivation, obtained similar results.With regard to self-assessment of course participation, the students on the Spanish version considered their participation to be higher than the Catalan version. This subjective perception is confirmed if we look at the number of participants on discussion forums on both courses.

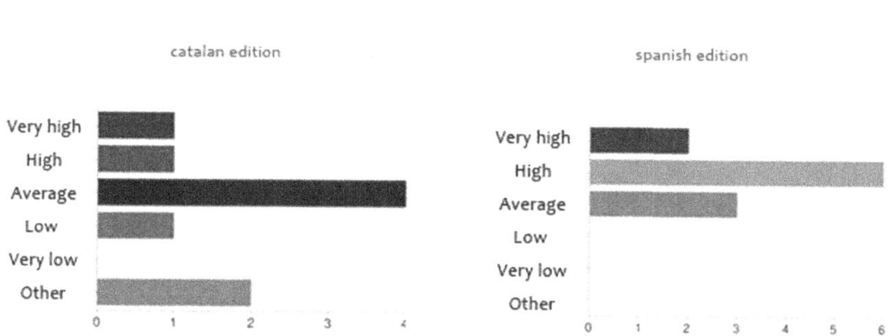

Fig. 1 Results item: *Assess your own participation in the course.*

Similarly, the students on the Spanish edition considered that they had spent more hours a day on course activities.

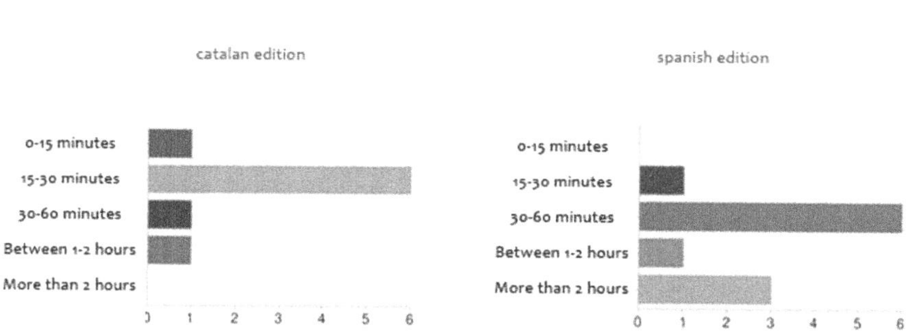

Fig. 2 Results item: *How many hours a day did you dedicate to the course?*

In the initial reflection phase on the design of a Facebook course, we considered whether it was necessary to have a teaching figure with a non-guiding role, or if we could simply do away with it. In short, we tried to respond to the question "up to what point can a student community self-manage itself from the start of the development of the learning process". Is this possible without a teaching figure? As we have mentioned in previous sections, our decision was to include an expert (in both course content and in on-line education), limiting the role to basically facilitating the community. Once the course had finished we asked the students if they thought that this teaching figure was really necessary for developing a course on Facebook. The answer is clear, and a large majority considers that this teaching role limited to facilitating the community was essential. With regard to Facebook as an actual platform, the majority of students concluded that the social dimension of the platform had facilitated their high level of participation greatly.

Finally, we also included a quantitative-type, overall rating that the students gave for the course.

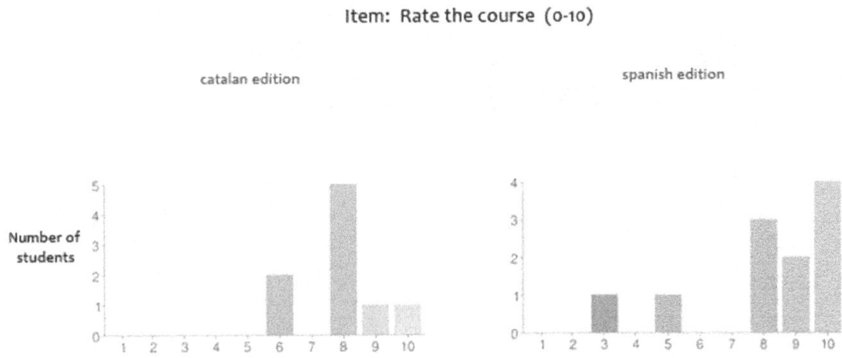

Fig. 3 Results item: *Rate the course (0-10).*

Although the quantitative evaluation presented provides general data as to the success of the activity, the open opinions given by the students and the qualitative analysis of how the course worked are in fact the main elements obtained for exploring the keys of the course in more detail.

With regard to the students' qualitative evaluations, undoubtedly the most common was the statement that it was a very enjoyable way of learning and therefore a clear source of motivation. However, some users mentioned that at the start of the course they felt unsure of the methodology, as they had never learnt in this way and they felt that it could be an unsuitable way to quickly assimilate content. It is precisely in relation to time where we find the main criticism of the course: "five weeks is not enough", was the main recommendation found on the evaluation questionnaires. A fact that is very relevant if we think that an initial adaptation period was necessary for the course methodology and that it was as of this methodological understanding when the students began to participate actively. Five weeks is clearly insufficient, even for habitual users of Facebook, if we want to make the most of the participation potential of an open method using this platform. With regard to participation, certain students pointed out that they didn't like the fact that the majority of their course colleagues who initially enrolled on the course ultimately did not participate. With regard, specifically, to the platform, some students mentioned the lack of notifications from the Facebook groups as a limitation, as they are obliged to continually visit the group if they want to be up-to-date with the discussion forums.

Continuing with the reflections at a qualitative level, we must also point out that the ethnographic approach enabled us to explore in more detail, the form of interrelation in the learning processes carried out on the platform. As we have previously mentioned, results, grouped in three categories, were obtained: content-type, reason for contributions and social or peer-to-peer learning.

With regard to type of content, in general, the content provided on the forums was informative. Despite being an open space, other contributions were not made to the course subjects and therefore we can conclude that the students did not misuse the open model and maintained an acceptable level of quality in their contributions. With regard to the emotional tone of the content, in general it was neutral, as what was communicated on the forums was normally information regarding the course's actual objectives. This fact did not prevent contributions being made that had a positive emotional tone, from both the lecturer and the students.

With regard to the reason for contributing, and in light of the analysis made, it is clear that the main reason for contributions from students was to respond to the subject opened by the lecturer. Even on a forum created by a student, the lecturer ended up focusing all of the subsequent contributions with one single contribution. Similarly, the inclusion of links in different comments by the students also always responded to the subject proposed by the lecturer. In this sense, the reputation a priori of the lecturer continues to give a hierarchical structure to contributions even on such an open environment as Facebook, strengthening the perception of quality of the experience but, to the contrary, distorting its spirit. It is also true that the environment favours alternative behaviours to this centralisation of interaction on subjects proposed by the lecturer and one group of students contributed links on their own initiative. These complemented the forum subject and in some cases also enabled comments on other students' contributions to be completed. Another reason for contribution by students was to flatter, show thanks or enthusiasm, which we could say provided a social relation to the learning.

With regard to social or peer-to-peer learning, in contributions from students we could observe explicit and implicit elements that denote this type of learning within forums, for either: a) achieving objectives proposed, based on doubts resolved between various students, b) providing links to help explain something or to inform about something related to the forum subject, c) contributing new information relating to the subject under discussion, or d) displaying critical opinions of some webs 2.0. With regard to the forums opened by the actual students on this subject, these did not differ from those opened by the lecturer. Somehow, the student who opens the forum acquires a certain lecturer role, simply for having had the initiative of opening a forum on the course subject and also by thanking the other students for their contribution, lecturer-style: *"María, José, thanks for sharing those links related to the course objective. I think it's a good idea to get to know the sites to have a better idea of how travel cyber-journalism works on 2.0"*.

Finally, with regard to learning outcomes, we have focused on two products: the blogs and the interaction created over their construction (observable in the forum posts of the Facebook group). We can not really compare this course with the 'standard' version we offer to our regular online students, since the learning outcomes and activities are different. The quantity and quality of interaction has been evaluated previously in this chapter as being very valuable for the learners that finished the course. Similarly, all the blogs created by students met the requirements expected by the facilitator (some are far better than we expected) and the quantity of resources and mash-ups in these blogs would make them useful for other travellers interested in visiting such countries or just interested in having some information about it. More specifically, learning has been produced individually through some reflection on the questions presented in the platform together with the process of searching for resources for mash-up development. Discussion and cooperative work for the blogs construction have been the group ways to produce learning.

7 Conclusions

The conclusions we have reached are related to the reflections on the results obtained regarding the new modes of interrelation for learning processes on the Facebook platform, in light of the experience carried out. In this section, we have also wanted to conclude recommendations of a practical nature for possible learning projects using Facebook.

First of all, we should mention that although this case experience dealt with the implementation of an open and horizontal type of methodology, this does not mean that the roles adopted by the different participants were completely equal: in this regard, the forums analysed made it clear that **contributions from the lecturer conditioned those of the students**. In other words, the majority of student contributions were motivated by a prior contribution from the lecturer. This prior contribution was either the opening of a forum or a contribution to a specific forum. In both cases, the subsequent contributions from students were an answer or comment to what the lecturer had said. Even on the forum opened by a student, the lecturer's contribution half way through with a question that slightly changed the forum's original subject, led to subsequent contributions from students being responses to the lecturer's participation, instead of being contributions in reference to the original subject of the forum.

With regard to the roles adopted by students, we can also conclude that these are not absolutely equal: quite the opposite, **the influence of active users was relevant to making the forum activity more dynamic**. On both the Spanish and the Catalan course there was a group of 3 or 4 students who participated more than the rest of the students. On the Spanish course, these users were those who participated by asking questions or querying something, who responded to these questions or queries, who on their own initiative provided links to websites to help explain something and those who opened forums. In other words, they were proactive students when contributing information or making the forums more dynamic, whereas, in the case of the Catalan course, the students who participated most on the forums were more focused on organising and promoting the study groups' learning process for the blog. These users did not make more contributions in terms of questions, responses or website links, as occurred in the case of the Spanish course.

Basing the project on a social network-type platform with an open methodology did not guarantee that a social learning would occur and we found more or less examples along these lines according to the subjects proposed on the forum. In other words, there were forums whose objective or subject favoured contributions from students that were closer to social learning and not limited to simply being a contribution. The most obvious example of this was on the forum called "2.0 Travel Blogs" in which the lecturer suggested as an objective the creation of a group or individual blog, and on the forum "A photo, a tool", where the lecturer suggested that the students include a representative photo of a journey and a photo management or edition web tool. On both forums, the students provided information and website links under their own initiative to help others create the blog (in the case of the '2.0 Travel Blog') and to explain something related to the forum subject (in the case of "A photo, a tool" forum). Also, on the '2.0 Travel Blog' the students created links to their blogs so that the other students and the lecturer could make comments and so that all of the students could see what each other had done.

Below we also detail practical-type recommendations when dealing with learning experiences on the Facebook platform. The first of these and perhaps the most important is that corresponding to the fact that there is an intrinsic trend towards transforming the learning activity, "open" by definition, into an activity with a more "facilitator-centred", traditional structure. Despite designing an activity for which horizontality should be the basis of participation and interaction in general, and carrying out actions so that the course works according to this methodology, there is pressure from the community, which expects the teaching facilitator to lay down more guidelines, propose new activities, clearly structure in time what needs to be done, etc. In short, it is a tendency towards returning to more traditional teaching models and to applying our teaching values, against which a specific strategy should also be designed.

In line with the previous point, it is advisable that the courses are not too short in duration. The introduction of an innovative methodology requires dedicating a certain amount of time to understanding it and putting it into practice. Based on our experience with students who were active users of Facebook, 2 to 3 weeks is the minimum amount of time necessary to guarantee that the open methodology is integrated. In this regard, an important aspect to consider is the course enrolment: should we choose to offer the course to any user, regardless as to whether or not they belong to our teaching institute, we cannot forget that the level of commitment when registering on any Facebook group is low. Specifically, in our experience, between 60 and 70% of users who initially signed up ultimately did not participate. Therefore, it is advisable to consider that at least half of the students who have enrolled have done so due to a type of "being involved" or "belonging to something" phenomenon, but when the moment to actively participate arrives, they will not respond in the same way.

A special section would be the analysis of Facebook as a platform for learning. If the structure based on groups offered by Facebook is used, we can conclude that we are dealing with a simple and relatively lineal interface. However, it is a closed structure and one which scarcely permits any level of customisation or personalisation. In fact, the lack of flexibility is a very significant limitation which is obviously balanced out by other positive aspects of Facebook. However, another aspect that seems even more arguable regarding the use of this social network is the fact that it is a platform that is not owned by us. In short, we have no control over Facebook. Any modification or changes in the Facebook policy will have a direct and unquestionable impact on the learning actions we develop. A specific example is that our pilot experience coincided with changes in the structure of discussion forums and groups, which our learning community suddenly suffered with no prior warning being able to be given by the course organisers. On a more positive note, we must point out that the use of Facebook's API has led to the development of thousands of specific applications which in many cases enable certain educational uses.

In general, and with participation being one of the key objectives of the instructional design, we must consider the difficulty of facilitating the participation of students using an open methodology, unguided by the lecturer. In this regard, we should not assume that the motivation and interest in the subject matter will lead the students to participate. At the same time, the initial orientation given by the lecturer is a key factor for encouraging this participation. Once again, we must remember that the non-compulsory nature of following the on-line course causes different actions. On one hand, it leads to a significant number of students who after having enrolled, don't ever participate in the activity. On the other, it arouses a lot of interest and encourages enrolment. These matters must be taken into account during the planning and design of later experiences. We also reiterate that time, which is a key factor, notably influences the encouragement of student participation. As it is not a traditional learning proposal, a period of adaptation for the participants to the course methodology, is necessary. We did not include this aspect in the course design and, as we have been able to establish, it is a very important point.

In order to improve the previously mentioned points, clarification of the methodology is essential, which will guide the learning activity from the start, so that participants and facilitator adapt to it and are completely aware of the situation in which they find themselves for the development of their own teaching-learning process.

With regard to virtual ethnography as an analysis technique for our experience, we can confirm that its use has been of great value when providing relevant information on the communicative process in relation to the learning carried out on the forums. Similarly, we have also been able to establish without a doubt, that the complementary nature of methodological techniques (quantitative and qualitative) has been relevant to enriching the analysis and main conclusions which have been made.

There is no doubt that the margin of design and use of the learning activity using platforms such as Facebook and similar, is enormous. A large part of our results depend on our actual university context and on the specific methodological decisions we take. One of the main lessons we have learnt from this experience is that values such as horizontality and openness are not simple changes to the educational design but imply a special effort to be made along many different lines in order to change a 1.0 paradigm which is more ingrained in our teaching minds than we had thought. The main piece of advice: let's not be fooled, this is an enormous change that goes far beyond the typical actions such as proposing open learning resources, holding small teaching-learning sessions or using 2.0 educational platforms.

References

1. Martin Márquez, J.M.: Uso de Facebook en E-learning (2009), http://sandyday.wordpress.com/2009/05/19/uso-de-facebook-en-e-learning/
2. Online College.com, 100 ways you should be using Facebook (2009), http://www.onlinecollege.org/2009/10/20/100-ways-you-should-be-using-facebook-in-your-classroom/
3. College Degree.com, The Facebook classroom. 25 Facebook Apps that are perfect for Online Education (2009), http://www.collegedegree.com/library/college-life/15-facebook-apps-perfect-for-online-education
4. Piscitelli, A.: El proyecto Facebook. Una herramienta de trabajo de la Cátedra de Procesamiento de Datos. Carrera de Comunicación, UBA (2009), http://www.proyectofacebook.com.ar/
5. De Haro, J.J.: Las redes sociales en Educación (2008), http://jjdeharo.blogspot.com/2008/11/la-redes-sociales-en-educacin.html
6. Anderson, P.: What is web 2.0? Ideas, technologies and implications for Education. JISC Reports (2007), http://www.jisc.ac.uk/media/documents/techwatch/tsw0701b.pdf
7. Bartolomé, A.: Web 2.0 and new learning paradigms. E-learning papers 8 (2008)
8. Downes, S.: Introducing edupunk (2008), http://www.downes.ca/cgi-bin/page.cgi?post=44760
9. Groom, J. (2008), http://edupunk.org/
10. Siemens, G.: Connectivism. A learning theory for de Digital Age (2004), http://www.elearnspace.org/Articles/connectivism.htm
11. Siemens, G.: Handbook of emerging Technologies for learning. University of Manitoba (2009), http://umanitoba.ca/learning_technologies/cetl/HETL.pdf
12. Dron, J.: Designing the indesignable: social software and control. Educational Technology & Society 10(3), 60–71 (2007)
13. Kirschner, P., Sweller, J., Clark, R.: Why minimal guidance during instruction does not work: An analysis of the failure of constructivist, discovery, problem-based, experiential, and inquiry-based teaching. Educational Psychologist 41(2), 75–86 (2006)
14. Newcastle&Durham Universities experiences (2009), http://pdfserve.informaworld.com/4089_788671212_917447096.pdf

15. Villatoro, J.M.: Uso de las redes sociales en formación online (2009), http://www.cibersociedad.net/congres2009/es/coms/uso-de-las-redes-sociales-en-la-formacion-online/776/
16. Udutech, Udutu Online Course Authoring (2008), http://www.udutu.com/products-ududuteach-and-udutulearn.html
17. Ellison, N.B., Steinfield, C., Lampe, C.: The benefits of Facebook "friends:" Social capital and college students' use of online social network sites. Journal of Computer-Mediated Communication 12(4), article 1 (2007)
18. Hine, C.: Etnografía virtual. Editorial UOC, Barcelona (2004)
19. Bielman, V., Putney, L., Strudler, N.: Constructing community in a postsecondary virtual classroom. Journal of Educational Computing Research 29(1), 119–144 (2003)
20. Browne, E.: Conversations in cyberspace: A study of online learning. Open Learning 18(3), 245–259 (2003)
21. Laurillard, D.: Rethinking University Teaching and Learning—a framework for the effective use of information technology, 2nd edn. Routledge, London (2002)
22. Siemens, G.: Connectivism. A learning theory for de Digital Age (2004), http://www.elearnspace.org/Articles/connectivism.htm
23. Chan, A.: The dynamics of motherhood performance: Hong kong's middle class working mothers on- and off-line. Sociological Research Online 13(4) (2009)
24. Goodsell, T., Williamson, O.: The case of the brick huggers: The practice of an online community. City Community 7(3), 251–271 (2008)
25. Skageby, J.: Semi-public end-user content contributions. A case-study of concerns and intentions in online photo-sharing. International Journal of Human-Computer Studies 66(4), 287–300 (2008)
26. Skageby, J.: Exploring qualitative sharing practices of social metadata: Expanding the attention economy. The Information Society 25(1), 60–72 (2009)
27. Lemai, N., Luba, T., Konrad, P., Brian, C.: Power relations in virtual communities: An ethnographic study. Electronic Commerce Research 6(1), 21–37 (2006)
28. Nicolas, D., Robert, J.M., Eric, N.: Virtual "Third places": A case study of sociability in massively multiplayer games. Computer Supported Cooperative Work 16(1-2), 129–166 (1989)
29. Moore, R., Gathman, E., Ducheneaut, N.: From 3D space to third place: The social life of small virtual spaces. Human Organization 68(2), 230–240 (2009)
30. Mey, G., Mruck, K. (eds.): Grounded Theory Reader (HSR-Supplement 19). ZHSF, Cologne (2007)

Development and Implementation of an Enterprise Learning Architecture for Collaborative Learning

Scott E. Grasman[1], Vicki Callaway[1], Can Saygin[2], Wooseung Jang[3], and Henrique Rozenfeld[4]

[1] Engineering Management and Systems Engineering, Missouri S&T
[2] Mechanical Engineering, University of Texas – San Antonio
[3] Industrial & Manufacturing Systems Engineering, University of Missouri – Columbia
[4] University of São Paulo, Brazil

Abstract. The "E-Warehouse" Framework is an innovative concept that provides an integrated learning environment for collaborative enterprise learning. It is based on an enabling computing foundation, learning pillars, and an integrating roof, which allows for cross-functional applications. Curriculum materials were developed to incorporate theory, homework, case studies, and other learning material by utilizing reference textbooks, journal articles, and other pertinent publications. These resources were converted into instructional material using multimedia tools in order to facilitate its effective dissemination and customization. Two software packages have been developed associated with E-Warehouse. The first allows the system administrators to create content modules, and the second package allows a user to customize the content. Additionally, a functional quiz application was developed, which allows a user to selection from a repository of questions, add their own questions, and grade the quizzes.

1 Introduction

Advances in supply chain management have had a tremendous impact on business and engineering schools, many of which are developing integrated and collaborative curricula. The growing importance of supply chain management and related concepts has caused an increase in the emphasis placed on integrated curricula. Although the curricula of top graduate business schools have been examined, and a framework for supply chain management material and pedagogy that stresses integration has been established (Johnson and Pyke, 2000, Pyke and Johnson, 2000), little emphasis has been placed on collaborative learning, in which knowledge can be created within a population and members can actively interact by sharing experiences (Mitnik at al. 2009). Collaborative learning redefines traditional student-teacher relationships in the classroom and involves joint intellectual effort by students or students and teachers (Smith and MacGregor, 1992).

T. Daradoumis et al. (Eds.): Technology-Enhanced Systems and Tools, SCI 350, pp. 47 – 67.
springerlink.com

Academia strives to provide students with a rewarding and effective educational experience, and these complex and multidisciplinary systems have given rise to a need for improved engineering curricula based on integrative real world scenarios, including collaborative team-based, open-ended projects. In fact, accreditation requirements, including ABET (2001), have called for a response to changing education needs (Besterfield et al., 2000). Many engineering programs have responded by developing new capstone courses in order to provide students with the ability to adapt to ever changing environments (Banios, 1991, Bond, 1995, Mertz, 1997, Newcomer, 1999, Safoutin et al., 2000, McKenzie et al., 2004, and Dunlap, 2005). Nevertheless, integrative learning that meets the needs of modern education is still lacking. It is necessary to develop integrative curricular that connect skills and knowledge from multiple sources and experiences, apply skills and practices in various setting, utilize diverse and even contradictory points of view, and understand issues and positions contextually (Kline, 1988).

In addition, several resources highlight the need to effectively use modern technology to gain more productive and rewarding undergraduate Sci¬ence, Mathematics, Engineering, and Technology (SME&T) education, including a National Science Foundation Symposium (1995). Computer-supported collaborative learning (CSCL) is a relatively new educational paradigm which uses technology in a learning environment to control and monitor interactions, to regulate tasks, rules, and roles, and to mediate the acquisition of new knowledge (Mitnik at al. 2009). For example, Mitnik at al. (2009) showed that using robots in the classroom to promote collaborative learning led to an increase in learning effectiveness of the activity and increase in the student's motivation. Another common approach to effectively use the modern technology is a Web-based e-learning environment (ELE). One of the most desired characteristics of ELEs is being adaptive and personalized (Brusilovsky and Peylo, 2003) so that students with different skills, background, preferences, and learning styles can use them. A contextualized ELE provides the learner with exactly the material he needs, and appropriate to his knowledge level and which makes sense in a special learning situation (Eyharabide et al., 2009)

In addition to the growth of information technology, the importance of hands-on practice and active learning has been highlighted in various resources (Seymour and Hewitt, 1994, Brock, 1993, and Prince and Felder, 2006). These factors, coupled with inadequate and insufficient real world experiences in undergraduate education, have become a major rea¬son for under-qualified and under-employed graduates. In the report of an Advisory Committee, under the auspices of the Education of Human Resources (EHR) Directorate of the National Science Foundation (George, 1996), the importance of hands-on practice is illustrated through the words of Professor Eugene Galanter (Director, Psycho-physics Laboratory, Columbia University):

"Insofar as every science depends on data for both theory and application; labora¬tory or field data collection experience is an absolute necessity. Adding up numbers from a textbook example is not the same as recording those numbers or qualitative observations based on one's effort. When students "own" their data, the experience becomes a personal event, rather than a contrived exercise."

Therefore, this chapter presents the development and implementation of a computer simulated enterprise management environment based on current research activities and real world scenarios that is intended to be integrated into current curricula, including capstone courses. The hands-on and active learning environment utilizes modern technology and provides the fundamental advantage of facilitating dynamic decision making and collaboration among decision makers.

2 E-Warehouse: A Framework For Learning

E-Warehouse is an innovative concept that provides a collaborative learning environment, and is based on a computing infrastructure, core business functions, and cross-functional applications in an integrated fashion. The enabling computing foundation for learning allows core business functions to be represented as learning pillars, which are built on the foundation and are made up of modular building blocks that represent knowledge, facilitate active learning, and provide assessment. E-Warehouse is completed with an integrating roof, which allows for cross-functional applications. The learning pillars have been packaged as a teaching tool for instruction of credit undergraduate and graduate courses.

The generic architecture utilizes an enabling computing foundation for learning (Figure 1).

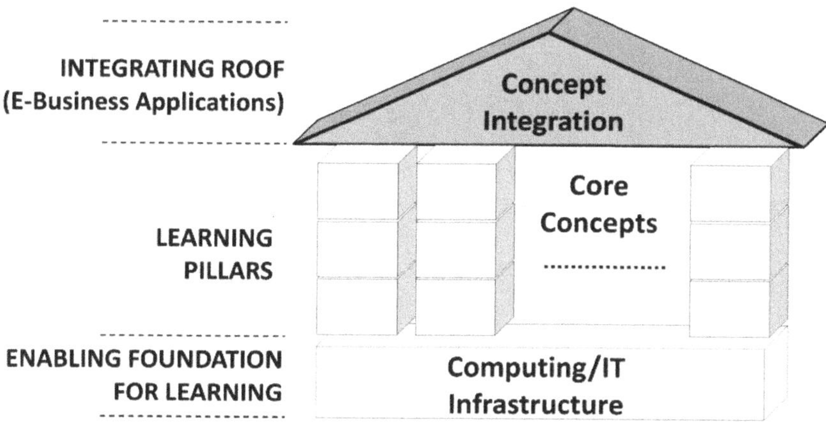

Fig. 1 Learning Warehouse.

Core concepts are represented as learning pillars, which are built on the foundation and are made up of modular building blocks that represent knowledge, facilitate active learning, and provide assessment. Core concepts are connected by the integrating roof.

Each learning pillar has a generic structure and, as shown in Figure 2, each pillar consists of three building blocks: Knowledge Block, Active Learning Block, and Assessment Block. The Knowledge Block contains figures, snapshots, and multimedia presentations on the core function that the pillar represents. Static examples, dynamic examples, and self-guided scenarios reside in the Active Learning Block. The Assessment Block includes homework, quizzes, and other assessment tools. Each block contains various levels of information that range from basic to advanced.

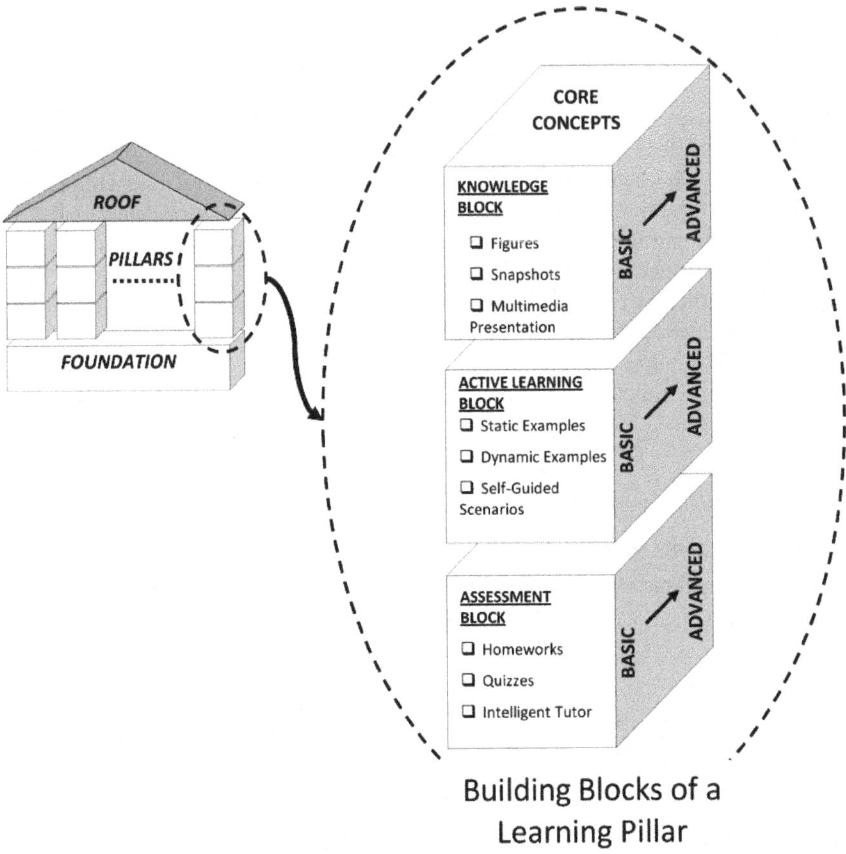

Fig. 2 Building Blocks of Learning Pillars.

Transportability, upgradeability, and adaptability are the basic characteristics of the E-Warehouse framework that allow it to be an effective collaborative environment. Transportability allows for easy dissemination, adaptability allows for customization, and upgradeability allows for sustained development. By creating these attributes within the E-Warehouse, the modules are easy to test, document, share with other universities, and keep current.

Existing platforms do not incorporate the theory and learning objectives normally needed in a classroom environment. Therefore, due to the lack of integrated coverage of theory and applications, the original format of imbedded scenarios is not effective and efficient for collaborative learning. E-Warehouse framework aims to facilitate effective collaborative learning by linking each process flow to the contents of its learning pillar using:

1. on-line documentation,
2. reference textbooks on the subject matter, and
3. articles and other pertinent publications.

As shown in Figure 3, the project team first focused on existing resources and tools.

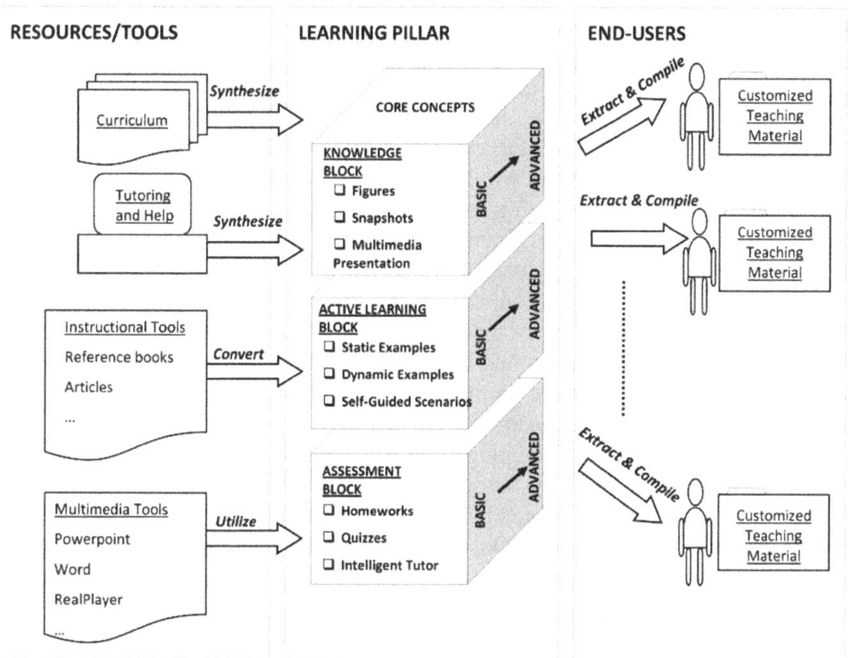

Fig. 3 Implementation of Learning Pillars: An End-Users' Perspective.

Theoretical concepts and methodologies can be compiled from reference books, articles, and online documentation, and converted into instructional material using multimedia tools in order to facilitate its effective dissemination, and associated with the appropriate blocks in the pillars. The outcome of this integrated effort leads to learning pillars on core functions. End-users can fully or in-part select the material available in the Knowledge, Active Learning, and Assessment blocks in order to customize materials tailored for their specific needs.

Thus, the E-warehouse framework establishes a potential foundation for increased mind share and use of integrated business processes for both educational and business communities. The concept provides several innovative features: (1) the generic architecture has the potential to lead to extended research efforts, (2) the HTML format of the educational material makes it easy to access and adapt by other faculty and institutions, and (3) it facilitates self-paced and interactive learning, easy navigation, and provides integrated coverage over the Internet. The modularity provided by HTML format also enables the developed curriculum to be kept current.

3 Implementation of Collaborative Environment

A main vision is a multi-disciplinary, distributed collaborative environment that integrates a variety of concepts (learning pillars). Individual students, teams, or classes can perform roles related to various concepts. Actions taken by one entity influence the decisions made by the other entities. Integration can occur by revolving curriculum around several cases arising from various business situations. For example, consider the following scenario:

A new system specified by Marketing students could be designed by Engineering Design students. The system would then be produced in an assembly cell operated by Manufacturing Engineering students, and Operations Management students would work with process planners to develop required policies. Accounting/Finance and Management students would participate in developing a business justification and planning and control mechanisms. Marketing students would be involved in developing a business and advertising/marketing plan. Information Systems and Accounting students would be involved in designing appropriate decision support and reporting systems to facilitate informed decision-making by the various functions. Computer Science and Computer Engineering students would be involved at several levels by developing computer applications and interfaces to support communications, control systems, data sharing and other integration needs.

Each entity could be represented by a class or, better yet, each entity could be resented by a cross-functional team. Once the initial scenario is established, events will unfolding over time resulting from the influence of the various collaborators. For example, a new partnership with an international company might be deemed necessary, and several questions may emerge that could be addressed by case studies.

This innovative approach creates a technology-based environment where students must react to the direct repercussions of decisions using appropriate tools in order to continuously adapt. Further, this approach allows for the creation of (1) a teaching environment that both encourages and supports collaborative learning and decision-making, (2) a set of role-based, pedagogical materials for use in introductory, core, and advanced concepts in business, information systems, and engineering, and (3) a data warehouse of simulated, "real-world" information over a series of years that can be used to stimulate research and the development of new ideas.

3.1 Curriculum Development

The learning environment provides a unique opportunity for innovative integrated and collaborative curricula. The environment provides the culmination of transforming curricula where classes are taught in isolation into a multidisciplinary collaborative environment (Figure 4), thus illustrating the need for integrative and dynamic curricula based on real-world scenarios. These scenarios utilize real-world and real-time information, and students have the opportunity to formulate solutions to real-world problems and gain the ability to apply classroom models.

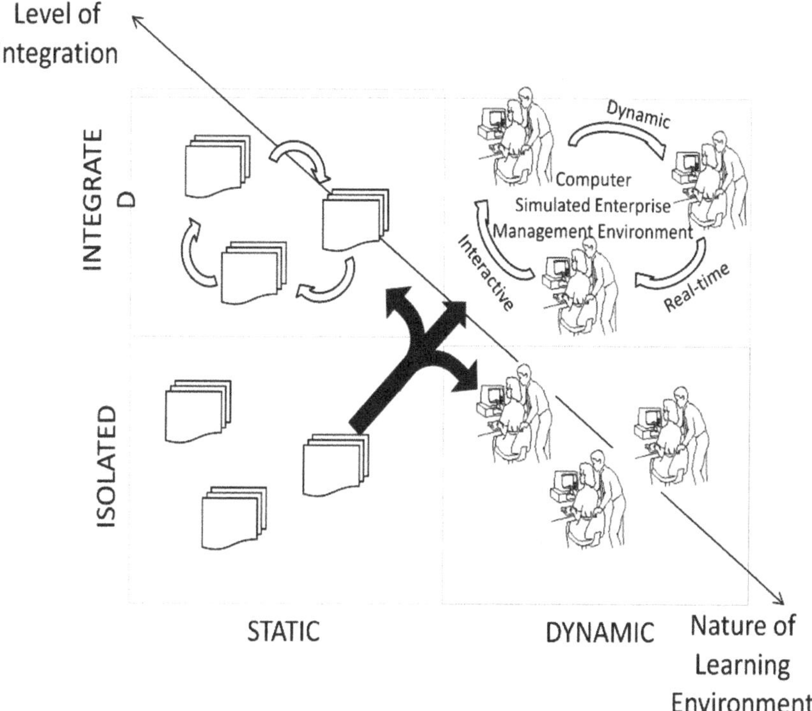

Fig. 4 Evolution of Curriculum

The learning environment may be replicated for implementation in a variety of classroom settings through dissemination of the pedagogical materials and data warehouse created through operation of the learning environment.

3.2 Pedagogical Materials and Data Warehouse

In addition to the fundamental advantages of facilitating dynamic decision-making and interaction among decision makers, the data collected during dynamic interaction can be used as a series of snapshots that be converted into learning module and used as static, but integrated, or isolated, but dynamic learning modules for integration into current courses. Throughout the curriculum, the pedagogical materials evolve from static problems that test the basic skill set and are appropriate for lower levels of learners, to dynamic open-ended case studies and simulated scenarios, which add depth and integrate all concepts, and are appropriate for higher level learners. These materials contain lecture notes, figures, snapshots, multimedia presentations, and self-guided scenarios, including a variety of simulated scenarios, and reflect definitions and concepts, and the ability to develop and apply models in decision making. Assessment materials include homework, quizzes, knowledge maps, and other assessment tools. Advanced classes use the scenario-based cases to learn the complexity and interdependence of concepts, and results of class and research projects (potentially industry sponsored) creates additional data that may be incorporated back into the learning pillars.

3.3 Assessment of Student Learning

Methods of obtaining evidence of student learning outcomes are based on objectives for student learning and a mix of qualitative and quantitative measures. In addition to the standard instruments of evaluation, such as exams, homework, and laboratory assignments, knowledge mapping is embedded in the learning pillars as evidence of learning outcomes. Knowledge mapping is a performance assessment that requires students to demonstrate their understanding of a content area by creating a network diagram, where nodes represent concepts and labeled arcs describe how concepts are related. Exercises that involve various tasks are assigned to the students, and students are asked to generate the knowledge map of the exercise prior to working in the collaborative environment. Upon completion of the exercise, the students are asked to reconstruct the knowledge map. The comparison of the two maps provides an indication of how students advance. By itself, reconstructing the map is a learning tool.

4 Sample Learning Context

The E-Warehouse Framework lays the foundation for sustainable development of learning modules, and will be illustrated by highlighting learning pillars for Supply Chain Management (SCM) and Product Lifecycle Management (PLM) based on the generic framework. These examples are based on curriculum materials in Supply Chain Management (Grasman and Jang, 2003) and Product Lifecycle Management (Rozenfeld, et al. 2003) that were developed based on a prototype framework (Saygin, 2003). The learning pillars were packaged as a teaching tool (Eller, 2003) used in the instruction of credit undergraduate and graduate courses, added as an innovative and integrative element to distance education programs, and been presented in workshops.

The learning modules are to be used to teach integrated concepts, collaborative product development and project management, and supply chain management. SCM and PLM curriculum materials were developed to incorporate theory, homework, case studies, and other learning relevant material. The interface between SCM and PLM occurs when the product development and design activity begins and requires integration of the entire development process; designers, suppliers, manufacturers, and customers, so enterprise engineering would no longer be a linear value chain, but a three-dimensional, collaborative community focused on a common goal.

4.1 Learning Pillars

These solutions provide a platform for curriculum development in the area of supply chain management. Supply Chain Management (SCM) addresses the planning, execution, and management of events that often interfere with supply chain excellence, enabling one to operate supply chain networks more efficiently and reach customers more effectively – often in real time. Coordination of financial, informational, and materials processes, as well as identifying processing exceptions, can help companies transform linear, sequential supply chains into adaptive supply chain networks that promote a distributed, dynamic environment. With the collaborative architecture, departments, business units, and companies gain greater visibility into inventory, planning, and scheduling, allowing them to anticipate problems sooner, adjust schedules and transactions quickly, and manage the extended supply chain more proactively. Furthermore, seamless integration with Product Lifecycle Management (PLM) ties suppliers into the design process, increasing quality and reducing time to market. Supply Chain Management creates value by enabling companies to reduce costs, increase revenue, and improve service to their customers, and offers all the capabilities one needs to design, build, and run the supply chain of the future.

4.2 Description of the SCM Pillar

The SCM Pillar consists of three blocks: the Knowledge, Learning, and Assessment. A brief description of the content and format of these blocks follows.

SCM Knowledge Block

The Knowledge Block (Figure 5) starts with an introduction and is structured according to the four functional areas of supply chain management, namely: Supply Chain Planning, Supply Chain Execution, Supply Chain Coordination, and Supply Chain Collaboration.

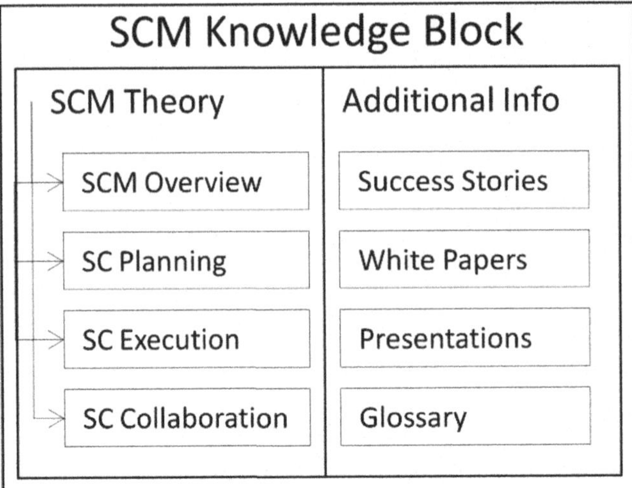

Fig. 5 SCM Knowledge Block.

In the Introduction, an overview of SCM is presented along with its benefits and an overview of its functional areas. The content of each functional area is based on several sources, supplemented by additional information that includes success stories, white papers and presentations. Finally, all the important terms used in the pillar are linked to a glossary.

SCM Learning Block

The SCM Learning Block (Figure 6) gives background information on all the scenarios and scenario-specific sections.

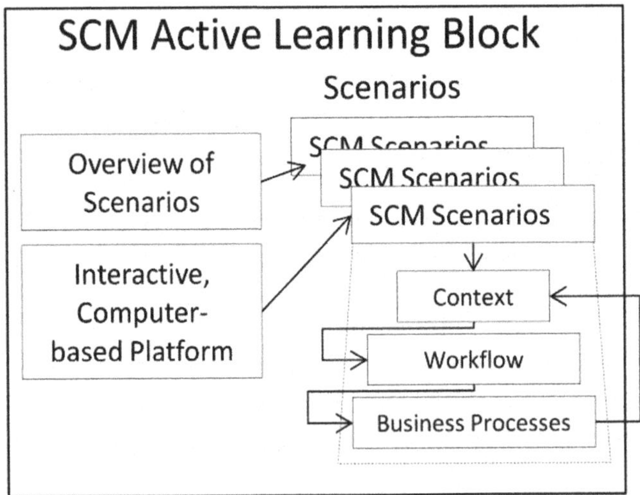

Fig. 6 SCM Learning Block.

The general sections consist of 1) an overview section introducing the scenarios, 2) a section introducing supply chain structure and other data, and 3) a section describing the various "workplaces" or menu structures and screens used to navigate through the SCM scenarios.

SCM Assessment Block

The Assessment Block (Figure 7) consists of a series of multiple-choice and true-false questions that users can answer online in a self-learning mode to assess the level of their understanding of the material. An online answer form is used to keep track of the user's answers and scores.

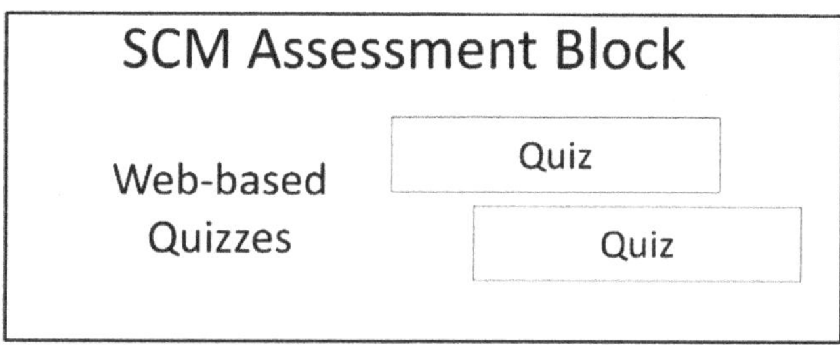

Fig. 7 SCM Assessment Block.

The integrating roof, which allows for cross-functional applications, is discussed later.

4.3 *Description of the PLM Pillar*

Product Lifecycle Management provides a flexible and adaptable approach that facilitates creativity by taking the idea of product innovation beyond the boundaries of traditional organizational constraints. PLM is a key Information Technology solution that improves management of collaborative and distributive product development projects, product data, asset maintenance, and quality and environment related issues. Despite the fact that PLM is viewed differently by different software vendors, there is a general agreement that its major benefit is in supporting the integration of product development in total supply chain. Providing students with a comprehensive view of PLM is a challenge that can be successfully addressed through the implementation of the E-Warehouse Framework.

The PLM learning pillar consists of three building blocks: knowledge block, active learning block, and assessment block. The knowledge block contains texts, figures, snapshots, and multimedia presentations on PLM core functions. The active learning block is basically a set of self-guided scenarios and hands-on work, while the assessment block includes homework, quizzes, and other assessment tools. In the remainder of this section we are focusing on one component of the PLM active learning block, which is related to scenarios, as an example of the work developed.

Figure 8 shows possible scenarios of the PLM active learning block.

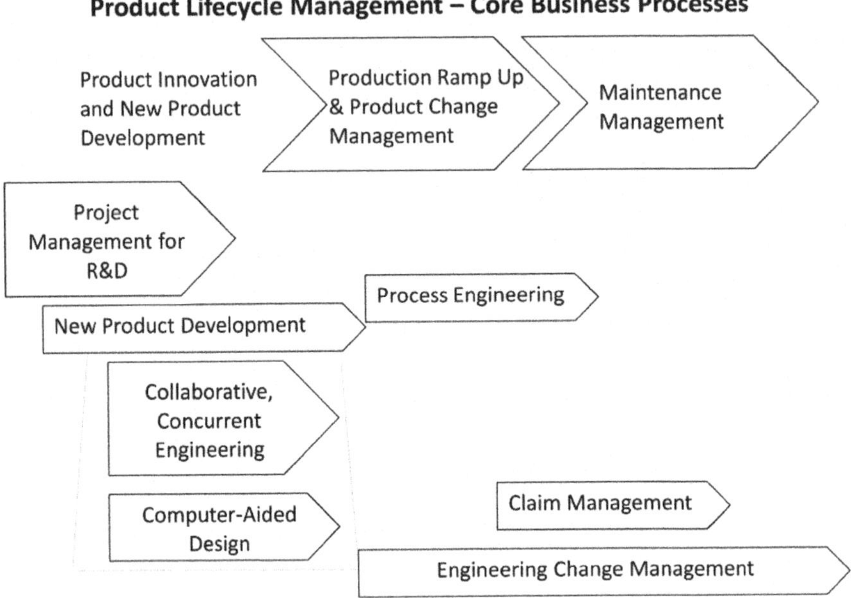

Fig. 8 Planned PLM Scenarios.

Each scenario includes an introduction, an overview, and a script. The introduction gives the context in which this scenario occurs, embracing the organizational structure of a company, the characters and their roles in the company and this scenario, the product and its bill of material (BOM) and the situation the student is going to face in the scenario. The overview presents the scenario problem statement, the previous activities that might have been carried out before this situation occurs, a short description of the scenario and finally the results to be expected at the completion of the scenario in a graphical form. Finally the scenario script guides the student in the hands-on work.

Students can use a scenario in two ways: First, by just studying it and going over the steps and copies of the screens shots presented without doing any hands-on work. Alternatively, he or she may elect to do the actual steps on available software packages that may be installed as a component landscape that includes a set of hands-on scenarios on PLM and other business functions. The student can consult, at anytime, the overview of a scenario to recall the context, the objectives and the description of the scenario and can also select the graphical workflow of the scenario to better locate the step he or she is performing within the whole process flow, so that he or she doesn't lose the whole picture of the process. A user can also access the theory in the knowledge block, where the main concepts and functions associated with the scenario are available.

4.4 Integration with Other Learning Pillars: E-Warehouse Roof: SCM / PLM Integration

A working environment is created by integrated applications shown in Figure 9. The end-to-end processes drive innovation and fast market positioning. Note that a stand-alone learning pillar for Customer Relationship Management is not presented; rather these concepts make be included in the SCM pillar.

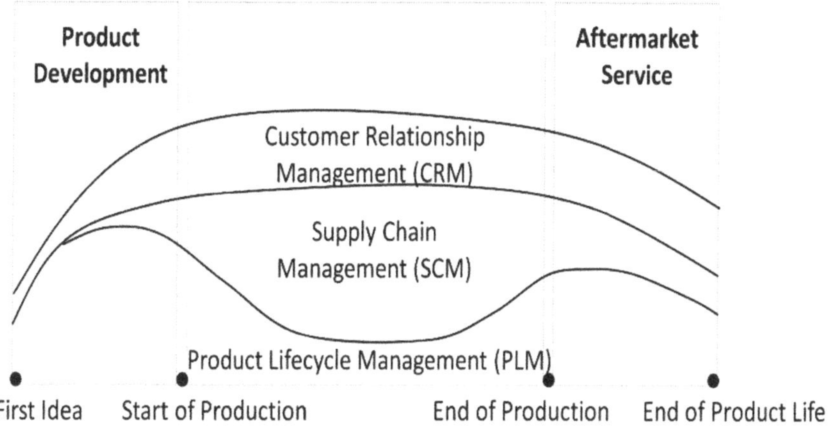

Fig. 9 Integrated Applications.

In the product development phase, the PLM processes are mainly applied, but the CRM processes are also crucial to guarantee that the customer requirements are being considered. The integration with SCM is more related to the product capacity analysis and with the SRM (supplier relationship management) to tie with the suppliers in the collaborative development and to support the procurement of material, facilities, equipment and services.

Figure 10 presents the SCM-related business processes presented together with the PLM processes.

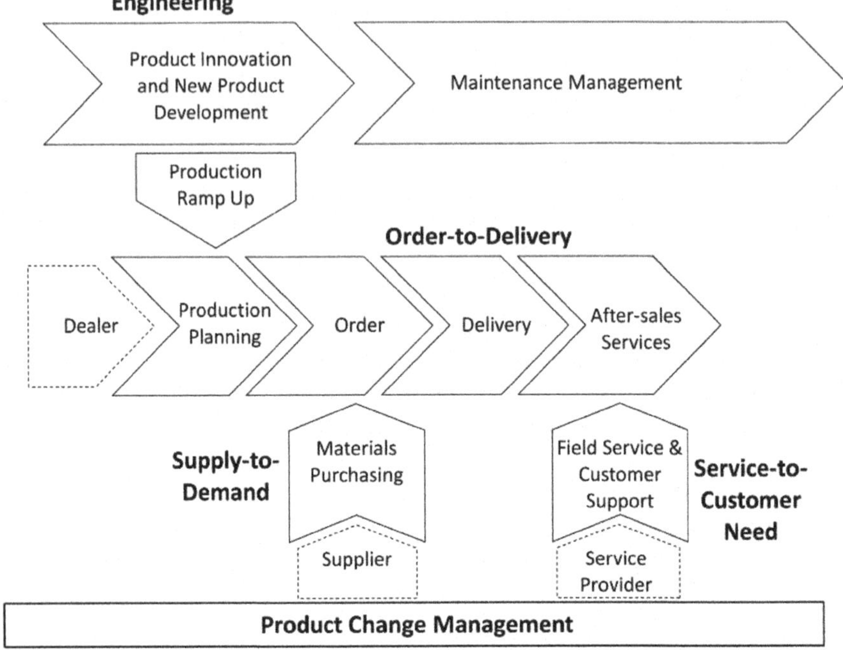

Fig. 10 Integration of SCM and PLM.

5 Management Environment for E-Warehouse

This section provides information about the development of software for the administration and dissemination of the E-Warehouse Learning Framework The goal of the software development team was to develop software that would allow the administrators to create "learning pillars". These pillars should be transportable, adaptable, and upgradeable. There are two software packages associated with this project. The first allows the learning pillar system administrators to create content modules called learning pillars. The second package allows an instructor wishing to use the learning pillars to customize the content and distribute it to the student.

The pillars are part of an integrating Learning Warehouse with three building blocks as illustrated earlier in Figure 2. The Knowledge Block contains self-paced learning modules. The Active Learning Block contains examples and interactive scenarios for active learning. The Assessment Block contains knowledge assessments. The challenge was to translate this structure into a software package that administrators could easily disseminate and upgrade, and that instructors could easily customize to use as a learning tool.

5.1 Design

The first step in the design was to translate the "pillar" into a file structure that could be easily distributed via the Internet, or on CD. It was decided that html pages with the file structure shown in Figure 11 would be appropriate. The Additional Materials folder could also contain any other file type. The entire file structure would be compressed and offered as a download from a website. This structure would constitute a "learning pillar".

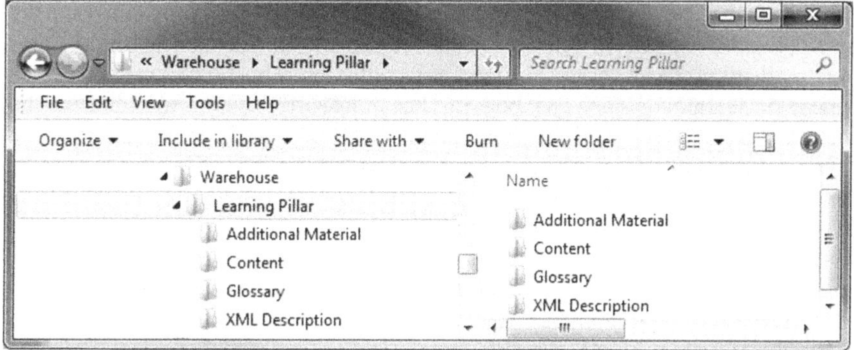

Fig. 11 Pillar File Structure.

The second step in the design process was to create a way that administrators could easily create the pillar structure, and upgrade the information in the pillars. It was determined that the administration tool should be able to do three functions. First, it should automatically create the basic folder structure when the pillar is created. Second, it should parse the information for terms and link them to a glossary. Third, the administration tool should parse the files and create links to content within the file structure. It is important to note that only html pages are currently parsed.

The modules required a separate administration tool for customization by the instructor. To protect the integrity of the system it was decided that the instructor could only add content to the Additional Files folder. This insured that the main structure of the information would not be compromised. The instructor should be able to parse any new html documents for glossary terms and be able to choose the information that would be used in the class. The software would then create a syllabus that contained links to that material in the structure. The entire pillar, including the syllabus, could then be uploaded to a web page or copied to a CD for dissemination to the students.

Several software languages were explored as platforms for the development of the tools. These included Java, Visual Basic, and C++. The Visual Basic .NET platform was then explored and was determined that a platform specific application would be acceptable for this version of the software. The conversion to Active

Server Page (ASP), which is not platform specific, is being explored for future versions.

5.2 Creating a Learning Pillar Using AdminTool

The two software packages created are referred to as the AdminTool and SyllabusCreator. Each contains a downloadable executable setup file that interacts with the individual "learning pillar" content modules. AdminTool and SyllabusCreator both have Installation Wizards that help you complete the installation process within a couple of mouse-clicks. Both applications need not be installed on the same computer; however, SyllabusCreator requires pillars created by the AdminTool application. Once the AdminTool is installed on a host computer the development of learning pillars can begin. The following section will discuss the creation of a pillar.

When AdminTool is started for the first time, the Source Directory is chosen. The Source Directory is the directory on your computer where the pillars created through the AdminTool are stored. Local and relative paths are allowed, with relative paths using the program directory as a reference. The Source Directory defaults to the program directory, if no directory is entered. The Source Directory can be set at any time after this, by selecting "Set Source Directory…" from the File menu, as displayed in Figure 12.

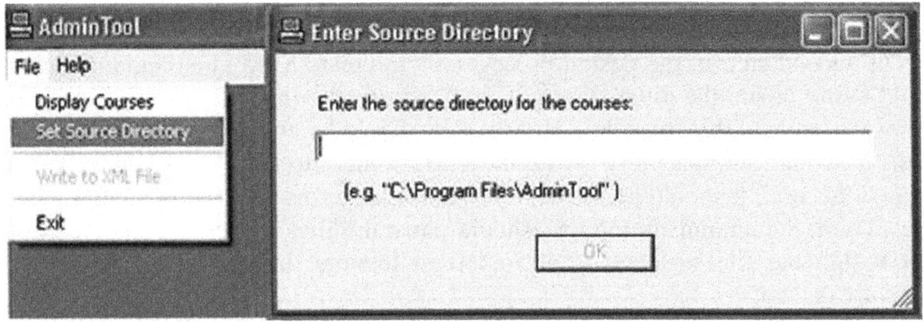

Fig. 12 Creating the Source Directory.

In the Courses display, seen in Figure 13, the courses currently residing in the Source Directory are displayed. The administrator can add or delete pillars from this view. The name chosen for the pillar must be unique, and should be descriptive.

To edit a course, the pillar is selected in the Courses display. This allows the administrator to add content to the course. To delete a course, simply select the course you want to delete, and click the "Delete Course" button. A confirmation is required to delete a course in order to help avert the possibility of unintentionally deleting one. When a course is deleted, all the contents of the course (including files) are deleted as well.

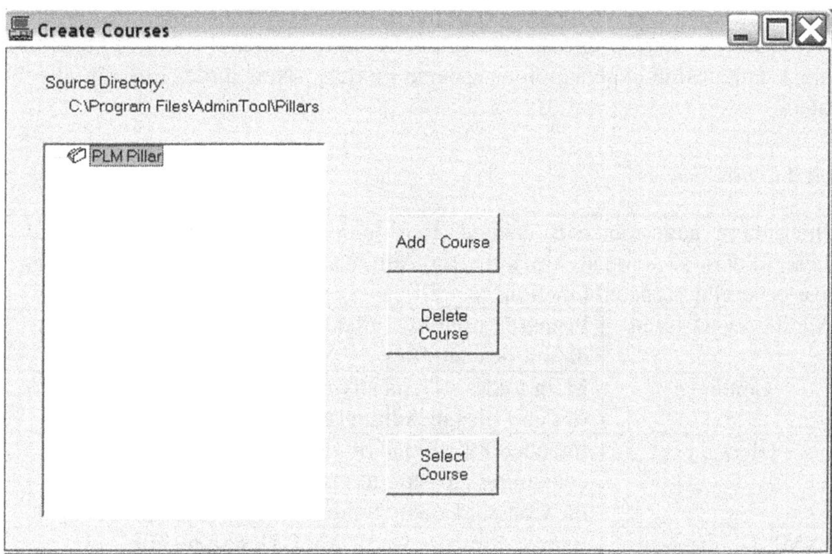

Fig. 13 Creating a Learning Pillar.

The Add/Remove content window as seen in Figure 14, displays files that a course currently contains, while displaying a "Windows Explorer"-type view of the computer's directories and files. The administrator may view a file if the application associated with it is available on the computer.

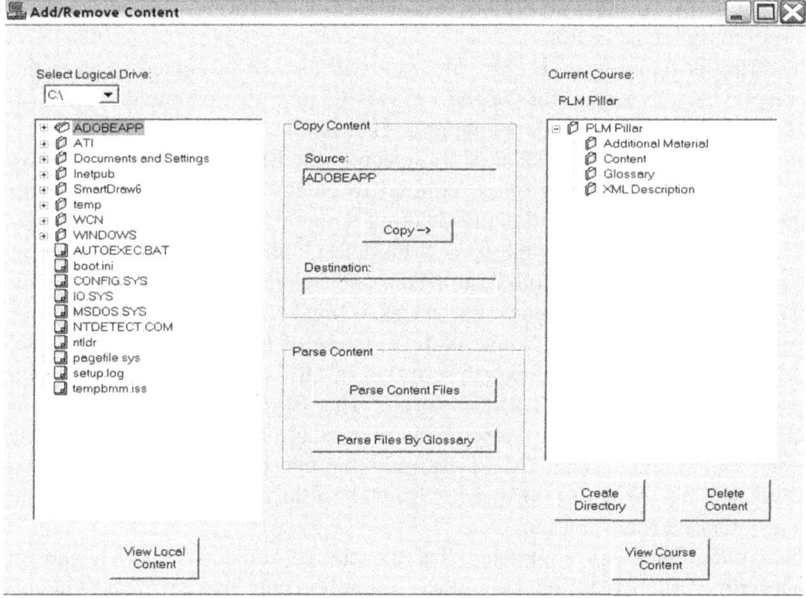

Fig. 14 Adding/Removing Content.

Content is added to a course by copying directories and files from the development location to the appropriate Course Folder designated in the Source Directory. Table 1 contains an explanation on Course Folders. New directories can also be created.

Table 1 Course Folders

Whenever a new course is created, four folders are automatically created. These folders are crucial structures for AdminTool and SyllabusCreator, and have general designated functions as well.	
Additional Material	Primarily used by SyllabusCreator to store additional directories and files.
Content	Main folder – Typically the default storage for directories and files in AdminTool.
Glossary	Intended for storage of glossary files – HTML files containing glossary terms mentioned within files in the Content Course Folder.
XML Description	Intended for storage of XML Descriptor file.

Parsing content files links HTML files in the Content Course Folder with other course files or glossary terms. This is accomplished by AdminTool parsing the source of each HTML file in the Content Course Folder for a particular string that contains the name of a course file, or for a glossary term. The format of the content file string is [F: filename.extension]. When the string is found by the AdminTool, it is replaced with a hyperlink to the indicated file. Hyperlinks are created with relative paths, so links do not become broken when courses are ported from AdminTool to SyllabusCreator.

If the string is found in an HTML file, you will also be prompted to enter the text to display as a hyperlink for the file. This is the text that people will click on to access the indicated file, while viewing the HTML file.

Content in a course is easily deleted by selecting the file or folder and choosing "Delete Content". A confirmation is required to delete content in order to help avert the possibility of unintentionally deleting something. An additional confirmation is needed to delete a default Course Folder. Course Folders are an integral part of a course's structure to both AdminTool and SyllabusCreator; thus, deleting a Course Folder will render a course useless to AdminTool or SyllabusCreator.

The final step in creating a course is the creation of the XML Descriptor file. The XML Descriptor file contains a description of all the material contained in a course, and the structure in which it is stored. This file is primarily intended for use with SyllabusCreator, but can also be distributed as a text-representation of the course and its structure. The XML Descriptor file can be created in one of two ways: by selecting "Write to XML file" from the File menu, or by clicking the "Create Content Package" button.

After a pillar has been completed (i.e. the course contains material, and an XML Descriptor file exists for the course), the pillar may be distributed via any medium (e.g. CD, FTP, Email). Courses may be compressed and decompressed without problems, as long as the internal structuring remains intact. Courses are stored in the Source Directory designated in AdminTool, as a directory. This directory is the structure that should be distributed.

5.3 Future Work

The current software packages are functional, but could be more user-friendly. Several upgrades to the software have been planned. A previously unmentioned quiz feature will be added to the software. A functional quiz application was developed separately, and will be integrated into both software packages. It will allow the administrator to select individual quiz questions to build exams and add their own questions. Currently, the syllabus is created with a list of files that follows the hierarchy of the file system. In the future, a customization feature would be added to allow the instructor to alter the file order on the syllabus. The instructor would also be able to add personal information and comments to the syllabus. Additionally, the glossary feature is currently separate and will be integrated in the next version. The feature will be upgraded to allow the administrators and instructors to add glossary terms.

6 Conclusions

The development of collaborative framework in a computer simulated environment and its integration into curriculum development will foster student development through active learning in the classroom and facilitate student projects based on current real world issues. This focus will improve the educational experience for the student, as well as motivation and retention of students. Further, the framework will improve students' abilities to apply their educational experience to their careers and consequently improve the knowledge base of firms. The framework will create a dynamic computer simulated environment utilizing available information technology and providing students with a hands-on and active learning environment. This approach will improve the undergraduate education experience by transforming a curriculum where classes are taught in isolation into a multidisciplinary integrated environment. Early in the education process, students will be asked to formulate real world problems so that they gain the ability to apply classroom models.

While formal lecture is important for presenting basic concepts and principles, much learning takes place when students apply lecture materials not only to homework assignments, but also in laboratory settings and real industrial settings. Creating integrated curricula offers the opportunity to provide students with a well-rounded education, including adequate theoretical knowledge, as well as real world examples and case studies. The E-Warehouse Framework will impact the curriculum through the integration of enterprise concepts into current courses and the development of a new capstone course, which will provide a more in-depth, integrated coverage of the basic concepts presented in earlier courses. An integrated curriculum, including real world scenarios and the opportunity to do research on current topics, will have a tremendous impact on students. Additionally, potential projects with industry will allow undergraduate students to gain industrial experience that will be beneficial to understanding the practical application of underlying theories, and provide additional benefits through increasing the interest in undergraduate research.

E-Warehouse provides tremendous opportunity for integrating current research topics in classical curricula, e.g., risk management and coordination, information technology and flow, global e-marketplaces, production planning and scheduling, and manufacturing. Expanding opportunities for a more active form of learning by students will encourage the interaction of undergraduate students with faculty. Exposing students to real world scenarios will spur greater interest in these activities. With proper support, students will achieve their educational goals. Integrating current research activities into the curriculum will lead to an effective learning environment. Fostering imaginative thought processes will allow students to derive creative solutions to both theoretical and practical problems, and will allow students to conceive new and better solutions to the issues that they are addressing. Providing students with a better integrated education will improve students' abilities to apply their educational experience to their careers and improve the knowledge base of firms, which in turn will increase the competitiveness of firms in the global marketplace.

References

1. ABET, Criteria for Accrediting Engineering Programs, Accreditation Board for Engineering and Technology (2001)
2. Banios, E.W.: Teaching Engineering Practices. In: Proceedings of the Annual Frontiers in Engineering Education Conference (1991)
3. Besterfield-Sacre, M., Shuman, L.J., Wolfe, H., Atman, C.J., McGourty, J., Miller, R.L., Olds, B.M., Rogers, G.M.: Defining the Outcomes: A Framework for EC-2000. IEEE Transactions on Education (43) (2000)
4. Bond, B.: The Difficult Part of Capstone Design Courses. In: Proceedings of the Annual Frontiers in Engineering Education Conference (1995)
5. Brock, W.E. (Chair): An American Imperative: Higher Expectations for Higher Education. In: Wingspread Group on Higher Education, Racine (1993)
6. Brusilovsky, P., Peylo, C.: Adaptive and intelligent Web-based educational systems. International Journal of Artificial Intelligence Education 13(2), 159–172 (2003)
7. Dunlap, J.C.: Problem-Based Learning and Self-Efficacy: How a Capstone Course Prepares Students for a Profession. Educational Technology Research and Development (53), 65–83 (2005)
8. Eyharabide, V., Gasparini, I., Schiaffino, S.N., Pimenta, M.S., Almandi, A.: Personalized e-learning environments: Considering students' contexts. Education and Technology for a Better World 302, 48–57 (2009)
9. George, M.D.(Chair): Shaping the Future: New Expectations for Undergraduate Education in Science, Mathematics, Engineering, and Technology (SME&T). In: Advisory Committee to the National Science Foundation, Directorate for Education and Human Resources, NSF, pp. 96–139

10. Grasman, S., Jang, W.: Supply Chain Management Learning Pillar. In: Proceedings of the ASEM National Conference, St. Louis (2003)
11. Holton, G.: Introductory Comments. In: National Research Council / NSF Symposium on Science, Mathematics, Engineering, and Technology Education, Boston (1995)
12. Johnson, M.E., Pyke, D.F.: A Framework for Teaching Supply Chain Management. Production and Operations Management 9 (2000)
13. Kline, P.: The Everyday Genius. Development of Integrative Learning (1988)
14. Liles, D.H., Johnson, M.E., Meade, L.: The Enterprise Engineering Discipline. The University of Texas at Arlington
15. McKenzie, L.J., Trevisan, M.S., Davis, D.C., Beyerlein, S.W.: Capstone Design Courses and Assessment: A National Study. In: Proceedings of the 2004 American Society of Engineering Education Annual Conference (2004)
16. Mertins, K., Jochem, R.: Architectures, Methods and Tools for Enterprise Engineering. International Journal of Production Economics (98), 179–188 (2005)
17. Mertz, R.L.: A Capstone Design Course. IEEE Transactions on Education (40) (1997)
18. Miller, J., Eller, V., Saygin, C., Grasman, S.E.: A Learning Pillar Management Environment for E-Warehouse. In: Proceedings of the ASEM National Conference, St. Louis (2003)
19. Mitnik, R., Recabarren, M., Nussbaum, M., Soto, A.: Collaborative Robotic Instruction: A Graph Teaching Experience. Computers & Education 53(2), 330–342 (2009)
20. National Science Foundation, Directorate for Education and Human Resources, Shaping the Future: New Expectations for Undergraduate Education in Science, Mathematics, Engineering, and Technology (SME&T), NSF 96-139
21. Newcomer, J.L.: Design: The Future of Engineering and Engineering Technology Education. In: Proceedings of the Annual Frontiers in Engineering Education Conference (1999)
22. Prince, M.J., Felder, R.M.: Inductive Teaching and Learning Methods: Definitions, Comparisons, and Research Bases. Journal of Engineering Education (95), 123–138 (2006)
23. Pyke, D.F., Johnson, M.E. (eds.): Supply Chain Management: Innovations for Education. POMS Series in Technology and Operations Management, vol. 2 (2000)
24. Rozenfeld, H., Najm, M., Saygin, C., Grasman, S.E.: Product Lifecycle Management in the E-Warehouse Learning Environment. In: Proceedings of the ASEM National Conference, St. Louis (2003)
25. Safoutin, M.J., Atman, C.J., Adams, R., Rutar, T., Kramlich, J.C., Fridley, J.L.: A Design Attribute Framework for Course Planning and Learning Assessment. IEEE Transactions on Education (43) (2000)
26. Saygin, C., Grasman, S.E., Najm, N., Jang, W., Rozenfeld, H.: E-Warehouse: A Framework for E-Business Learning. In: Proceedings of the ASEM National Conference, St. Louis (2003)
27. Seymour, E., Hewitt, N.: Talking About Leaving: Factors Contributing to High Attrition Rates Among Science, Mathematics, and Engineering Undergraduate Majors. Boulder, University of Colorado (1994)
28. Smith, B.L., MacGregor, J.T.: What is Collaborative Learning. National Center on Postsecondary Teaching, Learning, and Assessment at Pennsylvania State University (1992)

A Rational Framework for Student Interactions with Collaborative Educational Systems

Curtis Atkisson and Edward Brent

Ideaworks, Inc.
University of Missouri
100 W Briarwood Ln
Columbia, MO 65203
{catkisson,ebrent}@ideaworks.com

Abstract. The rise of a new class of collaboration tools should encourage us to examine parts of the collaborative process that may have been less valuable to examine in the past. Specifically, this research examines a computer-supported collaborative learning (CSCL) environment that makes possible new modalities for student-instructor collaboration. In particular, this environment makes possible time-shifted collaboration that allows students to collaborate interactively with instructors in real time with no noticeable delay, without requiring both individuals to be actively engaged at the same time. This learning environment makes it possible to examine both new ways in which students and instructors collaborate and to provide new evidence that addresses one of the fundamental problems faced by students – procrastination. Data routinely collected as part of this (CSCL) environment make it possible to empirically examine student behaviors and performance. This method of collaboration brings up an interesting dimension in education: A deadline can be assigned and the student can continue to collaborate with the instructor in a time-shifted manner right until the deadline. It becomes a natural question to ask how students alter their behavior as the time to deadline approaches. This paper empirically examines student behavior as time to deadline approaches and interprets that behavior using a rational framework based on Temporal Motivation Theory. Both qualitative and quantitative data are presented to highlight changes in student behavior and performance as time to deadline approaches.

1 Introduction

The increased use of technology in computer-supported collaborative learning (CSCL) environments is having a dramatic impact on how students interact with

T. Daradoumis et al. (Eds.): Technology-Enhanced Systems and Tools, SCI 350, pp. 69–91.
springerlink.com

instructors and material. Notable changes include the increasing ability to communicate directly with the instructor outside of class and office hours through email, chat, and discussion blogs; online posting of instructor lecture notes, videos and podcasts of lectures, and other instructor-created materials for review outside class; the use of student response systems to facilitate interaction and discussion in large lecture classes; and other changes in how students and instructors interact. These new learning environments provide both new opportunities for collaboration and rich data for assessing that collaboration.

This chapter addresses several themes of this book. It examines student learning strategies and performance in a CSCL environment. This is a web-based collaborative platform used in both distance learning and hybrid classes that combine classroom and distance learning modalities. Specifically, this CSCL environment is based on a system for automated grading of student essays, SAGrader™ (Brent, Carnahan, McCully, & Green, 2006b). Earlier work (Brent, Atkisson, & Green, 2010) examined how SAGrader™ provides time-shifted collaboration between students and instructors. This chapter uses Temporal Motivation Theory (TMT), based on a synthesis of theories by Steel and his colleagues (Piers Steel, 2007; P. Steel & Konig, 2006) to examine theoretical issues regarding adaptation methods and techniques for groups of learners in this environment. TMT provides a rational framework explicitly incorporating time into its predictions of student behavior. This provides a framework for assessing a fundamental problem for students – procrastination. Student activity and performance data routinely collected by this CSCL environment provide a unique opportunity to directly assess student procrastination and its consequences.

The CSCL environment studied here – SAGrader - automates the grading of essays, creating an environment in which the incremental cost for assessing student revisions is nearly free (Brent, Carnahan, & McCully, 2006a). This makes it cost-effective to permit students to revise their essays as often as they like. Thus, SAGrader is not just an assessment instrument. It is also a collaborative learning environment where students can submit essays, immediately see their grade and receive detailed, personalized feedback while the issues are still fresh in their minds, and have the opportunity to revise and resubmit their work. The result is a collaborative learning environment for e-learning that has the convenience of asynchronous collaboration and the power of synchronous collaboration. This method of collaboration in effect *time-shifts* instructor responses and assessments of student work. It allows the instructor to specify assistive knowledge to the student at a time convenient for the instructor. Then it allows the student to interact with the instructor-provided materials through assessments on the student's time frame.

SAGrader works by utilizing expert knowledge and computational linguistics. In this system, experts are able to specify the correct content that should be required in a paper. This content is easily specified so that thousands of valid ways of phrasing the content are accepted. SAGrader then uses principles of computational linguistics, such as parsing and fuzzy logic, to apply scores to a student's mentioning of the concept in the correct context. This system also uses computational linguistics to provide detailed, personalized feedback regarding a student's grammar. A further description of the system may be found in Brent and colleagues (2010).

Together the elements of time-shifted learning—motivation, information, feedback, and opportunity—create a uniquely powerful "*teachable moment*" analogous to the teachable moment for smoking cessation when someone is diagnosed with lung cancer (Gritz, 2006). Students just received their grade and are motivated to improve it. They see detailed feedback to guide that revision. They have just completed the previous draft and have the necessary information in their grasp. They have the opportunity to revise multiple times. Prior research by Brent and colleagues has shown that students take advantage of this learning opportunity to revise their work based on feedback from the program, often dramatically improving their performance by as much as two letter grades (2010).

SAGrader's transformation of the collaboration between teacher and student through time-shifted learning also provides a window into how students study and learn. In most courses the only information provided to instructors is the final product of student work: performance on an exam or a paper submitted to an assignment. All the hours of work, the false starts and restarts, the drafts and redrafts of what ultimately becomes the final paper are usually inaccessible to the instructor. Because SAGrader permits unlimited revisions by students and rewards each new submission with immediate feedback regarding the student's grade and suggestions for improvement, students have a strong incentive to make multiple submissions to the program as they work on their papers. The SAGrader program automatically tracks a wealth of information for each submission by students, including the number of submissions, when they occur, and how well students do on each submission. For each student it is possible to examine the track record left by their first submission and subsequent submissions through their final submission. This information provides an opportunity to examine how students manage their time as they work on papers and permits the examination of one of the fundamental issues students must address in learning and a fundamental issue in time-shifted collaboration – procrastination.

Procrastination is important. It is extremely common, with 80-95% of college students engaging in procrastination (O'Brien, 2002). Procrastination often leads to poorer performance (P. Steel, Brothen, & Wambach, 2001), greater dissatisfaction (Tice & Baumeister, 1997), financial costs such as increased taxes (Kasper, 2004), poorer health from failing to seek help earlier (White, Wearing, & Hill, 1994), and reduced retirement income from lack of saving behavior (O'Donoghue & Rabin, 1999). Most people who procrastinate would like to reduce it (O'Brien, 2002).

Procrastination is sometimes associated with increased cheating. Cheating on tests and plagiarism are positively associated with self-ratings of procrastination (Gerdy, 2004; Roig, 1995; Stover & Kelly, 2005) and inversely associated with two other inverse measures of procrastination - both the quality and quantity of study time (Kerkvliet, 1994; Norton, Tilley, Newstead, & Franklyn-Stokes, 2001). Some authors even distinguish between cheating that may result from procrastination (panic cheating) and cheating that is more deliberate (planned cheating).

> *Planned cheating may involve making crib sheets for tests, copying homework, or plagiarizing a paper; it occurs with full knowledge that it is wrong. Panic cheating, on the other hand, occurs during a test when the student finds herself at a loss for an answer. Although she did not plan to cheat, she looks at another student's paper and copies the answer. Although both types of cheating involve weighing costs and benefits, if social norms differ for planned and panic cheating, the subjective costs and benefits may be different for planned and panic cheating.* (Grijalva, Nowell, & Kerkvliet, 2006)

Procrastination has been the subject of considerable research. Steel (2007) summarized much of this research in a meta-analytic study of 216 separate works. Steel (2007)argues that much of the findings of this diverse research can be explained by a particular theory, Temporal Motivation Theory (TMT) to be described in more depth below. However, most prior studies have relied on surveys with scales measuring self-reported procrastination (Piers Steel, 2007). Such studies are subject to memory bias and the genuine possibility that people are not totally aware of how much and under what conditions they procrastinate. What are lacking are studies of behavior that directly measure how students conduct their learning activities over time. It is only by directly examining learning behaviors over time that procrastination can be measured.

Procrastination is often viewed negatively as the deliberate and often irrational delay of an intended course of action (Silver & Sabini, 1981). However, here procrastination is viewed as a rational act, deliberately chosen by actors from possible courses of action and behavior which can be examined from a rational framework (Piers Steel, 2007; Zarick & Stonebraker, 2009). Whether such procrastination ultimately leads to negative outcomes is regarded as an empirical question rather than a foregone conclusion. The research reported here examines empirically both the extent to which students procrastinate and ways in which this affects student behaviors and performance as time to deadline become shorter.

2 A Rational Model of Procrastination

Temporal Motivation Theory (TMT) is an attempt to synthesize theories of procrastination (Piers Steel, 2007). This theory is particularly appropriate for this study because it explicitly incorporates time and provides a basis for predicting how student behaviors may change as the deadline for an assignment approaches. In addition, the variables in the theory that predict the utility function are likely to be influenced by a wide range of other variables, including several variables available for examination in this study.

Temporal Motivation Theory models the utility of a course of action such as working on an assignment by the following equation:

Utility = EV/ΓD, where
E = Expectancy – the likelihood the assignment will be completed successfully
V = Value – how rewarding the assignment is
Γ = Sensitivity to delay – the extent to which the person is influenced by delay, and
D = Delay – how long the person typically must wait to receive the payout or outcome

This theory predicts utility will change as a function of each of these four parameters. Each of the parameters may be measured by specific variables that, in turn, should be related to utility and hence to procrastination. For example, assignments with more points should have greater value and hence higher utility, students who have low self-confidence, such as poor students, should have reduced expectancy of success for assignments and hence lower utility. Very difficult assignments should also carry with them lowered expectancies of success and hence lower utility. Students who are easily distracted or have trouble concentrating might be expected to be less sensitive to delay and hence be more likely to procrastinate.

Together these parameters and indices of them provide a very rich theory of motivation the full examination of which is far beyond the scope of this study. The focus of this study is on the impact of time on utility and the pattern of behaviors expected of students as time to deadline becomes shorter. As the time to deadline (Delay in this model) decreases, utility will increase in inverse proportion. Thus, utility plotted by time to deadline should increase in a geometric manner as time to deadline becomes smaller. That is, students will have greater utility for working on their homework as the time to deadline approaches.

Changes in utility are not measured directly in this study, but have clear implications for a number of measured behaviors by students, leading to several specific hypotheses to be tested. We predict results in three areas: increased activity, diminshed performance, and increased efforts to overcome low performance.

Increased activity. We hypothesize that, as time to deadline becomes shorter, this will lead to a dramatic rise of activity by students in an effort to successfully complete the assignment. This is the very essence of procrastination, as students put off the work to the point where performance often declines.

1. **Rate of submissions will increase geometrically as time to deadline becomes shorter.** As utility increases, students will increase their rate of submitting their papers. Hence, as the deadline approaches the rate of submission

should increase geometrically. This should be true for the first submissions, last submissions, and total submissions. One obvious reason to submit a revision is to improve one's grade. As available time narrows, it becomes more important to do it soon to make sure it is done before the deadline.

SAGrader is somewhat unusual because it provides nearly immediate feedback. This information should have some value to the student, further increasing the utility of submitting (based on both the probability of improving one's grade and the probability of receiving additional information that could help one make still more improvement). Less time will be taken between revisions. This will allow the individual more collaboration with the instructor in the form of earlier input knowledge. The feedback that a student receives is dependent on submitting a revised version of a previous work and allowing the program to compare that version against the ontological specification provided by the instructor. Because this will immediately reference the knowledge desired by the instructor, it would be rational to increase contact with this knowledge.

2. **Time since last submission will decrease geometrically as time to deadline becomes shorter.** As the rate of submissions increases, the corresponding time since the last submission (the difference in time between the current and last submission) will decline. Hence, as the deadline approaches the mean time since last submission should decrease geometrically. One obvious reason to submit a revision is to improve one's grade. As available time narrows, it becomes more important to do it soon to make sure it is done before the deadline.

Diminished performance. While TMT clearly predicts increased utility and hence increased activity as time to deadline becomes shorter, TMT does not predict whether this increased activity in a shorter time span will lead to greater or lesser success. We predict that students who procrastinate will experience increased cognitive load and stress leading to decreased performance in comparison to students who complete their work with plenty of time remaining.

3. **Final score will decrease as time to deadline becomes shorter.** As the deadline approaches, students who waited until near the deadline will have less time to work on the assignment and may have to settle for a lower final score, where if there had been more time they would try one or more additional revisions.
4. **Final number of submissions will be fewer when first submission occurs closer to deadline.** For students who begin later, we predict these students will be more likely to run out of time and be unable to complete as many submissions as students who begin earlier in the process. Above it is predicted students will attempt to compensate for this lack of time by revising more quickly (hypothesis 2). However, rushing revisions is unlikely to make up for all of the time and may lead to less effective revisions as indicated in hypothesis 4 below.

5. **Improvement in score per revision will decrease as time to deadline becomes shorter.** As the deadline approaches students will likely try to gain as much improvement as they can by speeding up revisions. However, rushing their work is likely to lead to poorer performance and smaller improvements in their score per revision.
6. **Total elapsed time spent on assignments (the time between the first submission and last submission) will decrease as time to deadline becomes shorter.** As time to deadline decreases, students who started late will need to compress their time spent on the assignment in order to complete work by the deadline. Conversely, students who begin early may take advantage of the added time to work longer and achieve a better grade. If, for example, a particular assignment would normally take around 2 hours for the average student to complete successfully, as time to deadline decreases, some students are likely to run out of time and have to work faster – and most likely less effectively – to complete their work.

Increased efforts to overcome low performance. As student performance is lowered by procrastination, students are more likely to take more extreme measures to increase their final score. These include increased challenges questioning the program's score and boundary-stretching behaviors in which student work looks suspiciously like that of other students or external sources.

7. **Student boundary stretching behavior, including copying from other students and plagiarism, will increase as time to deadline becomes shorter.** We predict similarity scores will increase as the time remaining until the deadline becomes shorter, particularly very near the deadline. As the deadline approaches, TMT predicts the utility of successfully completing the assignment increases dramatically and is likely to exceed the utility of maintaining academic integrity for more students, resulting in a number of students choosing to cheat, or at least stretch the boundary in a last-ditch attempt to improve their grade.
8. **The number of challenges will increase as time to deadline becomes shorter.** As the deadline looms and students find themselves unable to complete the assignment successfully before the deadline despite the increased utility of doing so, the utility of a good score is likely to exceed the threshold for complaining to the instructor by challenging the program's grade. Hence, we predict there will be more challenges the closer the deadline approaches.

3 The Impact of Procrastination on Collaboration

Procrastination is well established as an issue in traditional academic environments, and many academic and practical guides have been written on the subject. Procrastination in collaborative environments, however, is not as well established. The studies that have examined it find procrastination to be at least as bad a problem in collaborative environments, and potentially more of an issue. Discovering the unique impact of procrastination in collaborative environments seems to be a particularly important goal in collaborative research, and this paper addresses this issue.

Henry and LaFrance (2006) put together a primer on running group projects. They highlight 'Students tend to procrastinate' as a key blocker to having a successful group project. Last and colleagues (2000) detail a project that involved international collaboration between students and instructors. Both groups listed procrastination as a major hindrance, and not procrastinating was given as a key piece of advice for students working on this project in the future. These studies highlight the impact of procrastination on traditional collaborative environments.

Waite and colleagues (2004) have examined the impact of procrastination in a project that involved both synchronous and asynchronous collaboration. These students highlighted procrastination as a hindrance for asynchronous collaboration similar to the studies above. Furthermore, the students reported procrastination as a hindrance for synchronous collaboration. Students stated that procrastination often hindered collaboration because helping people catch up on material needed for the collaboration prevented discussion on the topic. This shows that procrastination impacts collaboration in both synchronous and asynchronous environments.

As a more explicit test of the effects of procrastination on personal versus collaborative assignments, Gafni and Gera (2010) had students complete a two-part assignment. In the first part, students were asked to write a comment on an article and incorporate topics covered in the class in their analysis. In the second part, students were asked to engage in commenting on other people's work. Gafni and Gera found that students completed the majority of their individual assignments on time. However, even though the students had a whole semester to comment on another student's work, the majority of comments took place in the last 21 days with some occurring on the last day (10 days after the deadline) that they would be accepted. Through observing student behaviors, the investigators identified this as indicating greater procrastination in the collaborative assignment compared to the personal assignment.

Procrastination has been found to influence collaborative efforts and projects. These studies have taken place in many types of collaborative environments, and a time-shifted environment is a natural extension. Furthermore, in each of these studies, procrastination was listed as a key way to avoid encountering problems. Together, this justifies an examination of the effect of procrastination on collaboration in a time-shifted environment.

4 Data

To test these hypotheses we conducted a field study in the classroom using a prospective panel study monitoring student learning strategies and outcomes in response to a series of writing assignments in a large introductory social science course at a Carnegie I Research University in the Midwest. Because this course is part of the core curriculum for the University it attracts a broad cross-section of students, representative of the student body. Two hundred eighty six students participated in the course, submitting writing in response to 22 different assignments, with an average of 3.3 submissions to each assignment and a total of 20,979 separate submissions.

Respondents were similar to the undergraduate student population at this university, with 58% female, 42% male, 2% Hispanics, 6% Asian/Pacific Islanders, 9% Blacks, and 84% Whites. Five percent spoke English as a second language. Most (63%) were freshmen, with 19% sophomores, 10% juniors, and 7% seniors. The most common majors were 24% in arts & sciences, 21% in the health professions, 16% in journalism, 15% in business, and 14% undecided, with the remainder in other majors. Their median high school GPA was 3.6 and their median ACT score was 25. Twenty-three percent were 18 years old, 47% were 19, 12% were 20, 16% were less than 25, and less than 2% were over 25.

4.2 Measurement

Data for this study are collected as part of the automated monitoring of student performance with SAGrader. The program is highly reliable with over 99.9% availability to users and built-in safeguards to assure that no student submissions are lost. All of these measures should be perfectly or nearly perfect in both reliability and validity. The specific indices used for each key variable in order of the hypotheses are described as follows.

- **Time to deadline.** This variable is obtained by subtracting the time of each submission in days, hours, and minutes, from the time of deadline. To facilitate graphing, this time is then grouped into 30-minute intervals.
- **Rate of submissions.** This variable is measured by the count of submissions during each 30-minute period.
- **Final score.** This is the score of the final submission for a student expressed as a percent of the possible score.
- **Time since last submission.** This variable is obtained by subtracting the time of each submission in days, hours, and minutes, from the time of the last submission. This number is measured to the nearest minute. For the first submission for each student there is no "last" submission and this variable is set to missing.
- **Final number of submissions.** This is the sequence number of the final submission submitted by a student.
- **Improvement in score per revision.** This variable is computed by subtracting the percent score of the most recent previous submission from the percent score for the current revision.
- **Total elapsed time spent on assignments.** This variable is computed by subtracting the time of the first submission for the student for a particular assignment from the time of the last submission by the same student for that assignment. If a student made only one submission, this score becomes zero.

- **Student boundary-stretching behavior** – SAGrader routinely compares submissions with all previous submissions by other students. A measure of the similarity between submissions is computed and expressed as a percent of the number of words in the current submission. For example, a similarity score of 36% would indicate 36% of the word phrases in the current submission are identical or nearly identical to those in previous submissions. Similarity scores can range from 0% to 100%. Similarities can occur in several ways, including when one student copies the work of another, when both copy phrases from the textbook, or when both copy from the same external source such as a web page on the Internet. Excessive copying of this sort is clearly plagiarism. The use of occasional phrases from the text may be more benign. The greater the similarity the greater the student is pushing the boundary and threatening to step over the line to clear cheating. The program does not make a judgment as to whether a certain level of similarity constitutes plagiarism. Rather, the program provides this information to instructors who make that determination themselves.
- **The number of challenges**. SAGrader has a built-in option permitting students to "challenge" their grade if they believe the program incorrectly scored an assignment. When students challenge the information is recorded and a message sent to the instructor who then reviews the student complaint and can override the program to provide a correct grade, revise the program so that it grades correctly, or determine that the program is correct and give the student helpful advice as to what they did wrong. Each challenge is associated with a particular submission and the number of challenges associated with submissions in each time period were examined.

5 Results

All of the hypotheses for this study predict particular patterns of behavior by students to change as time to deadline approaches. To test these hypotheses the appropriate data were graphed as a function of time to submission. Where appropriate a one-way analysis of variance was used to determine whether the changes over time were significant. In addition, for each hypothesis a particular case was selected from the data that illustrates the findings in a more concrete and easily understood fashion.

5.1 Increased Activity

We expect increased activity by students as time to deadline approaches. Specifically, we predicted higher rates of submissions and shortened time since last revisions. Data addressing these two hypotheses are summarized in Figure 1.

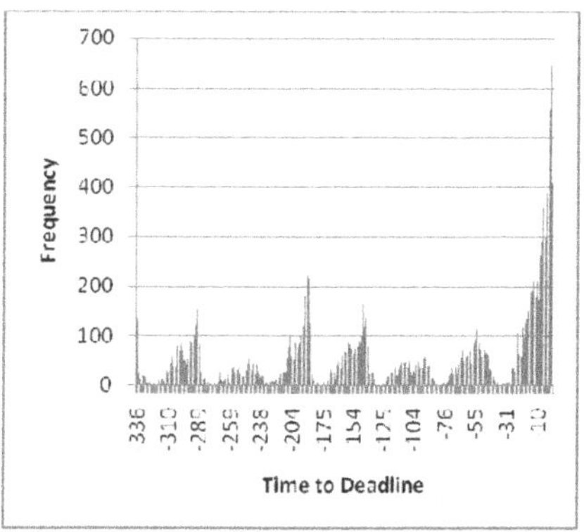

A. Rate of Submissions By Time to Deadline (Last Week)

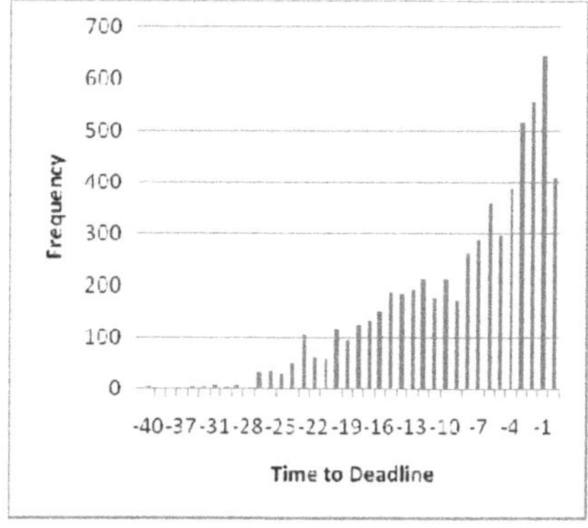

B. Rate of Submissions by Time to deadline (Last 20 Hours)

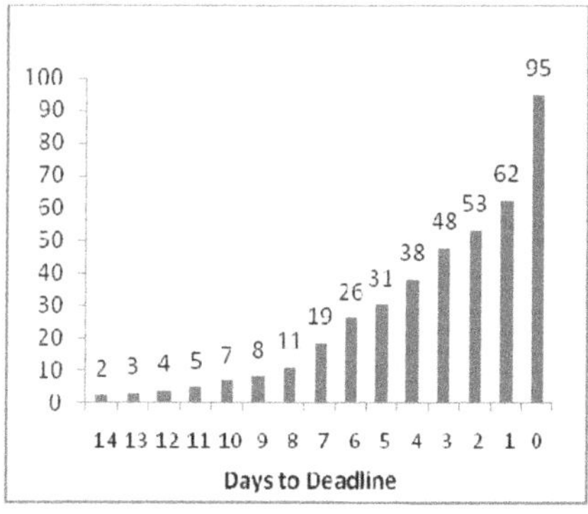

C. Cumulative Percent of First Submissions

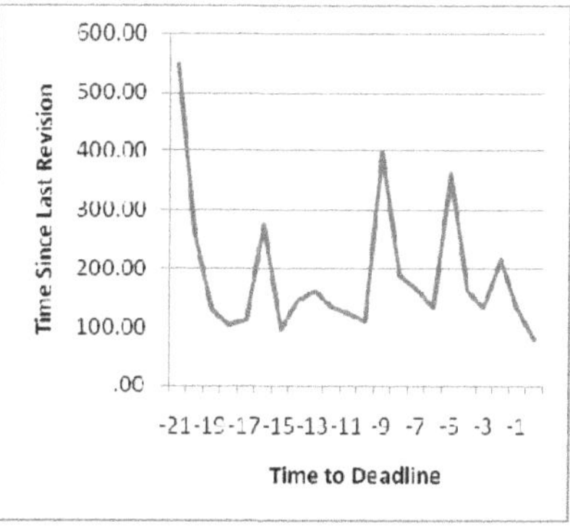

D. Mean Time Since Last Revision by Time to Deadline

Fig. 1 Increased Activity

5.1.1 Rate of Submissions as Time to Deadline Approaches

The first hypothesis is that the rate of submissions should increase along a hyperbolic curve as time to deadline gets smaller. To test this hypothesis, time was broken into 30-minute periods and the frequency of all submissions by students aggregated across all assignments were plotted over the week preceding the deadline. These results are displayed in Figure 1. In this figure, the displayed time-intervals were automatically generated by SPSS to facilitate graphing. A 24-hour period corresponds to a difference of 48 half-hour units on the X-axis of this figure.

The results in Figure 1A provide substantial support for Hypothesis 1. Clearly, the frequency of submissions during the day of the deadline (at the left in Figure 1A) is much higher than for prior days, and frequencies increase dramatically during that last day. In addition, the results in Figure 1A clearly display a cyclical pattern of submissions each day in which the frequency of student submissions begin at near zero around 3 or 4AM, steadily increases until it reaches a peak sometime in the evening, and falls off rapidly until around 3 or 4AM the next day. That daily cycle generally follows a pattern much like the hyperbolic curve shown the last day, increasing to a peak and then dropping off rather quickly. This pattern can also be interpreted in light of TMT if we recognize that there are daily deadlines for when students want to quit work for the day. While not as final and therefore not as highly motivating as the deadline for completing the assignment, the daily deadlines for quitting work produce similar fluctuations in the utility of working on the assignment and similar, but weaker changes in student behavior.

To further clarify the trend in submissions as the deadline approaches, Figure 1B displays the frequencies of submissions during each 30-minute period for just the 20 hours preceding the deadline for assignments. This graph clearly displays a hyperbolic curve with increasing frequencies of submissions as the deadline approaches up until the very last half-hour period in which submissions drop off moderately. To eliminate the confounding effects of daily cycling, subsequent hypotheses will generally be examined only for data from the last day before the deadline. Specifically, the last 11 hours in which there were sufficient numbers of submissions to provide accurate estimates will be broken into 22 half-hour periods for purposes of graphing and mean scores related to each hypothesis will be plotted over time for each half-hour period.

Figure 1C displays the cumulative distribution of first submissions by days to deadline. This graph too, clearly supports Hypothesis 1 and provides clear evidence that students procrastinate. Even though assignments are made available at least two weeks before they are due, fewer than half the first submissions are made by three days before the deadline. One-third of first submissions are made during the last day.

5.1.2 Time Since Last Submission (Speed of Revision)

The second hypothesis is that the time since last submission will decrease as time to deadline gets smaller. To examine this hypothesis, the mean time since last revision was plotted for each half-hour period over the last 11 hours before deadline. The results are displayed in Figure 1D. This mean was computed only for revisions (submissions during each time period for which there was at least one prior submission). These data do not support Hypothesis 2. A linear regression analysis produces a nonsignificant positive regression coefficient ($\beta = .002$, $t = .067$, $p = .946$, $R^2 = .000$). The time between revisions is smaller for submissions closer to the deadline than for submissions much earlier than the deadline. Curvilinear regression models were also examined, but their fit to the data was virtually the same as the linear model and the linear effect is easy to interpret so only the linear model results are presented here.

5.2 Diminished Performance

We expected to find diminished performance as time to deadline becomes short in the form of lower final scores, fewer revisions, less improvement in score per revision, and a reduction in the elapsed time spent on the assignment. Results for these four hypotheses are summarized in Figure 2.

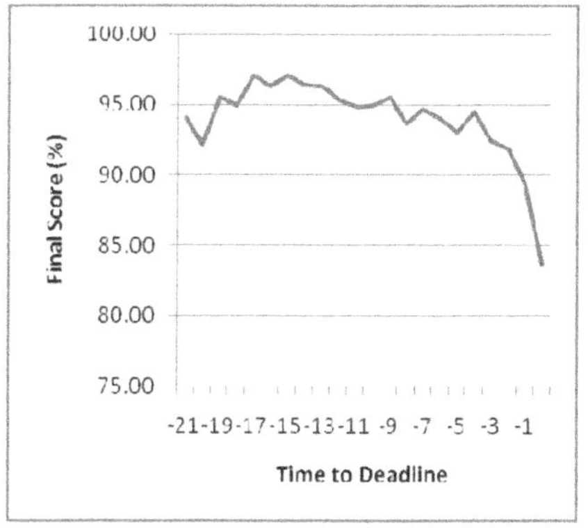

A. Mean Final Score (%) by Time to Deadline for First Submission

B. Final Number of Submissions by Time to Deadline of First Submission

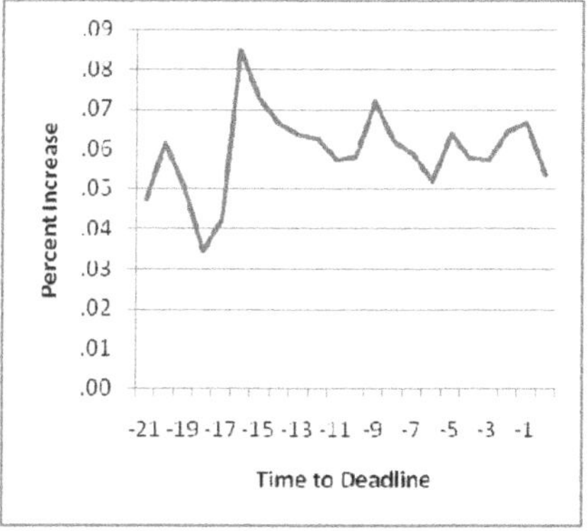

C. Mean Improvement in Score Per Revision by Time to Deadline

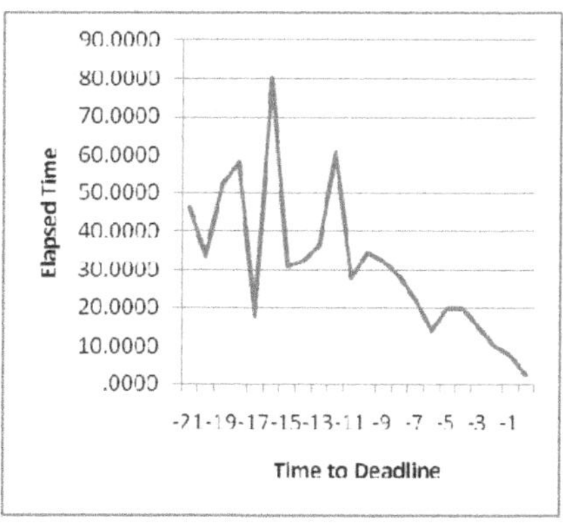

D. Mean Elapsed Time Spent on Assignment by time of First Submission

Fig. 2 Reduced performance as the deadline approaches

5.2.1 Final Score by Time of First Submission

The third hypothesis predicts mean final scores will decrease as time to deadline becomes shorter. This hypothesis is examined in Figure 2A where mean final percent score is plotted by time to deadline for first submissions. The expectation is that as students wait longer to begin their work they are more likely to run out of time and be forced to settle for a lower final score. These results clearly support this hypothesis. The mean final score for essays begun during the last half hour is less than 85%, while the mean final score for essays most time periods is well over 90%. A linear regression found a significant positive effect of time of first submission on the mean final essay score (β=.176, t=13.575, p<.000, R2=.031). There is a clear reduction in final essay score percent as the time of first submission gets closer to the deadline.

5.2.2 Final Number of Revisions by Time

The fourth hypothesis is that the final number of revisions will decrease as time to deadline becomes shorter. This hypothesis was examined by graphing the mean sequence number of the last submissions for each of 22 one-half hour time segments making up the last 11 hours of time before the deadline. The sequence number of the final submission for a student tells us the final number of submissions by that student. These results are displayed in Figure 2B. This graph shows a lot of variation and the trend is difficult to discern. However, a linear regression analysis finds a significant positive relationship between time to deadline at first submission and the final number of submissions ($\beta = .081$, $t = 5.685$, $p < .000$, $R^2 = .006$), indicating that students who wait until nearer the deadline to make their first submission are more likely to end up with a smaller number of submissions than students who begin earlier.

5.2.3 Improvement in Score Per Revision

Hypothesis 5 states that the improvement in score for each revision will decrease as time to deadline diminishes. This hypothesis is examined in Figure 2C where the mean percent increase in score is plotted against the time to deadline for all revisions. Improvement in score of course has no meaning for the first submission, so only revisions are included in this analysis. Results in Figure 2C do not support the prediction. There is considerable variation in these data making it hard to discern the trend visually. A linear regression analysis found no significant relationship between submission time and the percent increase in score per revision (β= -.005, t=-1.357, p=.721, R^2=.000). It is difficult to know why this hypothesis was not confirmed. But certainly one difficulty is separating students desperately trying to improve their score quickly as deadline approaches from other students who spent a long time on their revision and only submitted it just before deadline.

5.2.4 Elapsed Time Spent on Assignment

Hypothesis 6 predicts that the mean elapsed time spent on assignments from the first submission to the last will decrease as the time of initial submission gets closer to the deadline. This is a straightforward hypothesis based on the expectation that the longer students wait to begin on their assignment the more likely they are to run out of time and be forced to spend less elapsed time on the assignment.

In practice, however, this hypothesis is more difficult to test because many students would begin working on an assignment one day, perhaps make a few revisions, then put it aside and come back to it days or even weeks later. That pattern of intermittent work on the assignment clouds our understanding of the amount of time students actually spend on the assignment. Worse yet, very long gaps between work sessions unduly influence the mean elapsed time, hiding any trends. Since even the most complex of these assignments should not generally take more than a few hours to complete, mean elapsed times of days or more for assignments are clearly a distortion.

To compensate for this, we only examined students who both began and ended their work on the assignment during the last 11 hours. The question is, do students who begin closer to the deadline expend less elapsed time on the assignment than those who began earlier in the day. The mean elapsed time spent on analyses as a function of how long before the deadline was the students' first submissions is reported in Figure 2D. The tendency for students to spend less elapsed time on the assignments when their first submission was closer to the deadline is quite apparent. In addition, a linear regression analysis finds a significant relationship between elapsed time spent on assignments and the time before the deadline when the initial submission is made ($\beta = .227$, $t = 9.111$, $p < .000$, $R^2 = .051$).

5.3 *Efforts to Overcome Low Performance*

Our third set of hypotheses predict that, as student performance is lowered by procrastination, students are more likely to take more extreme measures to increase their final score. These include increased challenges questioning the program's score and boundary-stretching behaviors in which student work looks suspiciously like that of other students or external sources. Results pertaining to these hypotheses are summarized in Figure 3.

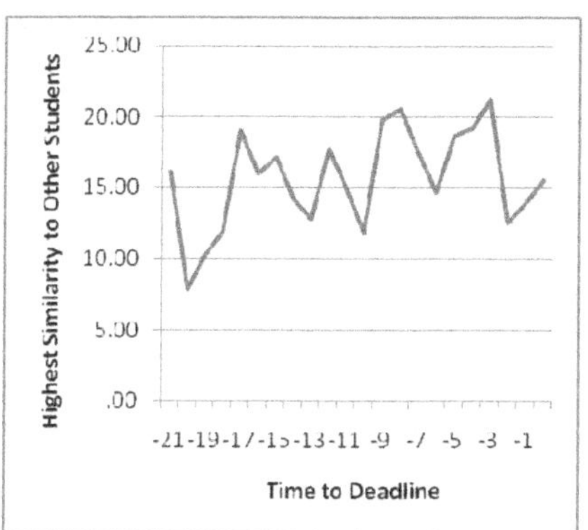

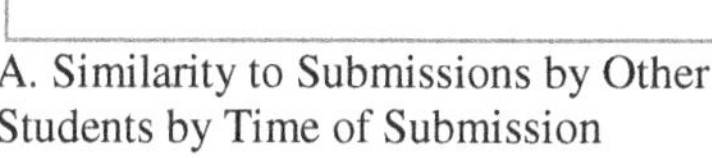
A. Similarity to Submissions by Other Students by Time of Submission

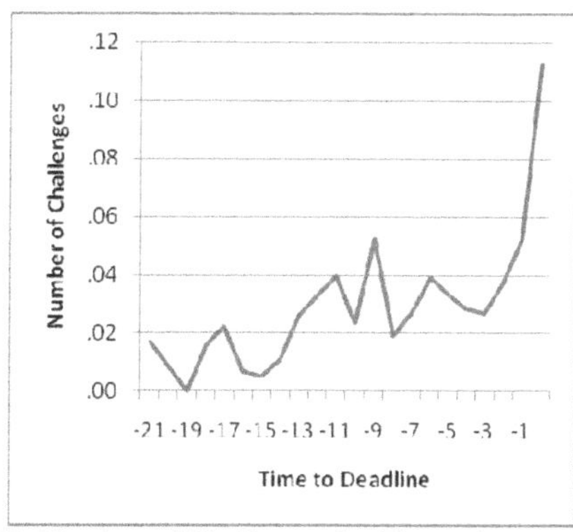

B. Number of Challenges by Time of Submission

Fig. 3 Increased efforts to overcome low performance

5.3.1 Increased Student Boundary-Stretching Behaviors

The seventh hypothesis predicts students will exhibit great boundary-stretching behavior as time to deadline becomes shorter. Specifically, we expect that as time to deadline gets shorter, students will display a higher mean percent match between their submissions and those of one or more other students. Some degree of similarity is to be expected in assignments where students are discussing the same issues. In some cases, they are even asked to cite passages from text files to illustrate important concepts. So low to moderate amounts of overlap with other students are legitimate. But obviously extremely high overlap is likely an indication of actual cheating. This hypothesis predicts that students will be more likely to push the boundaries and perhaps "borrow" some or all of the work of other students as time to deadline decreases.

This hypothesis can be examined by plotting the mean of the highest match between a submission by one student and submissions by other students as a function of time to deadline. Such a plot is provided in Figure 3A. These data do not support the hypothesis. As time to deadline becomes shorter, the mean match between student submissions and submissions by other students increases slightly. However, this trend is not significant as measured by a linear regression analysis, producing a nonsignificant negative regression coefficient ($\beta = -.016$, $t = -1.248$, $p = .212$, $R2 = .000$).

5.3.2 Number Challenges by Time

Finally, hypothesis 8 suggests that the number of challenges will increase as time to deadline becomes shorter. This hypothesis is examined by plotting the mean number of challenges by time to deadline for all submissions in Figure 3B. These results generally support the hypothesis. Over the period between 11 hours before the deadline and the deadline there is a clear upward trend in the mean number of challenges per submission. This means that students are more likely to challenge SAGrader's score on a submission if the submission occurs nearer the deadline. A regression analysis identified a significant negative relationship between the number of challenges and time to deadline ($\beta = -.052$, $t = -3.945$, $p < .000$, $R^2 = .003$).

5.4 A Few Illustrative Cases

To help make the implications of these findings more concrete, we examined in depth a few cases that illustrate a number of the key findings of the study. These cases are not selected at random, but were identified by scanning the submission records of 10 students making submissions during the last half hour before the deadline for a particular assignment. From those we selected three students, each of whom exemplify one or more important findings from the quantitative analysis of procrastination and its effects. The assignment chosen was worth twice as much as most assignments in the course and was made available to students more than two weeks before the deadline, as is true of all assignments in the course.

Case 1 is a male student. His first submission to this assignment is at 29 minutes before the deadline and receives a score of 77%. He makes his second submission 8 minutes later, receiving a score of 86%. His final submission is at one minute before the deadline and receives a score of 93%. At 17 minutes after the deadline, he challenges the program's grade on one concept. At twenty minutes after he challenges the program's grade on the second concept. At 26 minutes after the deadline the course TA gave him credit for the first challenge, and at 27 minutes after the deadline he was given credit for the second challenge. Case 1 clearly illustrates the tendency to procrastinate, first submitting to the assignment less than ½ hour before it was due, followed in rapid succession by two more submissions. He also illustrates the increased likelihood of challenging as time to deadline approaches by challenging two components of his grade. He does not illustrate an increased likelihood of cheating, the tendency for lower performance (his score is higher than the average final score on this assignment of 86% or the predicted (but not found) tendency to show lower improvement per submission as time to deadline approaches (he improves by 9% on his second submission and an additional 7% on his third submission).

Case 2 is a male student. He first submits to this assignment 72 minutes before it is due with a grade of 48%. He follows with four more submissions in rapid succession at 40 minutes before the deadline (his score is 62%), at 13 minutes before the deadline (his score is 70%), 8 minutes before the deadline (again a score of 70%), and finally at 5 minutes before the deadline with a final score of 70%. At 2-½ days after the deadline he challenges the program's grade on a particular section. Within an hour of that challenge, the course TA responded to his challenge pointing out how he was incorrect and steering him to the right section of the lectures to answer that part of the assignment. Case 2 illustrates many of the findings of the quantitative analysis. He clearly procrastinated with his activity only becoming apparent in the last hour before the deadline. He shows diminished performance, having a final score of 70% well below the average of 86% for this assignment; he makes many revisions in short order with only small improvements in score; and ends with 5 submissions, slightly less than the average 5.1 overall for this assignment. He also illustrates the tendency as the deadline approaches for increased challenges, but he does not illustrate the increased possibility of cheating.

Case 3 is a female student. Her first submission for this assignment occurs 35 minutes before the deadline. Her score is 97%. Seven minutes later she makes her second submission which is unchanged and receives the same score. At 21 minutes before the deadline she makes her final submission which includes one additional sentence and brings her score to 100%. Her initial score is a surprisingly high score on one of the most difficult assignments for the class. A later examination of SAGrader's tool to compare matches between student submissions revealed that 98% of her final submission matches the previous submissions of another student from an earlier semester. The instructor judged this to be cheating and SAGrader then generated a pdf file comparing the submissions of the two students highlighting identical passages. The student was given a zero on the assignment and was reported to the campus administration for possible disciplinary action. While her case does not illustrate most of the findings we would have expected near the deadline, she clearly cheated and that appears to have reduced her perceived need for the other behaviors normally expected near deadline.

6 Discussion and Conclusions

With one exception, these results provide consistent support in each of the three broad areas in which we make predictions about student behaviors.

First, as time to deadline gets shorter we see **increased activity** by students through increased rates of submissions and reduced time lag between submissions. Students submit more often and more quickly nearer the deadline. This provides strong and clear evidence of procrastination. The hyperbolic curve showing dramatic increases in submissions for the last day before deadline also supports the predicted pattern of increased activity based on the TMT perspective. The cumulative distribution of first submissions in which one third of first submissions do not occur until the last day before the deadline provides clear evidence of procrastination.

Second, three of the four hypotheses are supported regarding the predicted **decrease in performance** as time to deadline decreases. Students who wait until nearer the deadline to make their first submission are more likely to have a lower mean final essay score (hypothesis 3), make fewer submissions (hypothesis 4), and spend less elapsed time on assignments than students who began earlier (hypothesis 6). The single exception to this trend, hypothesis 5, which predicted the mean improvement in percent score per revision would become smaller as time to deadline became shorter, was not supported. That is, even near the deadline, students continue to achieve roughly the same level of improvement each time they revise, rather than a decline as expected. In general, these findings show, as expected, that students who procrastinate will spend less elapsed time on the assignment, make fewer submissions, and perform more poorly. That is, procrastination tends to lead to diminished performance.

Finally, as the deadline approaches students make increased **efforts to overcome low performance**, including more boundary-stretching behaviors and more challenges. As time to deadline decreases the mean similarity to submissions by other students and the mean number of challenges both increase significantly. This may reflect what Grijalva, Nowell, and Kerkvliet (2006) call panic cheating. Results are in many respects consistent with a panic response as the deadline approaches. Students wait too long to begin work, see their initial scores are not as high as they had hoped, begin submitting more often and with less time between submissions, realizing that they are running out of time and being forced to choose between lower performance and cheating or challenging. At some point, for at least some students, the utility of achieving a good final score may exceed the utility of maintaining academic integrity.

Together, these results paint a picture of student time management in which procrastination is rampant and its consequences significant. The strongest consequence of procrastination, accounting for 4.5% of the variance, was the most obvious one – a reduction in time students spent on assignments. The magnitude of other consequences of procrastination was fairly small, with low R^2 accounting for less than 1% of the variance. Essentially, waiting until late to begin work on the paper reduces the time available for the task, forcing students to reduce the elapsed time spent on the work and leading to lowered final scores despite their increased rate of revision. For some students this also leads them to pursue additional efforts to mitigate the damage, including challenging their grade or even cheating.

This study makes a significant contribution to understanding procrastination behaviors and their consequences. In most courses students working on a written assignment conduct most or all of their work on their own and that work is not monitored. As a result, much of our knowledge of how student behavior changes as time to deadline approaches is anecdotal. In contrast, SAGrader provides a unique opportunity to study how students engage in homework over time as a deadline approaches. The large number of writing assignments provides the opportunity for studying the student writing process in greater depth than in most courses. The opportunity to make multiple submissions and the availability of detailed feedback give students an incentive to revise their work until they achieve an acceptable grade. Students typically submit multiple revisions, opening up a window for studying the writing process in much greater depth than possible in most classes. This made it possible to track single students through multiple submissions, monitoring improvements in their papers through changes in their score, and examining their pattern of time-management through the timing of their submissions. These data have provided direct empirical evidence of procrastination and some of its important behavioral consequences.

Results of this study have practical implications for teaching and learning. These results show high levels of procrastination among these college students, with over a third of students not making their first submission to assignments until the last day. These results also show that among the consequences of procrastination are both lowered student performance and an increased incidence of cheating.

Admittedly, there are significant limitations to the information available in this learning environment. The timing and performance of students for each of multiple submissions are examined and aggregated across many different assignments. Left unmeasured are the details of when and for how much time students worked on their first draft. The process of writing takes time. There could have been hours of reading and writing that took place before students make their first submission. So the first evidence of work by students in the form of their first submission should naturally be expected to be more prevalent as the deadline nears. Nor does this study have access to the entire process of revision. This study tracks only each submission, not intermediate revisions of drafts of which there may be several between submissions.

Having acknowledged these limitations in the information available, it should also be said that there are factors that mitigate the problems it poses. Students have a clear incentive to submit early and often to receive feedback to improve their grades. In course evaluations students consistently mention the immediate detailed feedback as one of the greatest strengths of SAGrader. In fact, in many cases students submit only very cursory initial drafts clearly designed to elicit the detailed feedback from the program that can help them revise and improve their paper.

This study addresses only the effect of time delay in Temporal Motivation Theory. It does not examine variables that are likely to influence the other parameters in the model. Examples include demographic variables such as gender that is often cited as a factor influencing procrastination, or other variables that may influence perceived likelihood of success, such as high school GPA. Nor does this study examine ways in which those other variables interact with time delay to influence utility. Utility itself is also not measured directly in this study. Instead, behavioral indicators such as the rate of submissions are used to measure likely consequences of changed utilities. For these reasons, while this study clearly documents procrastination and its effects, it does not provide practical guidance for reducing procrastination.

The generalizability of results of this study can be assessed both with respect to the sample of data included and the learning environment created by the use of this essay grading program. These short writing assignments were assessed in an introductory social science course as part of a general education curriculum for students in a large public Midwestern university. The students in this course were representative of the undergraduate student population at this university but reflect less racial and ethnic diversity than the U.S. as a whole. Conceivably, this might limit the generalizability of findings somewhat.

The learning environment studied here is distinctive and offers advantages over other learning environments by permitting multiple submissions without penalty and providing immediate feedback to guide revision (Brent, et al., 2010). Yet it remains in many respects similar to homework assignments in other courses. As in many other courses, students are asked to write essays addressing specific issues covered in the course and to complete their work by a fixed deadline with significant penalties for being late (50% reduction the first week after the deadline and 100% reduction after that).

The SAGrader CSCL environment may differ in some important ways from other kinds of academic homework. One concern is that, because SAGrader permits students to revise their work many times and receive immediate feedback this provides a set of intermediate deadlines for each revision. That might actually discourage student procrastination. For example, Wesp (1986) found that students in a course with daily quizzes completed the course more rapidly and with higher grades than students with self-initiated quizzing. He argues the scheduled quizzes help students manage their time more effectively and reduce procrastination. The multiple submissions students make with SAGrader are not scheduled deadlines. But they are self-imposed deadlines students meet in order to check their work and receive feedback. Conceivably these deadlines give students added incentive to reduce procrastination. If so, then levels of procrastination found in this study may be lower than levels that might occur in settings that do not permit multiple revisions.

Some evidence in support of the importance of self-imposed deadlines for affecting student behavior is provided by the daily cycles of student submissions with large increases near the end of each day as displayed in Figure 1A. Those daily patterns resemble the effects of the final externally imposed deadline only with reduced magnitude. These are not formal deadlines imposed on students by the course, but are practical deadlines for students who have many demands on their time and need to find time for rest as well as study. Conceivably the deadlines individual students impose upon themselves for the submission of each revision may have effects similar to the daily deadlines of needing to find time to rest. Furthermore the presence of those daily "deadlines" are likely to affect work in many types of learning environments further reducing the uniqueness of this learning environment.

This study also helps us understand procrastination in the context of collaborative assignments. Time-shifted collaboration provides a unique environment in which students can interact with the instructor without the presence of the instructor. The effects of procrastination in this environment, while small, are significant. Students that procrastinate in seeking feedback receive lower scores. In a time-shifted environment, overcoming the issue of collaboration may require having intermediate assignments. One way to do this is to have assignments that focus on parts of a larger assignment, in a similar manner to scaffolding. This would allow students to receive feedback early, which could then be incorporated into their future assignments. In traditional collaboration, this might be accomplished, with little cost to the instructor, by having intermediate grading points in which students grade the work of each other. This would decrease procrastination on the part of the student and, by so doing, decrease procrastination on the part of the group. This would also expose members of the group to each other's work earlier in the process, allowing a better synthesis of the whole project to be achieved by each student. This same strategy would be effective in all collaborative contexts (e.g., business), and could be accomplished by setting many small deadlines within a larger collaborative project.

This research is only a beginning of direct studies of procrastination and its effects, particularly in collaborative (and especially in time-shifted) contexts. It has effectively demonstrated that we can empirically study these effects. But the large fluctuations seen in the trends over time and the relatively small (but significant) effects show that there is much left to be learned. Larger effects can be expected if future research can capture more of the underlying dynamics of procrastination. Other variables influencing the TMT model need to be examined in future studies. Among these, task characteristics such as difficulty of assignment and individual characteristics like past student GPA and gender seem among the most promising additional variables that should be considered. Other process variables need to be examined as well. For example, it seems likely that students making good progress improving their score on revisions will be encouraged to continue revising, while those who are making little progress may be more likely to give up. As additional variables are included in the research it should become feasible to more accurately estimate some of the key parameters of the TMT model such as the appropriate unit of measure for time delay, D, and the range of individual levels of susceptibility to time delay, Γ. Future research should examine how well these results hold up for other student populations and in other digital collaborative learning environments where data are available for tracking procrastination and its consequences. Of particular interest are environments where there may be less incentive and opportunity for intermediate deadlines. Finally, it would be interesting to conduct future research with added precision that monitors not only changes from one submission to the next but also the timing and nature of modifications that occur within each revision.

References

1. Brent, E., Atkisson, C., Green, N.: Time-Shifted Online Collaboration:Creating Teachable Moments through Automated Grading. IGI Publishing, Hershey (2010)
2. Brent, E., Carnahan, T., McCully, J.: Students Improve Learning by 20 Percentage Points with Essay Grading Program, SAGrader ™. Idea Works, Inc, Columbia (2006a)
3. Brent, E., Carnahan, T., McCully, J., Green, N.: SAGrader ™: A Computerized Essay Grading Program, Version 2. Idea Works, Inc., Columbia (2006b)
4. Gafni, R., Geri, N.: Time Management: Procrastination Tendency in Individual and Collaborative Tasks. Interdisciplinary Journal of Information, Knowledge, and Management 5 (2010)
5. Gerdy, K.: Law Student Plagiarism: Why It Happens, Where It's Found, and How to Find It. BYU Education 431 (2004)
6. Grijalva, T.C., Nowell, C., Kerkvliet, J.: Academic Honesty and Online Courses. College Student Journal 40(1), 180–185 (2006)
7. Gritz, E., et al.: Successes and Failures of the Teachable Moment: Smoking Cessation in Cancer Patients. Cancer 106(1), 17–27 (2006)
8. Henry, T.R., LaFrance, J.: The Changing Role Of Computing Education: Fostering Collaboration. Issues in Information Systems 7(1) (2006)
9. Kasper, G.: Tax procrastination: Survey finds 29% have yet to begin taxes (2004), `http://www.prweb.com/releases/2004/3/prweb114250.html` (retrieved May 25, 2010)
10. Kerkvliet, J.: Cheating by economics students: A comparison of survey results. Journal of Economic Education 25, 121–133 (1994)
11. Last, M.Z., Almstrum, V.L., Daniels, M., Erickson, C., Klein, B.: An international student/faculty collaboration: The Runestone Project. SIGCSE Bulletin 32(3), 128–131 (2000)
12. Norton, L.S., Tilley, A.J., Newstead, S.E., Franklyn-Stokes, A.: The pressures of assessment in undergraduate courses and their effect on student behaviors. Assessment & Evaluation in Higher Education 26, 268–284 (2001)
13. O'Brien, W.K.: Applying the Transtheoretical Model to Academic Procrastination. University of Houston (2002)
14. O'Donoghue, T., Rabin, M.: Incentives for procrastinators. Quarterly Journal of Economics 114, 769–816 (1999)
15. Roig, M., De Tommaso, L.: Are college cheating and plagiarism related to academic procrastination? Psychological Reports 77(2), 691–698 (1995)
16. Silver, M., Sabini, J.: Procrastinating. Journal for the Theory of Social Behavior 11, 207–221 (1981)
17. Steel, P.: The Nature of Procrastination: A Meta-Analytic and Theoretical Review of Quintessential Self-Regulatory Failure. Psychological Bulletin 133(1), 65–94 (2007)
18. Steel, P., Brothen, T., Wambach, C.: Procrastination and personality, performance, and mood. Personality and Individual Differences 30, 95–106 (2001)
19. Steel, P., Konig, C.J.: Integrating Theories of Motivation. Academy of Management Review 31, 889–913 (2006)
20. Stover, M., Kelly, K.: Institutional responses to plagiarism in online classes: Policy, prevention, and detection. Paper presented at the 18th Annual Conference on Distance Teaching and Learning (2005)

21. Tice, D.M., Baumeister, R.F.: Longitudinal study of procrastiation, performance, stress, and health: The costs and benefits of dawdling. Psychological Science 8, 454–458 (1997)
22. Waite, W.M., Jackson, M.H., Diwan, A., Leonardi, P.M.: Student Culture vs Group Work in Computer Science. SIGCSE Bulletin 36(1), 12–16 (2004)
23. Wesp, R.: Reducing Procrastination Through Required Course Involvement. Teaching of Psychology 13(3), 128–130 (1986)
24. White, V.M., Wearing, A.J., Hill, D.J.: Is the conflict model of decision making applicable to the decision to be screened for cervical cancer? A field study. Journal of Behavioral Decision Making 7, 57–72 (1994)
25. Zarick, L.M., Stonebraker, R.: I'll Do It Tomorrow: The Logic of Procrastination. College Teaching 57(4), 211–215 (2009)

Interaction Analysis as a Tool for Supporting Collaboration: An Overview

Georgios Kahrimanis, Nikolaos Avouris, and Vassilis Komis

Human-Computer Interaction Group, Department of Electrical and Computer Engineering, University of Patras, Greece

Abstract. This chapter constitutes an overview of logfile-based interaction analysis techniques that can be used for the support of Computer Supported Collaborative Learning (CSCL) activities. Interaction analysis is central in the study of CSCL activities, since in such activities through interactions between partners the state of evolving group knowledge is communicated. This interaction is facilitated by tools that allow logging of events that take place, capturing thus information about the content and the process of collaboration. Automated analysis techniques of this information can be developed. The objective of this analysis is often to support participants, in several ways: explicitly, by providing feedback to them in order to regulate their practices, or by making adaptive changes to some aspects of the collaborative setting; or implicitly, by making available to them representations of their activities. This chapter presents the most common approaches used in interaction analysis, while it particularly emphasizes recent innovative efforts to reap the advantages of machine learning techniques in order to overcome common shortcomings of previous approaches.

1 Introduction

Computer Supported Collaborative Learning (CSCL) constitutes a field of research and practice in the broader context of study and development of educational technologies. This research field is inspired by multiple research backgrounds, as it covers a wide range of activities and engages a multi-disciplinary community. In this context, an approach used extensively is the analysis of interaction (Jordan and Henderson, 1995).

CSCL constitutes a suitable field of applying analysis of interaction since, in collaborative learning, the state of evolving knowledge must be continuously displayed by collaborating participants with each other (Stahl 2002). Therefore, what one participant communicates with others is accessible to researchers, providing thus an objective source for analysis (Dillenbourg et al. 1995). Analysis is based on such observable interactions rather than measures of learning outcomes, models of students' mental representations, or internal cognitive processes, as is the

T. Daradoumis et al. (Eds.): Technology-Enhanced Systems and Tools, SCI 350, pp. 93 – 114.
springerlink.com

case with other paradigms of instructional technology (Koschmann, 1996). Moreover, the tools that mediate collaboration allow for the logging of events that capture aspects of the content and the process of interaction.

Based on this recorded information, automated or semi-automated analysis techniques can be developed that are used for supporting the collaborative process. This support can be provided in different ways: explicitly, by providing feedback to the participants in order to regulate their practices, or by making adaptive changes to some features of the collaborative setting; or implicitly, by making available to the participants representations of their activities. This support of the collaborative process may be important in many cases, as it can scaffold and enhance collaborative learning.

In order to successfully support and guide collaborative learning activities, and preferably in a dynamic, adaptive way, it is necessary that some knowledge of significant aspects of the process, as it evolves through time, is obtained. This is not a trivial task and in traditional settings depends on the knowledge, experience and intellect of human tutors that intervene to the process accordingly. However, interaction analysis for the study of collaborative processes and the technological collaborative facilities can offer possibilities for automatic evaluation of collaborative processes. Collaboration tools usually keep logs of events of the users' interaction and maintain them in suitably structured logfiles. These entries can then be manipulated and lead to targeted metrics that indicate meaningful aspects of collaboration, interaction, or learning, a process that is conceptualised and discussed in the framework presented later in this chapter.

This chapter constitutes an overview of logfile-based interaction analysis techniques that can be used for the support of CSCL activities. We start with a short description of general issues of analysis and evaluation of CSCL activities, followed by an introduction of a framework of the different stages that interaction analysis usually follows. The most important approaches in CSCL literature of automated interaction analysis based solely on logfile entries are then discussed, including cases where participants of the CSCL processes are forced to annotate parts of their interaction themselves. Such approaches were popular especially in the first years of the establishment of the research field, they have been, however, extensively criticised, the former because they may lead to "surface" metrics that lead to poor indications of collaborative practices, and the latter because they are likely to influence the collaborative process in ways not desired by their designers. The subsequent section is devoted to the most common interaction analysis techniques for which human intervention in the process of analysis is necessary, and that are nonetheless formalizable and suitable to be used for the support of CSCL processes. Such approaches can lead to analysis of collaboration on a deeper level, since they are based on subtler evaluations accessible to the human intellect that can not be conveyed by technologically feasible formalisations. However, such techniques are often arduous and time-consuming and cannot be used for the support of CSCL activities on a timely manner. Finally, the article concludes with thorough discussion of recent advances of automated interaction analysis that try to combine advantages of the two general aforementioned categories of interaction analysis techniques, while they aim at overcoming their shortcomings. These approaches use deeper-level evaluations of CSCL activities conducted by human analysts in order to train models of interaction analysis based on automated logfile captures. It is expected that the latter approaches would offer qualitatively advanced opportunities for the meaningful and efficient use of automated interaction analysis for supporting CSCL activities.

2 Analysis and Evaluation for the Support of Collaborative Learning Activities

CSCL covers a wide range of educational activities many of which are characterised by extended complexity. For this reason, the study of CSCL activities follows several approaches and traditions of research that can be discriminated in several ways. A major distinction that applies to the case of CSCL as well as to most research disciplines regards the distinction between *basic* and *applied research*. This distinction is determined by the objectives of a research study. In the first case, the goal of research is to gain insight into CSCL activities themselves in order to build new knowledge in the field, whether this is done by descriptive, qualitative studies of detailed episodes of collaboration, or by testing experimental hypotheses in order to understand the role of significant variables that influence and affect collaboration and its possible learning outcomes (Stahl et al. 2006). The first body of studies in CSCL research focused on the comparison between the efficiency of the new educational approach and traditional methods of instruction, in order to prove that the new approach was worth pursuing in terms of the learning benefits that it can offer and the efficient use of resources possibly spent in an institutional context (Dillenbourg et al. 1995). As the field was evolving, it became evident that success in the field of study was subject to multiple and extensively intermingled factors. Therefore, the next trend of basic research put more emphasis on the conditions under which the CSCL approach can be fruitful. A number of factors of different kinds can influence a CSCL process. The means of communication (synchronous vs. asynchronous collaboration), spatial constraints (co-located vs. distant collaboration), the structure of the activity, the profile and knowledge background of learners and the way they form collaborative groups are some of the factors that can shape the flow and the outcome of a CSCL process (Dillenbourg et al. 1995).

At another level, in addition to CSCL basic research proper, there is a strong need for development of efficient and effective analysis and evaluation techniques for collaborative activities, suitable for practical uses in real-world settings. We refer to this general approach as applied research. Evaluation can be discriminated from research in general or analysis in that it intends to lead to judgments on the activity, whereas research's focus is mainly to describe, explain, or predict. Moreover, analysis is descriptive, whereas evaluation is normative. Analysis is conceived as of lower level than research and can inform the latter without necessarily producing axiological judgments, although it may be influenced by some form of implicit values. The main general objectives of applied CSCL research are:

- to inform the design of new tools that mediate or analyze / evaluate interaction
- to inform new pedagogical and organizational designs of CSCL activities
- to provide teachers with the means of making assessments of students' performance (by evaluating not only the outcome of a CSCL activity, but the process through which learning gains may be achieved)
- to intervene to the collaborative process in ways that are deemed beneficial for participants

The tools that mediate or analyze collaboration are crucial for the shaping of CSCL and CSCL research respectively, as is the design of tasks that students are asked to engage with, and the shaping of the broader setting of a CSCL activity. In cases where this kind of objectives necessitates an evaluation approach, this can be of a formative or summative variety. Formative evaluation is conducted in some intermediate part of the process and is concerned with the improvement of the object of study, whereas summative evaluation takes place after the end of the studied phenomenon and intends to examine its overall effects. A specific case of the use of summative evaluation regards the need of assessing students participating in CSCL activities in some educational context in ways appropriate for this new educational approach. An example of formative evaluation relates to timely feedback that can be given to students of a CSCL process based on their collaborative practices.

The latter case relates to the goal of many CSCL analysis and evaluation studies that aim at monitoring the progress of the collaborative process and at allowing for timely adjustments to be made. An overview of such approaches constitutes the object of this article. The need for supporting collaborative processes arises from observations that effective CSCL activities need, in many cases, to be designed in such a way as to provide for adequate feedback that scaffolds the learning process. It has been found that simple participation in a collaborative activity does not guarantee that learners gain any benefits (e.g. Salomon and Globerson 1989), as collaborative learning activities can be fruitful, and preferable to more traditional approaches, under specific circumstances.

Interventions in collaborative processes can be made by tutors and supervisors, by the tools that mediate interaction, or by both. A categorization of CSCL tools regarding this issue has been proposed by Jermann et al. (2001) and Soller et al. (2005), distinguishing CSCL tools into monitoring, mirroring, and guiding tools.

Monitoring tools refer to the elementary facilities that a mediating CSCL tool must provide. The basic objective that such a tool must fulfill is the consistent transmission of one user's actions to all their partners. The tool must provide awareness (Dourish and Bellotti 1992; Rodden 1996; Gutwin and Greenberg 2002) of each user's actions, coordinate their actions, and ensure technologically seamless communication. Monitoring tools do not support any kind of analysis.

Mirroring tools or *meta-cognitive tools* extend the scope of CSCL tools by integrating analysis facilities. Such tools process data that are stored in logfiles they sustain, and supply the results of processing to collaborating participants and, possibly, to supervisors that may intervene in the process, and to researchers. This way, learners can use the results of analysis in order to assess the extent of collaboration of their group or their personal contribution to the process. Analysis data are thus reflected (or "mirrored") to the students, who are responsible for the interpretation of results and the adaptation of their practices so that they become beneficial for the whole process.

Guiding tools go a step further: they use analysis results, for advising or tutoring the students. Analysis results are not simply reflected to the users in order to be interpreted by them, but the system intervenes directly, trying to substitute or complement the role of a human tutor, and inevitably, evaluates the practices of the students at a given time. This way, the learning process is suitably adapted, based on the performance of the participants.

The support of CSCL activities that is of interest in this chapter refers to mirroring and guiding types of CSCL tools, as the first case of monitoring tools does not involve any kind of analysis or evaluation. The *meta-cognitive* character of mirroring tools concerns the existence of awareness about cognitive aspects of the collaborative process so that participants control and self-regulate their current practices in order to overcome perceived shortcomings (Brown 1987). It was originally perceived at the level of just an individual but this can also be generalized at the group level, based on conceptualizations of distributed cognition (Salomon 1995). Meta-cognition is supported by computer tools by automated interaction analysis processes, as will be discussed later in this chapter.

Guiding tools support the collaborative process in more interventionist ways and there can be several conceptualizations of their use. They can be thought of as scaffolding tools. In more conventional educational settings, scaffolding refers to targeted interventions by tutors and other educational agents that aim at changing the problem at hand so that the learner is able to perform tasks that would otherwise be out of their reach (Reiser 2002). There is a long history of theorizing related to the concept of scaffolding, from Vygotsky's work on the "zone of proximal development" (Vygotsky 1930/1978) to the concept of cognitive apprenticeship (Collins et al. 1989). Moreover, paradigms of instructional technology such as Intelligent Tutoring Systems (that eventually aim at substituting a human tutor with a computer-based, automated one) are extensively based on the concept of scaffolding (Shute and Ptsotka, 1995). Scaffolding usually targets at task-related issues in most single learner educational approaches, but in collaborative learning, scaffolding may also focus on the improvement of the practices of students that regard the process per se, their collaborative skills, their contribution to teamwork etc. It is thus adaptive in the sense that it behaves dynamically depending on knowledge of significant aspects of the collaborative process as it evolves through time. It may also refer to simple aspects such as the need for balanced interaction between the students in terms of contributions to the communicative process, or to subtler interventions that shape the whole educational design of the CSCL activity. In the latter case, the concept of scripts (Kollar et al. 2006) plays a crucial role. Scaffolding may concern changes of the whole design of a CSCL activity implementing therefore the case of adaptive scripts (Rummel and Weinberger 2008), and part of this dynamic behaviour can be based on tools that are informed by interaction analysis techniques. The least task-specific cases of scaffolding are of interest in this article that deals with the utilisation of interaction analysis techniques for that purpose.

3 A Framework for Interaction Analysis and Evaluation Techniques

The typical process of logfile-based interaction analysis can be formalized in a multiple stage process according to the representation of Fig. 1.

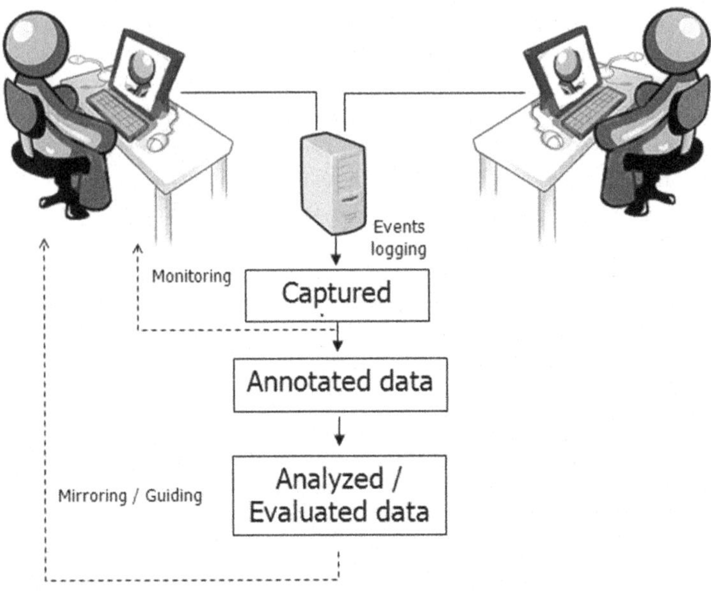

Fig. 1 A framework describing logfile-based interaction analysis

The CSCL mediating tool collects interaction data that refer to events captured and stores them in a logfile. Each recorded event is annotated according to a predefined typology, and related to the user who has generated it, the time when it occurred, and other aspects that convey additional information. Additional annotation may then be applied at another level. These annotated data are processed and analyzed so that meaningful results are obtained. The outcomes of this processing are then interpreted by the researcher or automatically by the tool, and can be used to reshape the collaborative process whether this regards automatic changes in the tool's behavior or the explicit provision of feedback to learners.

As stated above, in the first stage, CSCL mediating tools keep logs of interaction events that users generate with them and automatically assign them to categories. Of major importance is the typology used that describes the types of events logged. An example of such a typology is defined in OCAF (Object-oriented Collaboration Analysis Framework) (Avouris et al. 2003). This framework was created for the meta-description of data captured by CSCL tools in the case of collaboration through chat and shared workspace. In such cases, students collaborating in small groups create a joint model that constitutes a solution to a given problem. Typical events reported refer to the posting of chat messages and the creation and manipulation of objects in a shared workspace. OCAF serves for an integrated description of events generated through such means of communication. Workspace-related actions are automatically annotated according to predefined rules integrated in the tool's functionality. For example, the meta-description of the insertion of a new object in the shared workspace can be done straight after the action is recorded and propagated to all collaborating users. However, in the case of chat-related events, such automatic annotation is not always possible and usually demands the involvement of human annotators, as the meaning of natural language cannot be easily extracted by the machine. One way of bypassing this problem is to render the users responsible for annotating their own messages. This can be done explicitly, by necessitating that they associate each message they sent with a specific type (Barros and Verdejo 2000); or implicitly, by providing them with a set of "sentence openers" that are transparently related to specific categories (Dimitracopoulou and Petrou 2003). Yet the transfer of message annotation duties to participants of a learning process may significantly influence the activity under study, as it may inhibit fluent flow of interactions, shifting focus from cognitive to meta-cognitive tasks.

The last stage of interaction analysis, concerning the analysis and evaluation of annotated data is crucial for the process. It refers to processing of the original dataset that leads to metrics of interaction, informative of significant properties of the collaborative process. Alternatively, the analysis may be of a more interpretative nature, but this model mostly refers to qualitative, highly formalized interaction analysis that is bound to be more useful for the practical support of collaborative learning activities. This stage also involves the interpretation applied to the results of analysis, which is based on judgments about the collaborative process and possibly leads to guiding actions that reshape the process in ways desired by its designers. In the case of the use of metrics of interaction such interpretations may be based on solid rules regarding threshold values of metrics or other criteria that lead to decisions that the tool makes that determine the way it should intervene in the process.

Based on this general conceptualization, the rest of this chapter presents the main approaches to interaction analysis techniques, shown in fig. 2, that are relevant for the provision of support to the collaborative process in practical terms. Such techniques are distinguished in three main categories: fully automated techniques that only build on logfile entries of interaction events in a top-down manner, techniques that necessitate that human agents are engaged in analysis, and techniques that become fully automated only after the logfile-based metrics they use have been trained on deeper-level evaluations conducted by human analysts.

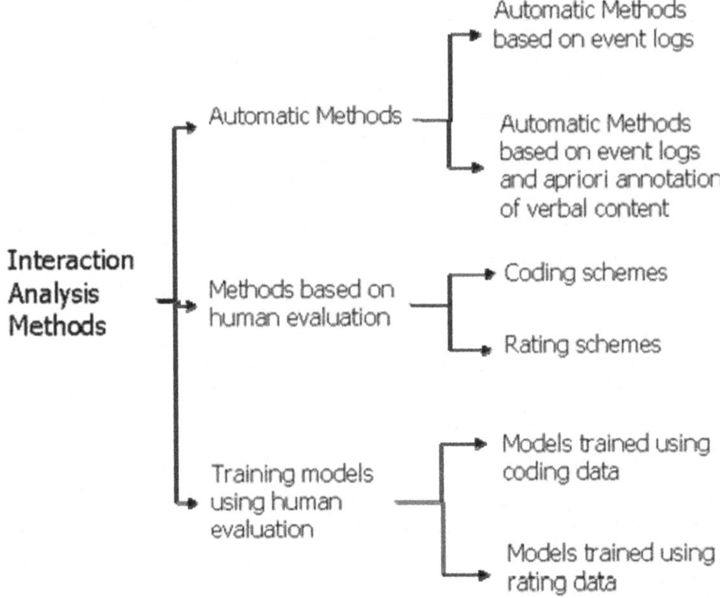

Fig. 2 Overview of interaction analysis techniques

4 Top-Down Automatic Interaction Analysis Techniques

4.1 Automated Interaction Analysis Based on Event Logs

In the first years of the development of the CSCL research field, analysis that was based on measures of automated logfile entries in a top-down manner was particularly popular. For example, numerous metrics indicating the symmetry in collaborative interactions (i.e. the balanced amount of contributions from all participants) have been developed in CSCL or other relevant research disciplines (Hiltz et al. 1989; Warschauer 1996; Constantino-Gonzalez and Suthers 2000; Fitze 2006; Jermann and Dillenbourg, 2008; Marshall et al. 2008; Buisine 2010). In other cases, more sophisticated metrics were proposed: e.g. in the frame of the Synergo analysis tools, Avouris et al. (2004) have developed a set of metrics that reflect interesting aspects of interaction, such as a Symmetry, a Balance, a History, and a "Collaboration factor". Schümmer et al. (2005) have similarly developed a metric that reflects the volume of interaction activity throughout a collaborative process, based on calculations of actions that are characterized by spatial or temporal proximity. Other studies concern calculations of the structure of threads in asynchronous discussions (Simoff 1999; Hewitt 2003), and associations between participants of a collaborative activity applying Social Network Analysis (SNA) (Scott 2000; Wasserman and Faust 1994) using measurements of event logs. SNA has gained wide popularity and several such studies have been conducted for asynchronous CSCL activities involving large groups or communities of participants interacting through file sharing systems (e.g. Martinez et al. 2003; Nurmela et al. 1999), asynchronous discussion fora (e.g. De Laat, 2002; Lipponen et al. 2001), or emailing systems (e.g Reffay and Chanier 2003).

Totally automated metrics, such as the ones discussed above, can be reflected back to the participants in order to inform them on their collaborative performance. Metrics of participation may constitute the input of suitably visualized metacognitive artifacts, such as the ones developed by Jermann and Dillenbourg (2008) and common visualizations of social network analysis, or even for explicitly guiding the collaborative process (Constantino-Gonzalez and Suthers 2000).

4.2 Automated Interaction Analysis Based on Event Logs and a Priori Annotations of Verbal Content

All the approaches mentioned in the previous section use event logs to calculate metrics of interaction. They do not involve any systematic analysis of the verbal content of interaction and do not involve human-made annotations. One way to enrich the information to be analyzed is to take into account to some extent verbal content of e.g. exchanged communication messages without resorting to the assistance of human evaluators. This can be done by enforcing the participants of the collaborative process to explicitly or implicitly annotate their verbal actions. For example, when using the DEGREE tool for asynchronous online discussion, participants have to associate each message they send with a specific predefined message type (Barros and Verdejo 2000). Moreover, the types of annotation available to participants are dynamically defined, based on types assigned to previous postings and predefined graphs of desired sequences of types of such contribution. In addition, several metrics are calculated and are integrated into a fuzzy reference procedure that produces ratings of collaboration. This way, the mediating tool guides the collaborative processes based on fully automated analysis of interaction.

If having participants of the collaborative process annotate messages they send themselves is considered to be too intrusive for the ecology of the collaborative process, a "milder" approach involves implicit ways of a priori annotating verbal interaction. Participants can be provided with a set of "sentence openers" when they want to post a message, which are transparently related to specific categories (Dimitracopoulou and Petrou 2003). This can be designed on a voluntary (e.g McManus and Aiken 1995; Baker and Lund 1997) or a mandatory basis (e.g. Robertson et al. 1998; Soller et al. 2002).

This extra information resulting from annotations of verbal content of interaction constitutes a richer source for automated analysis than simple logfile counts. Therefore, several tools and studies conducted reaped the advantages of the application of more sophisticated artificial intelligence techniques for automated analysis. Such techniques regard finite state machines (McManus and Aiken 1995; Inaba and Okamoto 1997), fuzzy inferencing (Barros and Verdejo 1999), rule learning (Katz et al. 1999), decision trees (Constantino-Gonzalez and Suthers 2000), plan recognition (Muehlenbrock and Hoppe 1999), and Hidden Markov Models (Soller 2002; Soller and Lesgold 2003) (Jermann et al. 2001). The purpose for all these advanced calculations was that collaborative systems would provide timely feedback to the collaborative process.

5 Interaction Analysis Techniques with the Aid of Human Evaluators

As stated in the previous section, it is not always possible to fully automate the annotation and analysis process described in Figure 1, and obtain meaningful results. The difficulty of formalizing verbal content of interaction, or the side-effects of obligating participants to annotate verbal content themselves (explicitly or implicitly), often requires the involvement of human agents in the annotation and analysis process. Of course, this approach misses the opportunities for totally automated and timely analysis of interaction. This section focuses on two approaches of interaction analysis techniques that necessitate the interference of human agents in analysis: the application of *coding schemes* (referred as *content analysis* in many cases) involves human intervention in the stage of annotation, whereas the application of *rating schemes* (or rating scales) skips the annotation stage and renders human raters responsible for overall evaluation. Both approaches are considered as techniques that produce outcomes that can be useful from several methodological standpoints. Moreover, they are formalisable and closer to previous quantitative approaches than other deeper-level qualitative analysis approaches.

5.1 Coding schemes

As discussed, the verbal content of messages and postings, standing at the core of most CSCL interaction, cannot easily be manipulated and categorized in automated ways. Message content is highly contextual and elliptical, while the structuring of subsequent messages is of increased complexity when compared to face-to-face interactions (Garcia and Jacobs 1999; Herring 1999; O'Neil and Martin, 2003; Suthers et al. 2003). Therefore, formalizations of CSCL verbal content that can render analysis automatable cannot easily lead to useful results that take deeper aspects of collaboration into account. Moreover, the alternative approach of forcing participants of a CSCL session to use sentence openers can influence the process in not desirable ways.

It is therefore necessary that in many circumstances, human agents apply appropriate, additional annotations to the recorded data. Often, this involves the application of theoretically derived coding schemes such as the one developed by Gunawardena et al. (1997). This technique is generally known as *content analysis,* which is defined as "a systematic, replicable technique for compressing many words of text into fewer content categories based on explicit rules of coding" (Krippendorf 1980; Weber 1990). Its main goal is to extract from the complexity of exchanged messages in a CSCL process indicators related to basic aspects of interaction, collaboration, or learning.

There is a diversity of such indicators in the CSCL literature, depending on the specific research objectives and research theory. A first approach in the field by Henri (1992) dealt with indicators of cognitive and meta-cognitive skills. Newman et al. (1997) examined indicators of critical or deep thinking in contrast to surface thinking (Garrison 1991). Later studies followed a socio-constructivist framework for the study of knowledge co-construction (Gunawardena et al. 1997; Veerman and Veldhuis-Diermanse 2001).

A set of indicators forms a protocol of annotation of dialogues that is accompanied by an established theoretical framework. Such a scheme should be easily applicable by appropriately trained researchers. Some of the most influential coding schemes have been proposed by Gunawardena et al. (1997) for studying CSCL activities in asynchronous discussion fora, Garrison et al. (2000) in similar settings but covering in addition aspects of tutor participation in the CSCL process, and Baker et al. (2003) in synchronous CSCL activities with the use of an argument graph tool with an integrated chat tool.

The unit of analysis of the coding process can vary according the theoretical underpinnings of the schemes and the specificities of the interaction media used. The most common choices refer to the *message* (or *event* in another medium if applicable), the *thematic unit*, and the *illocutionary act*. In verbal content analysis, these types of units correspond to aspects of the syntactic structure of a message, its thematic content (in a less objectively defined way), and the structuring of dialogue according to *speech act theory* (Howell-Richardson and Mellar 1996; Searle 1979) respectively.

Since annotations in content analysis are applied by human agents, some extent of subjectivity in the assignments of units of content into categories is unavoidable. It is therefore necessary that reliability is assured by involving several suitably trained researchers in the annotation of the same content in parallel. A high level of concordance between coders constitutes an indication that the process is reliable. Several measures have been established for the testing of inter-coder reliability, such as a simple percent agreement, Holsti's measure (Holsti 1969), Scott's pi (Scott 1955), Cohen's kappa (Cohen 1960), and Krippendorff's alpha (Krippendorf 1980). A threshold has been empirically established in the research community for each measure, for the results to be considered acceptable.

The results of the application of a coding scheme to data of CSCL activities can be used in many ways. They may be used in a qualitative manner, serving just for reflecting aspects of collaboration and condensing related information, describing communication and interaction. However, the use of content analysis that is most relevant to the scope of this chapter, and can reap the benefits of technology to support a collaborative process, is to quantify attributes of verbal interaction that serve for further automated analysis.

Informed interventions in the collaborative process in this case are still possible in asynchronous longer-term activities, or series of synchronous collaborative sessions, that last long enough so that the time-demanding analysis of the data is feasible.

5.2 Rating schemes

A *rating scheme* or a rating scale is "a measuring instrument that requires the rater or observer to assign the rated object to categories or continua that have numerals assigned to them" (Kerlinger and Lee 2000, p. 736, cited in Meier 2005). Rating schemes are discriminated from coding schemes in that they are used to make a judgments on a larger set of data at a time, and are based on the knowledge and critical skill of the human agent that applies them, whereas coding schemes usually demand from the coder to neutralize the process by following strictly defined rubrics (Kerlinger and Lee 2000). Rating approaches are thus normative, and refer to evaluation in the stricter sense of the term, whereas content analysis approaches are usually descriptive and regard analysis, unless further statements related to their place in a research process are made. A rating approach can either cover all the stages of the logfile-based interaction analysis framework of Figure 1, or it can cover just the stages of annotation and analysis, since interpretations, may be based on further elaborations.

Rating scales may be intuitive, without any strict theoretical grounding, or concept-oriented (Langer and Schulz von Thun 1974). Concept-oriented rating schemes require precise definitions of the concepts that determine the rating grades and provide information on the means of correctly applying the process (Guilford 1954). Other facilities such as the use of anchoring examples or handbooks that provide guidelines for the correct conduct of the rating process are also deemed necessary (Meier 2005).

However, even if the rating process is done in a rigorous and systematic way, it still relies on judgments of human agents that cannot be totally objective. Therefore reliability testing of rating processes is even more important than in the application of coding schemes. In this case, reliability refers not only to the extent of exact agreement between the grades different raters apply, but to how close they are in the range of the scale. The most commonly used measures of inter-rater reliability are *intra-class correlation* (ICC; Shrout and Fleiss 1979) that measures the explained variance based on the ANOVA-model, *adjusted ICC* that, in addition, discards any differences in raters' mean values, *Cronbach's alpha* (Cronbach 1951), *Spearman's rho* and *Kendall's tau* (Wasserman 2006) as correlation factors that can also give an interesting approximation of concordance. Thresholds of acceptable concordance for each measure have been proposed as empirical conventions (e.g. ICC scores higher than .7 are considered to signify good inter-rater reliability; Wirtz and Caspar 2002, cited in Meier 2005; 0.6 is considered acceptable for Cronbach's alpha; George and Mallery 2003).

Several studies have applied the rating scheme technique in the CSCL field. Järvelä and Häkkinen (2003) developed a concept-oriented scale for assessing the level of perspective taking (Selman 1980) in asynchronous online discussions. Meier et al. (2007) developed a rating scheme for the multi-dimensional assessment of collaboration quality in synchronous interdisciplinary problem-solving through videocoenfencing systems. This scheme was also adapted in order to be suitable for another CSCL setting (Kahrimanis et al. 2009), without sacrificing its core conceptual rationale and operational properties. The latter version of the scheme was used in a pilot study that involved the provision of adaptive feedback to collaborating dyads. Students received feedback from tutors in dimensions of collaboration in which they had poor performance in prior similar sessions (Meier et al. 2008).

6 Trained Automatic Interaction Analysis Based on Human Evaluations

As discussed above, interaction analysis techniques sometimes fall short of providing empirically meaningful indications of important aspects of collaboration without the intervention of human evaluators. Measures of automatically logged events cannot often account for deeper level aspects of collaboration.

On the other hand, human-based evaluations are usually arduous and time consuming, especially in the case of content analysis, and miss the advantage of supporting the collaborative process in real time.

One way to proceed to new, qualitatively different automated interaction analysis tools is to use human evaluations as an external point of reference to the values that automatic interaction analysis leads to, and estimate metrics of interaction in a way that they can reflect aspects of collaboration proved to be meaningful by human analysis. The necessary precondition for pursuing such an approach is that the results of human analysis are formalizable, as is the case with coding and rating schemes.

6.1 Automated Interaction Analysis Trained on Coded Data

Recent advances in CSCL research regard efforts to support the coding process of content analysis in automated ways. In contrast to aforementioned approaches to annotate verbal interaction during the ongoing collaborative process (e.g by forcing participants to annotate their actions), the aim in this case is to provide trustworthy automated content analysis without any unintended influence on the collaborative activity itself.

A technically simple approach to that problem is to define *keywords* or *key phrases* that are linked to specific categories of a coding scheme. This approach is followed in the work of Law et al (2007), who have developed an analysis tool that facilitates the process of content analysis by highlighting specific predefined keywords or assigning preliminary codes to segments of data that are supposed to be eventually annotated by human analysts. Erkens and Janssen (2008) follow a similar rationale using *discourse markers* or *clue phrases*, which are used for segmenting and mapping dialogue content into predefined categories. Both approaches constitute encouraging attempts to automate the content analysis procedure (which can otherwise be extremely tedious). They are, however, tightly related to specific a priori defined coding schemes, and more importantly, the rationale of annotation is defined in a top-down way, significantly influenced by aspects of the technical manipulation of dialogue content.

An alternative approach, able to overcome some of these shortcomings, was proposed by Rosé et al (2008) who took advantage of recent advances in text classification technology in computational linguistics to apply machine learning techniques to a large CSCL corpus that had been analyzed by human coders using a theory-based multidimensional coding scheme (Weinberger and Fischer 2006). In this way, annotations applied automatically are not determined by a priori defined rules, but are trained on empirical annotations by human evaluators.

The study involved approximately 750 university students that mostly collaborated in groups of three through a discussion forum. Their task was to apply theoretical concepts from Attribution Theory (Weiner 1985) to specific case problems, while following (in some cases) a predefined script for collaboration that emphasized mutual feedback between participants (Weinberger et al. 2005).

The resultant dataset, comprising 250 discussions in the forums, was object to content analysis. Appropriately trained coders categorized each segment using the coding scheme (Weinberger and Fischer 2006). The unit of analysis for assigning categories to segments of dialogue was not defined by strict linguistic structural properties, but was related to the information conveyed in dialogue (closer to a thematic unit), following the approach of Weinberger and Fischer's (2006) coding scheme.

A part of the whole coded corpus consisting of 1250 coded segments was used to train machine learning algorithms that learnt rules and applied them automatically to segments of data that had not been annotated by human evaluators. The algorithms were based on mappings between a set of input features and a set of output categories. Input features included punctuation marks, unigrams and bigrams (single or pairs of words), part-of-speech bigrams (pairs of grammatical categories), line length counts, etc., while appropriate practices for other technical aspects of verbal content such as the omission of rare words or the grouping of similar words (stemming) were also applied. Starting from an already defined segmentation by human evaluators, researchers pursued two basic approaches to the development of machine learning models: a feature based approach, such as the one described above, and an algorithmic approach. Results were encouraging ranging from very good for certain dimensions of the scheme, to more problematic scores for other dimensions (Rosé et al 2008). The work resulted also in the development of the TagHelper application, which can be used for content analysis type evaluation approaches for other CSCL settings as well.

Although this significant research work can be thought of as being still in a an evolving phase, encouraging results obtained so far have initiated a new thread of automated interaction analysis tools, which, if suitably improved, can lead to automated support of CSCL processes that can stand on comparable performance to human agents. Provided that estimation scores are further improved, the development of specified meta-cognitive aids available in real time, the provision of targeted timely feedback, and the handling of large datasets would be possible following the discussed here approaches.

6.2 Automated Interaction Analysis Trained on Rated Data

Automated interaction analysis using coding schemes (or automated content analysis), stands in parallel to following a similar approach when using rating schemes. The advantages of automated techniques involving training models using human assessments can be pursued for this alternative method of evaluation as well. Still, the differences between coding and rating methods, as discussed above, necessitate that a different approach is followed.

Kahrimanis et al. (2010) have developed an innovative approach that aims at automatically rating collaboration quality in a way similar to the evaluation conducted by human raters in previous studies (Meier et al. 2007; Kahrimanis et al. 2009). The goal is that automated metrics of interaction that are calculated based on events stored in logfiles, are trained by collaboration quality ratings applied by human agents.

A prerequisite for the training of automated models of collaboration quality is that a large dataset of evaluation data is gathered from a large number of collaborative activities of similar characteristics. Therefore, numerous collaborative activities were arranged (Kahrimanis et al. 2010). Students collaborated in dyads trying to solve an elementary problem of computer algorithms using a diagrammatic representation. Collaboration took place with the use of a chat tool and a shared workspace where diagrams can be built. Students collaborated synchronously for sessions that lasted from around 60 minutes. The dataset gathered comprised 228 collaborating dyads. All instances of collaboration were evaluated by two raters using the rating scheme approach reported in Kahrimanis et al. (2009).Each collaborative session was rated for each dimension of collaboration quality defined by the rating scheme of Kahrimanis et al. (2009). Inter-rater reliability was ensured using approximately 1/3 of the whole dataset for that purpose.

After the application of the ratings, a set of automated metrics of interaction had to be defined and implemented in order to provide the technical basis on which automated estimations of ratings of collaboration quality would be based. The metrics designed and developed reshaped and augmented a metric set previously implemented based on logfile entries annotated with a typology that follows the OCAF model. Four categories of events were defined: *chat messages*, *main actions in the workspace*, *overall actions in the workspace* (including actions in the workspace of secondary importance as well, such as the movement or resizing of existent objects), and *overall events* (including all categories of events captured). Eight types of metrics were then defined, each one of them applied to each category of events: *number of* [], *rate of* [], *symmetry of* [], *alternations in* [], *rate of alternations in* [], *mean response time in* [], *median response time in* [], and *number of* [] *gaps per minute* (e.g. *number of [chat messages]*). 4 additional metrics were also added: *number of words per message*, number *of question marks*, *symmetry of text changes*, *number of objects altered more than X times*. So the final set consisted of 36 metrics. Metrics were kept relatively simple: since the aim of the study was that the metrics' usefulness for indicating collaboration quality would be tested empirically, it was deemed that the use of too sophisticated metrics was premature for this case.

A correlation analysis that was conducted led to encouraging results. Chat-based metrics were highly correlated with all dimensions of collaboration quality. The highest correlations were found for communicational and information processing dimensions (Kahrimanis et al. 2010). The most valuable chat-based metrics for indicating collaboration quality were the *number of chat messages*, the *alternation of chat messages* and the *mean response time in chat messages*. A notable exception was the *symmetry of chat messages* which did not correlate with any of the rating scheme's dimensions. Regarding workspace-based metrics, the most notable findings relate to *symmetry in main actions* or *overall workspace actions*, which are positive indicators of the quality of the commitment of students to the task, and the number of workspace-related actions, which is a negative indicator of collaboration quality on most of its dimensions. The latter finding indicates that too much activity in the workspace is usually related to bad coordination and redundant actions in the workspace.

Scores of correlation reported suggest that models can be developed, that are able to estimate collaboration quality based on automatic metrics of collaboration with relative success. For example, the highest correlation score reported between one metric and a dimension of collaboration quality is .427 for Kendall's τ metric of correlation and .552 for Spearman's ρ metric of correlation ($p<.001$ for both cases). More importantly, correlation scores of similar level are reported for many cases, something that suggests that it is likely that models built can indicate collaboration quality with a high score. More generally, if such an approach leads to good estimation rates some practical applications would be available. Timely feedback could be given to participants targeting specific dimensions of collaboration where it is found that they have problems in a similar way to a pilot study conducted by Meier et al. (2008), but using automated means.

7 Conclusions

This chapter presented an overview of logfile-based interaction analysis techniques as a tool for supporting processes of Computer Supported Collaborative Learning. It included a discussion on common practices of supporting collaborative processes in ways that are deemed beneficial for the participants. A major categorization was made between interaction analysis techniques that are fully automated and based on event logs in a top-down manner in order to mirror or guide the collaborative process, and interaction analysis techniques that require the interference of human evaluators. It was claimed that while the former approaches have the practical advantage that the support of the collaborative process based on them can be timely and totally automated, they often lead to indications that are extensively based on surface representations of interaction. Similarly, the latter approaches may cover more meaningful aspects of collaboration accessible to the human intellect but they miss many of the practical advantages of automated ones. Therefore, special emphasis was then given to a recent thread of interaction analysis techniques that aim towards automated interaction analysis provided that the metrics and other technical aspects of automated calculations have been appropriately trained by deeper-level evaluations made by human agents. It is believed that the latter types of approaches are the most fruitful for reaping the benefits of the technology of collaborative tools in order to achieve automated support for the collaborative process that stands at similar levels of sophistication with other human-based evaluation approaches, rather than resorting to shallow criteria of evaluation.

References

Avouris, N., Dimitracopoulou, A., Komis, V.: On analysis of collaborative problem solving: An object oriented approach. Computers in Human Behavior 19(2), 147–167 (2003)

Avouris, N., Margaritis, M., Komis, V.: Modelling interaction during small-group synchronous problem-solving activities: The Synergo approach. In: 2nd Int. Workshop on Designing Computational Models of Collaborative Learning Interaction, ITS 2004, Maceio, Brasil (September 2004)

Baker, M.J., Lund, K.: Promoting reflective interactions in a CSCL environment. Journal of Computer Assisted Learning 13, 175–193 (1997)

Baker, M.J., Quignard, M., Lund, K., Séjourné, A.: Computer-supported collaborative learning in the space of debate. In: Wasson, B., Ludvigsen, S., Hoppe, U. (eds.) Designing for Change in Networked Learning Environments, pp. 11–20. Kluwer Academic Publishers, Dordrecht (2003)

Barros, M., Verdejo, M.: Analysing student interaction processes in order to improve collaboration. The DEGREE approach. International Journal of Artificial Intelligence in Education 11, 221–241 (2000)

Brown, A.: Metacognition, Executive Control, Self regulation, and other more mysterious mechanisms. In: Weinert, F.E., Kluwe, R.H. (eds.) Metacognition, motivation and understanding. Laurence Erlbaum Associates, Hillsdale (1987)

Buisine, S.: Quantitative assessment of collaboration. International Reports on Socio-Informatics 7, 32–39 (2010)

Cohen, J.: A coefficient of agreement for nominal scales. Educational and Psychological Measurement 20, 37–46 (1960)

Collins, A., Brown, J.S., Holum, A.: Cognitive apprenticeship: making thinking visible. American Educator 15(3), 6–11,38–39 (1991)

Constantino-Gonzalez, M., Suthers, D.: A Coached Collaborative Learning Environment for Entity-Relationship Modeling. In: Gauthier, G., VanLehn, K., Frasson, C. (eds.) ITS 2000. LNCS, vol. 1839, pp. 324–333. Springer, Heidelberg (2000)

Cronbach, L.J.: Coefficient alpha and the internal structure of tests. Psychometrika 16(3), 297–334 (1951)

De Laat, M.: Network and content analysis in an online community discourse. In: Stahl, G. (ed.) Proceedings of CSCL 2002 Computer Support for Collaborative Learning: Foundations for a CSCL Community., pp. 625–626. Lawrence Erlbaum Associates, Hillsdale (2002)

Dillenbourg, P., Baker, M., Blaye, A., O'Malley, C.: The evolution of research on collaborative learning. In: Reimann, P., Spada, H. (eds.) Learning in Humans and Machines, pp. 189–211. Springer, Berlin (1995)

Dimitracopoulou, A., Petrou, A.: Advanced Collaborative Learning Systems for young students: Design issues and current trends on new cognitive and metacognitive tools. THEMES in Education International Journal (2003)

Dourish, P., Bellotti, V.: Awareness and Coordination in Shared Workspaces. In: Turner, J., Kraut, R.E. (eds.) CSCW 1992: Proceedings of the Conference on Computer-Supported Cooperative Work, Toronto, Canada, pp. 107–114. ACM Press, New York (1992)

Erkens, G., Janssen, J.: Automatic coding of communication in collaboration protocols. International Journal of Computer Supported Collaborative Learning 3(4), 447–470 (2008)

Garcia, A., Jacobs, J.: The eyes of the beholder: Understanding the turn talking system in quasi-synchronous computer mediated communication. Research on Language and Social Interaction 32(4), 337–367 (1999)

Garrison, D.R.: Critical thinking and adult education: A conceptual model for developing critical thinking in adult learners. International Journal of Lifelong Education 10(4), 287–303 (1991)

Garrison, D.R., Anderson, T., Archer, W.: Critical thinking in a text-based environment: computer conferencing in higher education. Internet and Higher Education 11, 1–14 (2000)

George, D., Mallery, P.: SPSS for Windows Step by Step: A Simple Guide and Reference. 11.0 Update. Allyn & Bacon, Boston (2003)

Guilford, J.P.: Psychometric methods. McGraw-Hill, New York (1954)

Gunawardena, C.N., Lowe, C.A., Anderson, T.: Analysis of global online debate and the development of an interaction analysis model for examining social construction of knowledge in computer conferencing. Journal of Educational Computer Research 17(4), 397–431 (1997)

Gutwin, C., Greenberg, S.: A Descriptive Framework of Workspace Awareness for Real-Time Groupware. Computer-Supported Cooperative Work 3-4, 411–446 (2002)

Henri, F.: Computer conferencing and content analysis. In: Kaye, A. (ed.) Collaborative Learning Through Computer Conferencing: The Najaden Papers, pp. 117–136. Springer, New York (1992)

Herring, S.C.: Interactional coherence in CMC. Journal of Computer-Mediated Communication 4(4) (1999)

Hewitt, J.: How habitual online practices affect the development of asynchronous discussion threads. Journal of Educational Computing Research 28, 31–45 (2003)

Hiltz, S.R., Turoff, M., Johnson, K.: Experiments in group decision making, 3: Disinhibition, deindividuation, and group process in pen name and real name computer conferences. Journal of Decision Support Systems 5, 217–232 (1989)

Holsti, O.R.: Content Analysis for the Social Sciences and Humanities. Addison-Wesley, Reading (1969)

Howell-Richardson, C., Mellar, H.: A methodology for the analysis of patterns of participation within computer mediated communication courses. Instructional Science 24, 47–69 (1996)

Inaba, A., Okamoto, T.: Negotiation Process Model for Intelligent Discussion Coordinating System on CSCL Environment. In: Proceedings of the 8th World Conference on Artificial Intelligence in Education, Kobe, Japan, pp. 175–182 (1997)

Järvelä, S., Häkkinen, P.: The levels of web-based discussions: Using perspective-taking theory as an analytical tool. In: van Oostendorp, H. (ed.) Cognition in a Digital World, pp. 77–95. Lawrence Erlbaum Associates, Mahwah (2003)

Jermann, P., Soller, A., Muehlenbrock, M.: From mirroring to guiding: A review of state of the art technology for supporting collaborative learning. In: Dillenbourg, P., Eurelings, A., Hakkarainen, K. (eds.) Proceedings of Euro CSCL 2001: European Perspectives on Computer-supported Collaborative Learning, Maastricht, The Netherlands, University of Maastricht, pp. 324–331 (2001)

Jermann, P., Dillenbourg, P.: Group mirrors to support interaction regulation in collaborative problem solving. Computers and Education 51(1), 279–296 (2008)

Jordan, B., Henderson, A.: Interaction Analysis: Foundations and Practice. Journal of the Learning Sciences 4(1), 39–103,1532–7809 (1995)

Kahrimanis, G., Meier, A., Chounta, I.A., Voyiatzaki, E., Spada, H., Rummel, N., Avouris, N.: Assessing collaboration quality in synchronous CSCL problem-solving activities: Adaptation and empirical evaluation of a rating scheme. In: Cress, U., Dimitrova, V., Specht, M. (eds.) EC-TEL 2009. LNCS, vol. 5794, pp. 267–272. Springer, Heidelberg (2009)

Kahrimanis, G., Chouda, I.A., Avouris, N.: Study of correlations between logfile-based metrics of interaction and the quality of synchronous collaboration. International Reports on Socio-Informatics (IRSI) 7(1), 24–31 (2010)

Katz, S., Aronis, J., Creitz, C.: Modelling pedagogical interactions with machine learning. In: Proceedings of the Ninth International Conference on Artificial Intelligence in Education, LeMans, France, pp. 543–550 (1999)

Kerlinger, F.N., Lee, H.B.: Foundations of behavioral research. Harcourt College Publishers, New York (2000)

Kollar, I., Fischer, F., Hesse, F.W.: Collaboration scripts–a conceptual analysis. Educational Psychology Review 18, 159–182 (2006)

Koschmann, T.: Paradigm shifts and instructional technology: An introduction. In: Koschmann, T. (ed.) CSCL: Theory and practice of an emerging paradigm, pp. 1–23. Lawrence Erlbaum Associates, Mahwah (1996)

Krippendorf, K.: Content analysis: An introduction to its methodology. Sage Publications, Beverly Hills (1980)

Langer, I., Schulz von Thun, F.: Messung komplexer Merkmale in Psychologie und Pädagogik - Ratingverfahren. Reinhardt, München (1974)

Law, N., Yuen, J., Huang, R., Li, Y., Pan, N.: A learnable content & participation analysis toolkit for assessing CSCL learning outcomes and processes. In: Chinn, C.A., Erkens, G., Puntambekar, S. (eds.) Mice, minds, and society: The Computer Supported Collaborative Learning (CSCL) Conference 2007, vol. 8, pp. 408–417. International Society of the Learning Sciences, New Brunswick (2007)

Lipponen, L.: Exploring foundations for computer-supported collaborative learning. In: Stahl, G. (ed.) Computer-supported collaborative learning: Foundations for a CSCL community. Proceedings of the Computer-Supported Collaborative Learning 2002 Conference, pp. 72–81. Erlbaum, Mahwah (2002)

Marshall, P., Hornecker, E., Morris, R., Dalton, N.S., Rogers, Y.: When the fingers do the talking: A study of group participation with varying constraints to a tabletop interface. In: Proceedings of IEEE International Workshop on Horizontal Interactive Human-Computer System, pp. 37–44 (2008)

Martinez, A., Dimitriadis, Y., Gomez, E., Rubia, B., de la Fuente, P.: Combining qualitative and social network analysis for the study of classroom social interactions. Computers and Education, special issue on Documenting Collaborative Interactions: Issues and Approaches 41(4), 353–368 (2003)

McManus, M., Aiken, R.: Monitoring computer-based problem solving. Journal of Artificial Intelligence in Education 6(4), 307–336 (1995)

Meier, A.: Evaluating Collaboration: A Rating Scheme for Assessing the Quality of Net-Based, Interdisciplinary, Collaborative Problem Solving. Unpublished diploma thesis, Institut für Psychologie, Albert-Ludwigs-Universität Freiburg im Breisgau, Freiburg, Germany (2005)

Meier, A., Spada, H., Rummel, N.: A rating scheme for assessing the quality of computer-supported collaboration processes. International Journal of Computer-Supported Collaborative Learning 2, 63–86 (2007)

Meier, A., Voyiatzaki, E., Kahrimanis, G., Rummel, N., Spada, H., Avouris, N.: Teaching students how to improve their collaboration: Assessing collaboration quality and providing adaptive feedback in a CSCL setting. In: Rummel, N., Weinberger, A. (eds.) New Challenges in CSCL: Towards adaptive script support. Proceedings of the Eighth International Conference of the Learning Sciences (ICLS 2008), Utrecht. International Society of the Learning Sciences, June 2008, vol. 3, pp. 338–345 (2008)

Muehlebrock, M., Hoppe, H.U.: A collaboration monitor for shared workspaces. In: Proceedings of the International Conference on Artificial Intelligence in Education (AIED 2001), San Antonio, TX, pp. 154–165 (2001)

Newman, D.R., Johnson, C., Webb, B., Cochrane, C.: Evaluating the quality of learning in Computer Supported Co-operative Learning. Journal of the American Society for Information Science 48(6), 484–495 (1997)

Nurmela, K., Lehtinen, E., Palonen, T.: Evaluating CSCL log files by social network analysis. In: Hoadley, C. (ed.) Proceedings of Computer Supported Collaborative Learning 1999, pp. 434–444. Erlbaum, Mahwah (1999)

O'Neill, J., Martin, D.: Text chat in action. In: Proceedings of the 2003 International ACM SIGGROUP Conference on Supporting Group Work, Sanibel Island, Florida, USA, November 9-12 (2003)

Reffay, C., Chanier, T.: How social networl analysis can help to measure cohesion in collaborative distancelearning. In: Proceedings of Computer Support for Collaborative Learning: Designing for change in Networked Environments, CSCL 2003, pp. 343–352. Kluwer Academic Publishers, Amsterdam (2003)

Reiser, B.J.: Why scaffolding should sometimes make tasks more difficult for learners. In: Proceedings of the Computer Support for Collaborative Learning (CSCL 2002), Boulder, Colorado, USA, pp. 255–264 (2002)

Robertson, J., Good, J., Pain, H.: BetterBlether: The Design and Evaluation of a Discussion Tool for Education. International Journal of Artificial Intelligence in Education 9, 219–236 (1998)

Rodden, T.: Populating the Application: A Model of Awareness for Cooperative Applications. In: Ackermann, M.S. (ed.) Proc. of Conference on Computer Supported Cooperative Work (CSCW 1996), pp. 87–96. ACM Press, Boston (1996)

Rosé, C., Wang, Y.C., Cui, Y., Arguello, J., Stegmann, K., Weinberger, A., Fischer, F.: Analyzing collaborative learning processes automatically: Exploiting the advances of computational linguistics in computer-supported collaborative learning. International Journal of Computer Supported Collaborative Learning 3(3), 237–271 (2008)

Rummel, N., Weinberger, A.: New Challenges in CSCL: Towards adaptive script support. In: Proceedings of the Eighth International Conference of the Learning Sciences (ICLS 2008), Utrecht. International Society of the Learning Sciences, June 2008, vol. 3, pp. 338–345 (2008)

Salomon, G., Globerson, T.: When teams do not function the way they ought to. International Journal of Educational Research 13, 89–98 (1989)

Salomon, G.: Distributed Cognitions: Psychological and educational considerations. Cambridge University Press, Cambridge (1995)

Schümmer, T., Strijbos, J.-W., Berkel, T.: A new direction for log file analysis in CSCL: Experiences with a spatio-temporal metric. In: Koschmann, T., Suthers, D., Chan, T.W. (eds.) Computer Supported Collaborative Learning 2005: The Next 10 Years, pp. 567–576. Erlbaum, Mahwah (2005)

Scott, J.: Social Network Analysis: A Handbook, 2nd edn. Sage, London (2000)

Scott, W.: Reliability of content analysis: The case of nominal scale coding. Public Opinion Quarterly 17, 321–325 (1955)

Searle, J.: Expression and Meaning: Studies in the Theory of Speech Acts. Cambridge University Press, Cambridge (1979)

Selman, R.L.: The growth of interpersonal understanding. Academic Press, New York (1980)

Shrout, P.E., Fleiss, J.L.: Intraclass Correlations: Uses in Assessing Rater Reliability. Psychological Bulletin 2, 420–428 (1979)

Shute, V.J., Psotka, J.: Intelligent Tutoring Systems: Past, Present, and Future. In: Jonassen, D. (ed.) Handbook of Research on Educational Communications and Technology. Scholastic Publications (1995)

Simoff, S.: Monitoring and Evaluation in Collaborative Learning Environments. In: Proceedings of the Computer Support for Collaborative Learning (CSCL) 1999 Conference, Stanford University, Palo Alto (1999), http://www.ciltkn.org/cscl99/A83/A83.html

Soller, A., Wiebe, J., Lesgold, A.: A machine learning approach to assessing knowledge sharing during collaborative learning activities. In: Proceedings of Computer- Support for Collaborative Learning 2002, Boulder, CO, pp. 128–137 (2002)

Soller, A., Lesgold, A.: A computational approach to analyzing online knowledge sharing interaction. In: Proceedings of Artificial Intelligence in Education 2003, Sydney, Australia, pp. 253–260 (2003)

Soller, A., Martínez, A., Jermann, P., Muehlenbrock, M.: From mirroring to guiding: A review of the state of the art technology for supporting collaborative learning. Int. J. on Artificial Intelligence in Education 15, 261–290 (2005)

Stahl, G.: Rediscovering CSCL. In: Koschmann, T., Hall, R., Miyake, N. (eds.) CSCL 2: Carrying forward the conversation, pp. 169–181. Lawrence Erlbaum Associates, Hillsdale (2002)

Stahl, G., Koschmann, T., Suthers, D.: Computer-supported collaborative learning: An historical perspective. In: Sawyer, R.K. (ed.) Cambridge Handbook of the Learning Sciences, pp. 409–426. Cambridge University Press, Cambridge (2006)

Suthers, D., Hundhausen, C.: An empirical study of the effects of representational guidance on collaborative learning. Journal of the Learning Sciences 12(2), 183–219 (2003)

Veerman, A., Veldhuis-Diermanse, E.: Collaborative learning through computer-mediated communication in academic education. In: Dillenbourg, P., Eurelings, A., Hakkarainen, K. (eds.) European Perspectives on Computer-Supported Collaborative Learning. Proceedings of the first European Conference on CSCL. McLuhan Institute, University of Maastricht, Maastricht, Netherlands (2001)

Vygotsky, L.: Mind in society. Harvard University Press, Cambridge (1930/1978)

Warschauer, M.: Comparing face-to-face and electronic communication in the second language classroom. CALICO Journal 13(2), 7–26 (1996)

Wasserman, S., Faust, K.: Social network analysis: Methods and applications. Cambridge University Press, Cambridge (1994)

Wasserman, L.: All of Nonparametric Statistics. Springer, Berlin (2006)

Weber, R.P.: Basic content analysis. University of Iowa, Iowa City (1990)

Weinberger, A., Reiserer, M., Ertl, B., Fischer, F., Mandl, H.: Facilitating collaborative knowledge construction in computer-mediated learning environments with cooperation scripts. In: Bromme, R., Hesse, F., Spada, H. (eds.) Barriers and Biases in Computer-Mediated Knowledge Communication–and How They May Be Overcome. Kluwer Academic Publisher, Dordrecht (2005)

Weinberger, A., Fischer, F.: A framework to analyze argumentative knowledge construction in computer-supported collaborative learning. Computers & Education 46(1), 71–95 (2006)

Weiner, B.: An attributional theory of achievement motivation and emotion. Psychological Review 92, 548–573 (1985)

Wirtz, M., Caspar, F.: Beurteilerübereinstimmung und Beurteilerreliabilität. [Inter-rater agreement and inter-rater reliability]. Verlag für Psychologie, Göttingen (2002)

CLFP Intrinsic Constraints-Based Group Management of Blended Learning Situations

M. Pérez-Sanagustín, J. Burgos, D. Hernández-Leo, and J. Blat

ICT Department, Universitat Pompeu Fabra, C/Roc Boronat 138,
08018 Barcelona, Spain
{mar.perez,davinia.hernandez,javier.burgos}@upf.edu.com,
josep.blat@upf.edu.com

Abstract. When applying a Collaborative Learning Flow Pattern (CLFP) to structure sequences of activities in real contexts, one of the tasks is to organize groups of students according to the constraints imposed by the pattern. Sometimes, unexpected events occurring at runtime force this pre-defined distribution to be changed. In such situations, an adjustment of the group structures to be adapted to the new context is needed. If the collaborative pattern is complex, this group redefinition might be difficult and time consuming to be carried out in real time. In this context, technology can help on notifying the teacher which incompatibilities between the actual context and the constraints imposed by the pattern. This chapter presents a flexible solution for supporting teachers in the group organization profiting from the intrinsic constraints defined by a CLFPs codified in IMS Learning Design. A prototype of a web-based tool for the TAPPS and Jigsaw CLFPs and the preliminary results of a controlled user study are also presented as a first step towards flexible technological systems to support grouping tasks in this context.

Keywords: Constraints, Flexibility, CLFP, Group management, IMS LD.

1 Introduction

Scripts are the computational solution proposed in the Computer-Supported Collaborative Learning (CSCL) field to guide and support potentially fruitful interactions in terms of learning benefits. Scripting a learning process means shaping interactions without spoiling the natural richness of free collaboration in order to produce situations of effective learning [3, 4]. However, when applying a script to a blended learning scenario - where online, technology supported and face to face (f2f) activities are combined in a given space - some unpredictable situations arising from the context force the scripts' constraints to be re-defined *on the fly*. One of the main aspects usually affected by this contextual variability is the group organization and the role distribution along the script's phases. When these situations occur, it is necessary to re-distribute groups of participants and roles in a

T. Daradoumis et al. (Eds.): Technology-Enhanced Systems and Tools, SCI 350, pp. 115–133.
springerlink.com

flexible manner to adjust the script to the actual situation without violating its principles; i.e. the constraints that structure the collaboration. Different solutions and tools have been developed to provide support to collaborative practices [6, 9, 14]. Nevertheless, these systems are still too rigid to capture the unexpected changes occurring in educational contexts and, in particular, in blended learning contexts. Specialized and interoperable tools are needed for supporting these flexibility demands.

This work proposes a flexible solution for managing groups of students according to the variability of the context and the intrinsic constraints stipulated by Collaborative Learning Flow Patterns (CLFPs) codified with the IMS LD specification. CLFPs capture the essence of well-known techniques for structuring the flow of learning activities to potentially produce effective learning from collaborative situations [5, 7]. Whereas, the IMS Learning Design (IMS LD) specification allows its formalization into a computer-interpretable design. Taking as a basis a constrain-based framework proposed by Dillenbourg and Tchounikine we analyze the flexibility requirements of two representative examples of complex CL (Collaborative Learning) activities: the TAPPs and Jigsaw CLFPs. With the results of the analysis we implement a Web-based prototype for flexibly supporting the group management both examples.

Section 2 discusses the concept of flexibility, presents some of the existing approaches for supporting the group management that inspired this work and gives an overview of the solution proposed. Section 3 presents the results of studying the intrinsic constrains for the TAPPS and the Jigsaw CLFPs and their representation in IMS LD. Section 4 explains the web-based prototype and its architecture. Finally, section 5 and 6 report the preliminary results obtained from a controlled user study, the main conclusions and future work.

2 Flexible Solutions for Supporting CSCL Scripts

Using a script means to structure the learning flow and organize groups of students to constrain collaborative interactions. If these constraints are too strong, the script can spoil the natural richness of free collaboration; whereas if the constraints are too weak, the expected interactions might not be produced [2, 3, 4]. Consequently, the design of technological settings for supporting CSCL scripts must be sufficiently flexible for dealing with the main dimensions that arise from these two aspects. It must help to structure collaboration, but should also support some variability when applied into a real context. This section reviews some of the studies that inspired this work. In one hand, we discuss the concept of flexibility adopted as a basis for the solution proposed. On the other hand, we go through some approaches developed for supporting the group management in collaborative practices and highlight their limitations. Finally, we introduce our proposal for supporting teachers in the group organization and adaptation that will be developed in the next sections.

2.1 Flexibility as Disjunction of Intrinsic and Extrinsic Constraints

Dillenbourg and Tchounikine (2007) support the idea that, due to the unpredictability of the script during the enactment phase, the teacher and the student must be able to modify some script features. Based on this, they propose a conceptual constraint-based framework that defines flexibility in terms of intrinsic and extrinsic constraints [4]. The intrinsic constraints arise from the principles from which the script has been generated and must be respected in order to get a fruitful collaboration. The extrinsic constraints arise from those elements induced by the technology of contextual factors (limitations in the number of students, evaluation elements ...). The dissociation of constraints proposed marks the boundaries of flexibility for the teacher and students, and provides the basis for a computational platform of interaction. This platform should be sufficiently flexible to maintain interaction patterns in the space of extrinsic constraints, without violating the intrinsic constraints in each of the phases of the script development process (edition, instantiation and enactment). As a conclusion, Dillenbourg and Tchounikine propose addressing the operationalization of CSCL scripts by handling multiple representations of the same script: the script to be executed; the current interaction patterns or emergent organization of teams; the intrinsic and extrinsic constraints that result respectively from the pedagogical design; and from the decision and the visual representations of the script for the students and teachers.

In this work, we adopt the dissociation between intrinsic and extrinsic constraints proposed in this constraint-based framework for delimiting our notion of flexibility and the scope of this work.

2.2 Limitations in Supporting Group Management in Collaborative Blended Learning Scenarios

Several approaches have been developed for technologically supporting the group management in collaborative learning. However, and despite of their potential for solving some aspects of collaborative tasks, they lack on facing some of the problems arising when enacting collaborative learning flows in blended learning scenarios. Here we classify, describe and analyze some of these approaches under the idea of flexibility introduced in the previous section.

Specialized grouping tools

A study by Ounnas proposes a framework for learner group formation, based upon satisfying the constraints of the teacher by reasoning over semantic data about the potential participants [13, 14]. As a technological support based on this framework, Ounnas proposes a tool that enables forming groups of students according to a set of constraints defined by the user and the semantic data that characterize the potential students participating in the activity. The result is a simple and powerful solution for easily allocating all students in groups. In the same line, an study by Hwang et al [9] proposes a genetic algorithm as a basis for an assistant system for organizing efficient cooperative groups that fit the learning objectives set by the instructor.

Despite of the potential of these approaches, they propose solutions for supporting the group organization for a particular activity and not for sequences of activities following a learning flow such as those defined by scripts. Thus, these solutions do not consider the relations established within group members from a set of interrelated activities, i.e. group formation according to the students roles in previous activities. Moreover, these applications do not assist the teacher in understanding the adaptation needs that emerge from the contextual situations and their relation with the intrinsic script constrains.

Specialized grouping tools conforming with IMS LD

One of the best-established modeling languages that are used to develop applications in educational contexts is IMS Learning Design (IMS LD) [10, 11, 12]. This specification enables the computational representation of learning flows according to a wide range of pedagogies in online learning. These computational learning flows are defined in different phases: learning flows are typically determined according to the educational objectives at design time, particularized to the specific learning situation at instantiation time and delivered to the participants as an activity to perform at enactment time. In CSCL, different approaches conform to IMS LD have been developed to support one of these phases. These computational representations are suitable to be interpreted by a compliant system as a way of alleviating teacher and learner management tasks.

As a support for the design time, Hernández-Leo *et al* propose an authoring tool for the edition of designs based on Collaborative Learning Flow Patterns conforming to IMS LD [8]. These patterns represent the techniques used to structure the flow of types of learning activities involved in collaborative learning situations. As a result, this tool provides the educator with a computational learning flow suitable to be interpreted by a system conformig to IMS LD that organizes groups of students within an activity sequence during the edition time, but not during the enactment. Therefore, no changes on group organizations are possible with this tool.

For the instantiation phase, Hernández-Gonzalo *et al* propose an IMS LD compliant tool called iCollage [6]. This is a graphical tool for the particularization of role/group structures aiming at facilitating the creation of instances and population of groups. One interesting innovation that this tool features is that groups can be defined during the instantiation phase instead of during edition, allowing the user to adapt group structures to the real contextual situation. However, this tool only provides graphical support for the group population according to the previous structures determined during the script edition. Thus, it fails to allow modifications during the script enactment, in which the extrinsic constraints can force changes in the structure planned during the edition process.

Finally, Zarroandia *et al* proposes a mechanism for the introduction of small variations in the original IMS LD learning flow during the enactment [16]. This tool allows changing some aspects of the activity such as the title, the resources associated or the structure of the learning flow. Nevertheless, the group hierarchies and the roles defined during the edition phase cannot be changed during the enactment.

The main problem of these approaches is that they treat separately the edition from the instantiation and enactment phases. This means that the group structures planned during the edition cannot be adjusted to the contextual situations during the enactment.

2.3 Considering the Intrinsic Constraints of Two IMS LD CLFPs

This work proposes a solution for flexibly managing groups of students according to Collaborative Learning Flow Pattern (CLFP) principles when applied to blended learning contexts. For the proposal we adopt: (1) the constrain-based framework proposed by Dillenbourg and Tchounikine as a basis for understanding the flexibility requirements that arise from collaborative learning practices and (2) the IMS LD specification as the de facto standard for our implementations for assuring the interoperability with the current developments and an easier integration with the existing tooling conform to this specification.

The solution is based on a conceptual model developed by the authors in a previous work. This model proposes four factors conditioning the group management in blended learning scenarios [15]: the Pedagogical Method (the activity workflow that defines the groups and role distributions), the Participants (potential and actual people participating in the activity), the History (the unexpected events fruits from the context) and the Space (elements of the space involved in the activity). The first three factors proposed in the model are the basis for identifying the main aspects to be considered when analyzing the requirement of a system for supporting the group management. The Space factor will be considered in future studies. As the Pedagogical Method factor we adopt a CLFP codified with the IMS Learning Design specification.

For addressing the flexibility requirements of the group organization we analyze two particular CLFPs, Jigsaw and TAPPS (Thinking Aloud Pair Problem Solving) by dissociating the constraints intrinsic to the pedagogical design of the script from those induced by the contextual factors. From the analysis, we extract a set of constraints for each of the CLFPs and map them with some of the elements of their IMS LD codification. This mapping leads to a formal representation of the educational flexibility requirements. The results define the foundations of a technological architecture based on a notification system for facilitating the adaptation of the CLFPs to the unexpected events arising from the learning context by preserving their main rationale. In the following sections, the analysis of the constraints, their mapping with the IMS LD and the web-based prototype resulting from this proposal are detailed.

3 Flexibility Constraints for TAPPS and Jigsaw

To study the flexibility requirements for the group management in the Jigsaw and TAPPs CLFPs we follow the definitions given in [7]. We adopt the main indications regarding the group composition and the role distribution along phases for extracting the intrinsic constraints. The aim at selecting these concrete CLFPs is to consider two CLFPs with different levels of complexity in order to understand the effectiveness of using technology for supporting these practices. This section presents the 1) description of both CLFPs, 2) the analysis of the intrinsic constraints regarding the group management, 3) the notification messages proposed in case that these constraints are violated for guiding the users through the best grouping solution according to the actual circumstances and 4) the mapping of the IMS LD codification and these intrinsic constraints.

3.1 Jigsaw and TAPPS CLFPs

The Jigsaw CLFP organizes a complex learning flow for a context in which several small groups are facing the study of a lot of information for the resolution of the same problem [7]. The activity flow is structured in three phases: i) a first phase in which an individual or initial group studies a particular subproblem, ii) a second phase in which the students that are involved in the same problem are grouped in *Expert groups* for exchanging ideas, and iii) a third phase in which the students are grouped in *Jigsaw groups* formed by one expert in each subproblem to solve the whole problem. It is based on the principle that to solve a complex divisible task collaboratively promotes three main educational benefits: positive interdependence, discussion and individual accountability.

Table 1 analyzes the intrinsic constraints for the Jigsaw pattern. The intrinsic constraint (a) is related with the minimum number of students with a different expert role necessary for applying this pattern. Since the main script principle is based on the division of the task, applying this script requires, at least, having an enough number of students to define two different expert roles. Otherwise, the system should notify the teacher that the script could not be applied. The constraint (b) regards with the difference between the number of potential (E) and actual (E') students. A non equilibrated number of students per expert group can lead to an inconsistency when forming jigsaw groups, such as having a jigsaw group without one of the expert roles. For that reason, a variation on the number of expected students should be notified to advice the teacher that s/he should adjust their jigsaw groups in the next phases. Constraints (c), (d) and (e) have to do with the requirements of jigsaw groups (J). This CLFP defines that the appropriate number of students per jigsaw group is within 4 to 5. Although the script could be applied with three students per group, the system notifies the teacher that the restrictions imposed by the CLFP are not accomplished (notifications (c) and (d) in the table). Finally, in case of having jigsaw groups without one expert of each type, the teacher is advised that it is necessary to re-adjust the jigsaw groups for reaching the expected learning objectives defined by the script.

Table 1 Intrinsic constraints of CLFP Jigsaw

Intrinsic Constraints	Violations	Notification of the system
a) # E >=2	#E=1	Not.: You need at least 2 different expert groups for applying this pattern.
b) EG must be formed by the same # of students. The EGs must be equilibrated.	#E≠#E'	Not.: Be careful when creating the Jigsaw groups in the next phase. You have a non equilibrated group of students in each EG.
c)#J in JG <=max size JG (by default)	#J in JG > max size JG	Not.: The number of students in Jigsaw groups is different than the one stipulated by the CLFP
d)#J>min size JG (by default)	#J<min size JG	Not.: The number of students in Jigsaw groups is different than the one stipulated by the CLFP.
e) JG are formed by at least one E from each topic	JG<#E de un EG diferent	Not.: Your jigsaw groups don't contain members of the different expert groups. Please, review the proposed distribution and adapt your groups to this restriction.

E/ E'=# (potential/actual) students with Expert role J= # students with Jigsaw role, EG=Expert Group, JG=Jigsaw Group; T=total students.

The TAPPs CLFPs gives the organization for a context in which several students are paired and given a series of problems [7]. Each member of the pair is given a role of Problem Solver and Listener that switches for each problem. The Problem Solver reads aloud and talks through the solution of the problem. The Listener follows the problem solver's steps, catches the errors and asks questions for guiding the problem solver to the solution.

Table 2 analyzes the intrinsic constraints of the TAPPs pattern. Constraints (a) and (b) regard with the number of students (T) and the roles distribution. Since the script proposes working in pairs, if the number of students is odd, the system should notify the teacher that it is necessary to create a group of three persons and distribute the roles of listener and problem's solver accordingly. A group of three must have only one problem solver at once per phase. Constraint (c) is related with the role of the students. In case of having pairs, the role between listener and problem solver switches each phase. However, if there is a group of three, the teacher should control that one of the students in this group repeats the same role in two consecutive phases.

Table 2 Intrinsic constraints of the TAPPS CLFP

Phase 1: Individual or initial group		
Intrinsic Constraints	**Violations**	**Notification of the system**
a)T is pair	T is odd	Distribute the students in pairs and locate the orphan student in one of the groups and assign him the listener role. Not.: The number of students is odd and we propose you to do one group of three persons.
b)In a P there should be, at least, one L and one PS.	There are groups of three persons.	Not.: You have one group of three. Pay attention for the role distribution in this group. Be sure that there is only one problem solver at once per phase.
c) The P switch roles each phase. In case of having a group of three one student plays the same role in two consecutive phases (N, N+1)	If P>3, 1 PS and 2 Listeners.	Not.: You have one group of three. Be sure that one of the members in this group plays the same role of listeners in two consecutive phases.

T= total students in class; P=pair, L=Listener; PS=Problem's solver.

3.2 *Representing the Intrinsic Constraints with IMS LD*

We take as a starting point two CLFPs codified as a Unit of Learning (UoL) in IMS LD that we created with Collage [5, 8] and Recourse [17]. For the UoLs' definitions, we follow the guidelines specified in [7] and we configure them as the minimum units needed for representing the CLFPs in IMS LD. A UoL is composed by a set of resources and an *xml* file called *manifest* that relates them. We benefit from the *manifest* definition for extracting the intrinsic constraints defined in tables 1 and 2 of the previous section.

The component `<imsld:roles>` defines the *hierarchy* of the groups by setting the different roles that will be involved in the activity (Fig. 1). By default, IMS LD distinguishes between two types of roles: learners and stuff. Another attribute defines the minimum (`min-persons`) and the maximum (`max-persons`) number of persons playing the same role. This corresponds to the *size* of the groups and gives implicit information about the *amount of groups*. The last element is `create-new`.

When it is set to "allowed" indicates that it is possible to create occurrences of groups of the same type, i.e. groups of people with the same role.

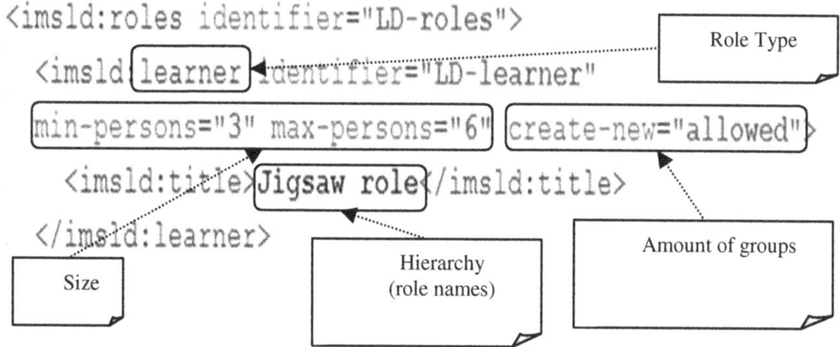

Fig. 1 IMS LD elements of the *manifest* defining the characteristics of the roles and groups

The learning flow with its activities and the activity-dependent-associations or *dynamic formation* are defined in the <imsld:method> (Fig. 2). This section defines a set of <imsld:act>. Each act refers to a sequence of activities defined in the <imsld:activities>, in which are also described the roles taking part in each activity (<imsld:role-part>).

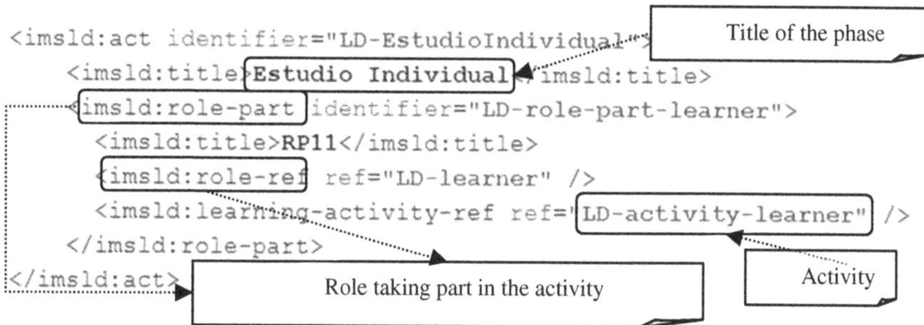

Fig. 2 IMS LD elements from the *manifest* defining the sequences of activities and the activity dependent associations.

4 Supporting Flexibility for Group Management: A Web-Based Tool

We present here below a prototype as a first effort for supporting group management in blended learning scenarios where CLFPs are applied. This prototype has been designed for the two particular CLPFs Jigsaw and TAPPs taking as a basis the analysis and representation of the intrinsic constraints presented in sections 2 and 3.

4.1 A Web-Based Application

We developed a web-based application that distinguishes between a view for the teacher and a view for the learner. The teacher's view includes functionalities for allowing the management of Participants' factor manually or automatically. When using the automatic distribution the system provides always the best possible distribution trying to respect as much as possible the intrinsic constraints. However, the teacher has always the flexibility to change the group distribution proposed (without changing the number of phases or the roles' definition). In case that one constraint is violated, the teacher will be notified but will be always free of leaving the organization as desired. The students view only shows the general group distribution for each phase and the position of the student accessing the system highlighted in another color. The student cannot change any configuration but access to the information stored about his role in other phases.

4.2 The Architecture

As a basis for the architecture we use three of the factors conditioning the group management to blended learning scenarios defined in [15]: the Pedagogical Method, the Participants and the History (Fig. 3). The Pedagogical Method defines the learning flow of the collaborative activity and it is represented here by a CLFP codified in a UoL conforming to IMS LD. Concretely, the flow of activities and their associations are represented by the elements described in section II, which are parsed from the *manifest* and codified as the intrinsic constraints in the system according to the tables 1 and 2. The Participants factor is directly associated, in one hand, to the list of potential students that the teacher can upload to the system during the preparation of the group distribution and, on the other hand, to the actual students during the development of the activity. Finally, the History factor stores the information about the group distribution and the new group configurations that occur during the activity development. The unexpected events affecting the group composition are stored as extrinsic constraints. A constraints' controller is always listening to the system for notifying the user if any of the intrinsic constraints have been violated. In this case, it will propose an optimal distribution of the participants according to the Pedagogical and the History factors. The system will always propose an alternative, except when the actual number of participant's configuration makes it impossible to satisfy them. In such cases, the system proposes the best alternative or recommends using other CLFPs for this learning scenario. Fig. 3 shows a general picture of the main elements of the system.

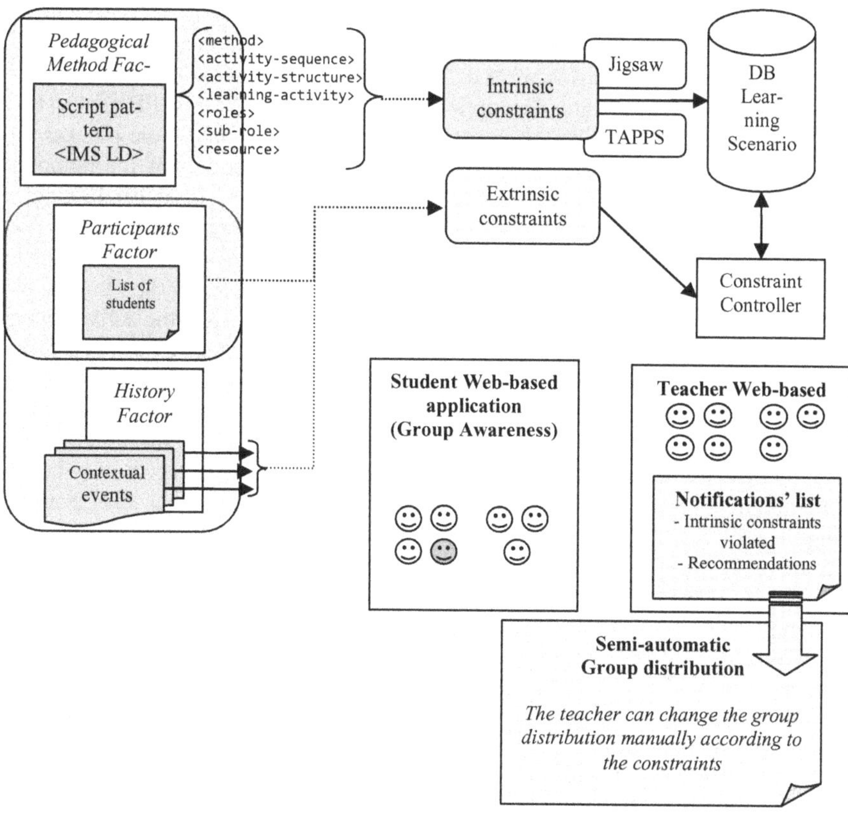

Fig. 3 Schema of the architecture underlying the prototype. The three factors are represented: the pedagogical method, the participants and the History

5 Preliminary User Study Evaluation

With the aim at obtaining the first evaluation results of the prototype we conducted a preliminary controlled user study. The study focuses on understanding the effectiveness of a tool for flexibly supporting the group management in front of a manual process and indicating in which situations this approach is useful. The main questions of interests were: 1) Do the users find helpful to have a semi-automatic tool for the group management in collaborative activities? 2) Is the tool flexible enough to freely adjust the groups to the unexpected situations? 3) Does the tool support correctly the whole process and in which situations?

5.1 Description of the User Study

For the user study we prepared two different scenarios: one for the Jigsaw and the other for the TAPPS. Both scenarios described a CLFP in the context of an e-Learning course of 13 students. The task of the teacher consisted in organizing the students in groups according to the restriction imposed by the collaborative activity proposed. The scenarios were delivered in a document containing an introduction to the context and the description phase by phase of the CLFP pattern that should be applied. For analyzing the strategies used during the whole process we proposed two different tasks: (1) prepare the group distribution of the potential students from a list according to the requirements of the activity before the class and (2) adapt the groups previously defined to a set of unexpected situations that were described in the scenario as a simulation of the type of events occurring in real educational contexts (i.e. one of the potential students leave the class at the second phase of the activity or a new student joins the class when the activity have already started). In all cases, the restrictions imposed by the CLFP needed to be accomplished. Since the focus of the study was to understand if the tool facilitates the group management in comparison with a manual process we asked the users to perform the two tasks twice, firstly by hand and secondly using the tool. Therefore, the evaluation process was divided in 3 phases: (1) familiarization with the CLFP and the context, (2) group management by hand and (3) group management using the tool.

5 university teachers with 1 to 8 years teaching experience participated in the controlled use case. 2 of them were experts in CSCL practices whereas the other 3 had never prepared a collaborative activity following a CLFP. We assigned the Jigsaw scenario to the 2 experienced users and to 1 inexpert and the TAPPS for the remaining 2. This distribution was focused on comparing the usefulness of the solution in relation to the complexity of the collaborative activity. After a brief explanation of the activity the users started the exercise by performing the group distribution manually. In the second phase, we devoted 5 minutes explaining the main functionalities of the tool and the users repeated the exercise using the tool. Since the objective of the evaluation was to understand the whole process and not the design or usability of the prototype, the users were allowed to ask about the functionalities during the experience. Fig. 4 shows the picture of two of the participants of the experience during the two different phases. Two different researchers were recording the observations on how the participants planned their group distributions and their spontaneous comments. During the whole process the users were guided through the different situations by a template with a set of steps. For each step they were asked to explain the strategies followed for the group management and their final students' distribution. All the resulting strategies and distributions were collected. Finally, the users answered a test with close and open questions in which they compared both, the manual and the technologically-supported processes. Table 3 summarizes the different data sources considered in the evaluation.

Due to the characteristics of the user study and the objectives of the evaluation, we followed a mixed evaluation method combining and triangulating [1] the qualitative and the quantitative data obtained from the different sources in Table 3. As the objective of the evaluation was focused on the process, the qualitative results were used as the main reference for understanding the strategies of the users for solving the unexpected situations and to identify the necessities emerging from this type of practices.

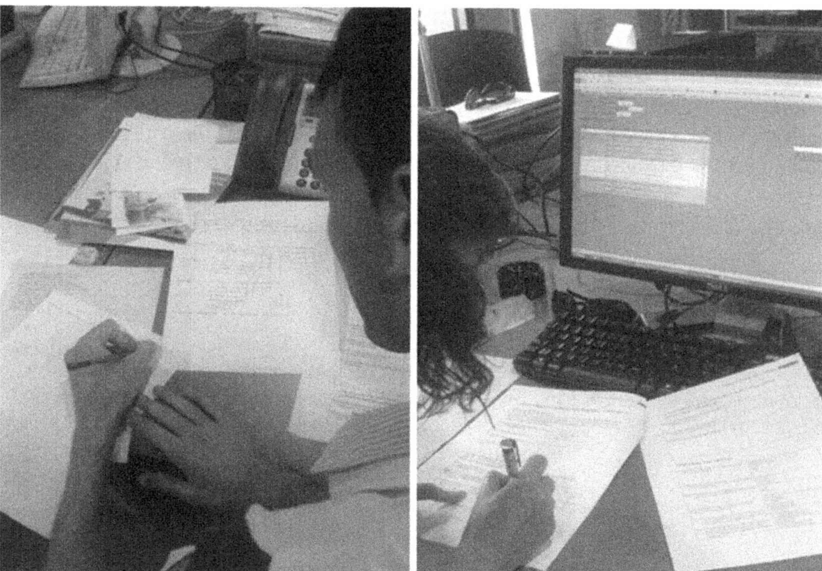

Fig. 4 Teachers participating in the experience. The picture on the left shows the phase in which the activity is carried out by hand and the one on the right corresponds to the phase carried out with the application.

Table 3 Data sources for the evaluation

Data source	Type of data	Labels
Process and outcomes described by users in a template	Qualitative descriptions and draws.	[Quest-JigsawX]
	Qualitative comments and opinions.	[Quest-TAPPSX]
		Where X is the number of the user, from 1 to 5
	Screenshots of the students' distribution resulting from the whole process step by step.	[ToolDistribution-JigsawX] [ToolDistributionTAPPSX]
Observations	Record of direct observations during the experience by 2 different researchers.	[Observer1] [Observer2]
Final questionnaire	Quantitative ratings and qualitative opinions comparing the manual and the technologically-supported process.	[Quest-comparison-JigsawX] [Quest-comparison-TAPPSX]

5.2 Results

To have a general view of the results we answered the main questions of interests by joining the results from the final questionnaire of the Jigsaw and the TAPPS scenarios (Table 4). A detailed analysis of these general results with the qualitative data permits extracting a generic picture of the tool's effectiveness in front of

Table 4 Questions of interest and main results achieved in the user study.

Questions	Results
1) Do the users find helpful to have a semi-automatic tool for the group management in collaborative activities?	*The 3 users that performed the Jigsaw scenario spent an average of 10 minutes less doing the exercise with the tool than by hand. Whereas the users performing the TAPPS scenario spent 5 minutes less in average by hand. Nevertheless, the two TAPPS' users commented that it would be very useful in case of having a bigger number of students, like 30 or 50. More time devoted for familiarization with the tool would decrease the average time spent in the semi-automatic management.
	*4 of the users preferred managing the groups using the tool instead of doing it by hand. The user that preferred doing it by hand commented that, in case of having more students s/he would have chosen the tool.
	*All the participants considered the tool very useful for managing groups. They mainly highlighted the automatic group distribution functionality and the visualization of full group organization in which the students are labeled with the name of the group they belong to.
2) Is the tool enough flexible to freely adapt the groups to the unexpected situations?	*All participants found the tool flexible or very flexible for reorganizing the groups according to the contextual situation. One of the users considered necessary to include the possibility of creating groups whenever s/he wanted (the tool only included the possibility of creating a new group in the first phase of the activity).
3) Does the tool support correctly the whole process and in which situations?	*All participants doing the Jigsaw scenario found that the notifications provided by the tool when a constraint was violated helped them to understand the errors that they need to solve in order to continue the activity correctly. From the users performing the TAPPS scenario one considered the notification system helpful whereas the other one marked that it did not helped him at all. Nevertheless, this last user answered in a previous question that it was helpful to understand the restrictions imposed by the CLFP.
	*All participants used the History of the students for confirming that the distribution proposed by the system was correct and to check the role of the students that they needed to re-allocate for adapting the groups to the real context. Only one user from the TAPPS scenario considered the History not very helpful, however, from her/his comments and the observations, it arises that s/he used it for controlling the role of the students.

the manual distribution, understanding how helpful is this approach for the users and which the missing requirements are.

Results of question 1 in table 4 show that the users found the tool a good support for managing big groups of students in complex collaborative tasks and for having general visualization of the full group distribution. The users performing the more complex activity (Jigsaw) had a better perception of the tool than those doing the simple one (TAPPS). This supports the idea that such type of solutions are helpful in case of having activities with many constraints to be accomplished and a big number of students to organize. As one of the users performing the easiest task said "*In small groups of people with few changes it's easier by hand. You don't need to form the groups with the tool. However, for big groups it would be useful.*" [Quest-Tapps2]. The draws of the users as outcomes from the manual part (see Fig. 5) also evidence the utility of having a graphical support showing the general group distribution.

With regards to the flexibility of the tool (question 2) for managing groups, the results show that all the users freely change their planned distribution according to the necessities required by the unexpected events. However, they missed the possibility of creating groups at any phase: "*I would like to have the possibility of creating new groups*" [Quest-Tapps2].

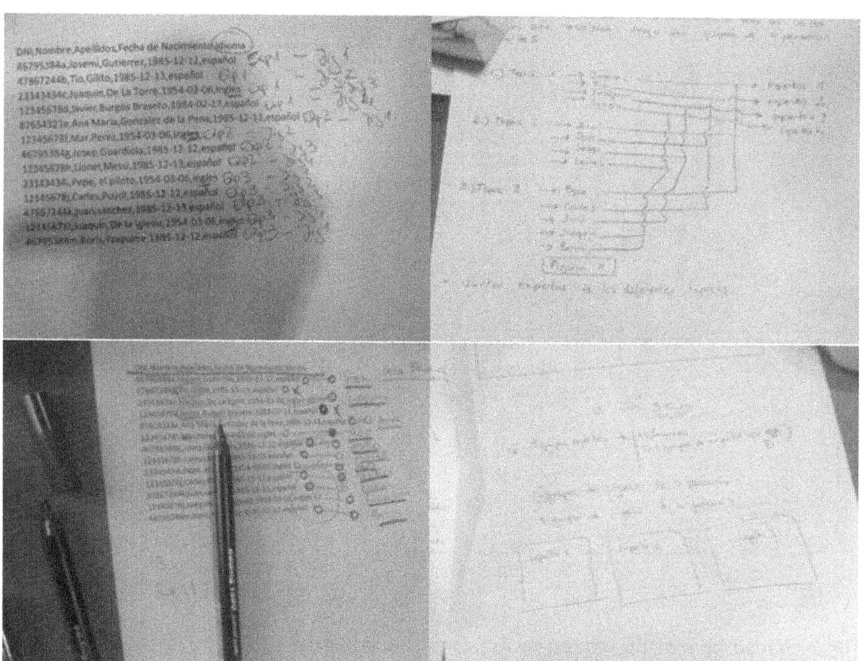

Fig. 5 Draws for organizing the group structures in the Jigsaw scenario.

Finally, the notification and the History of the students serve as a support in the whole process (question 3). All the users re-organized the groups following the notifications provided by the tool and using them as a guide for understanding the constraints that were not fulfilled in their group structure (see Fig. 6 for an example of a screenshot of the process). They used as well the History for checking their final distribution and the list of the students available, thus the potential students that were missing in some of the phases: *"I found it very useful to have the list of the students available (although deleted from the activity)"* [Quest-Jigsaw1]. One of the more interesting results was that all users agree with the necessity of adding a button for automatically providing in each phase the best group distribution according to the CLFPs' restrictions.

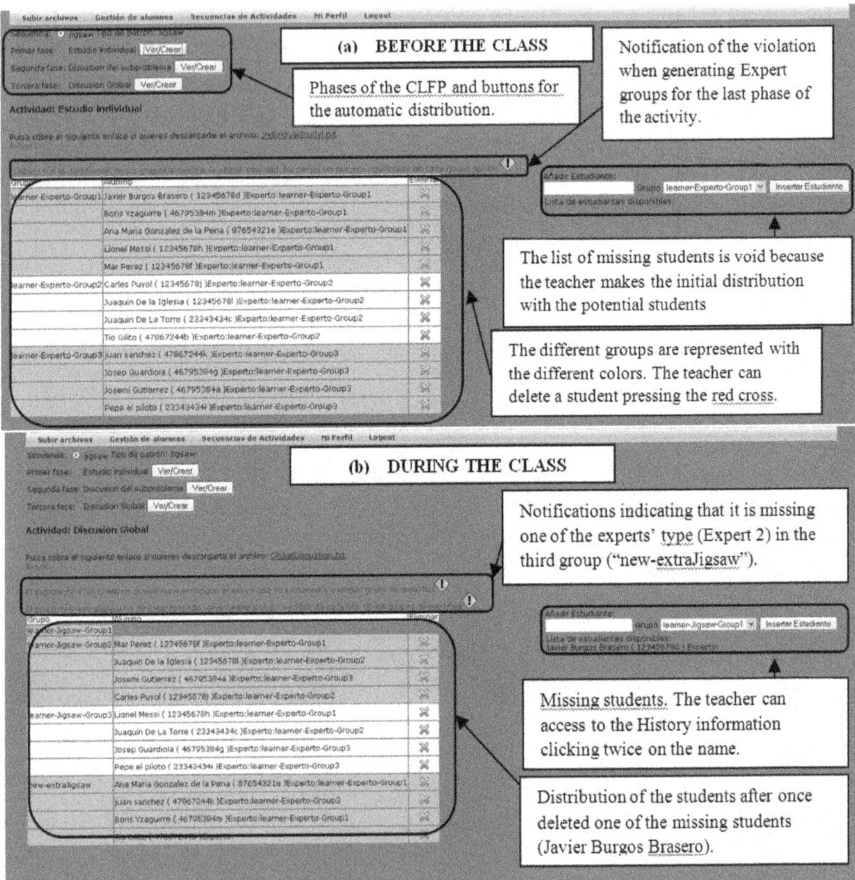

Fig. 6 Screenshots of the prototype. (a) Group distribution before the class. The system proposes the best distribution when clicking on the buttons next to each phase. (b) One of the students is missing and the final distribution is incorrect according to the CLFP's intrinsic constraints. The teacher manually deletes the missing students and attends to the notifications of violations for the final distribution.

Some other suggestions for improving the usability of the tool were proposed: 1) change the way that the notifications are showed to the user: "*I found the notifications useful just to be sure that everything is ok. However, I will put the warning in yellow and not in red because it seems an error instead of a notification* [Quest-Tabbs2]", 2) use more intuitive systems for manipulating the user in the list and change them from one group to another: "*It would be useful to have a drag&drop functionality to locate the students in the different groups* [Quest-Jigsaw1]".

6.2 Future Developments

Future developments are planned to improve the web-based prototype. The first improvement consists on adding the functionalities suggested by the users. According to the users' suggestions, we have already incorporated in the tool an automatic re-distribution button. This new functionality provides the teacher with the best students' distribution according to the intrinsic constraints and the contextual circumstances. We have also changed the color of the notifications from red to yellow for making them less aggressive for the user.

We also consider extending the tool by providing more sophisticated and formal mechanisms for proposing the best group organization fulfilling the constraints. This requires a further study of the intrinsic constraints for producing a hierarchy ordered depending on whether they are strong (i. e. the number of students is not enough for applying the script) or weak (i. e. although the students' distribution does not fulfill the requirements of the script, the activity can continue without affecting to the final learning outcomes). This classification of constraints would allow at providing more accurate suggestions to the user.

Currently, as an extension of the prototype presented, we are working on functionality for enabling the specification of the Space as a conditioning factor in the design and enactment of the scripting processes. This extension relates with the work of the authors in which the Space, understood as the place *where* the learning activity occurs and *which* elements compose it, is considered as a factor influencing the how the groups are distributed for in the design and the enactment of collaborative learning flows [15]. Thus, depending on the characteristics of the physical space where the activity is carried out (with places for working or groups or not), the movement of the students when applying a Jigsaw CLFPs will be possible or not. This physical arrangement will affect on the way students are grouped for the expert groups.

7 Conclusion and Future Work

This work presents a web-based prototype as a solution for flexibly supporting teachers in organizing the groups during the edition according to the principles stipulated by the Jigsaw and TAPPS CLFPs and guiding their re-distribution when unexpected situations occur. The preliminary evaluation results from a controlled user study show that such type of solution is useful mainly in two cases: 1) when performing complex collaborative learning activity in which there are many constraints to control and 2) when preparing activities with a big number of students. The evaluation also evidences that the introduction of a notification system and the History of the students is a good mechanism for guiding the users along the best solution for solving the non-fulfilled constraints. Although a more exhausted evaluation is needed, these preliminary results demonstrate that to consider the intrinsic constraints and the history of the activity facilitates the adjustment of the pre-defined groups to the variability of the context.

As next steps, we aim at performing an evaluation of the tool in a real learning scenario for studying how the notification system and the usability can be improved. We also plan to study the intrinsic constraints of new CLFPs to have a more extensive variety of collaborative situations to enact. The tool could be also tested as a solution to guide students in collaborative projects during longer periods such as a semester. The results from the planned evaluation will serve as a basis for improving the notification system by introducing a more sophisticated mechanism for guiding the user in the group adjustments according to the solutions adopted by other practitioners.

Acknowledgement

This work was supported in part by the Spanish Ministry of Science and Innovation in the Learn3 project TIN2008-05163/TSI.

References

[1] Creswell, J.W.: Research design: Qualitative, quantitative, and mixed method approaches, 2nd edn. Sage Publications, Thousand Oaks (2003)

[2] Dillenbourg, P.: Overscripting CSCL: The risks of blending collaborative learning with instructional design. In: Kirschner, P.A. (ed.) Three worlds of CSCL. Can we Support CSCL?, pp. 61–91. Open Universiteit Nederland (2002)

[3] Dillenbourg, P., Fischer, P.: Basics of Computer-Supported Collaborative Learning. Zeitschrift für Berufs- und Wirtschaftspadagogik 21, 111–130 (2007)

[4] Dillenbourg, P., Tchounikine, P.: Flexibility in macro-scripts for computer-supported collaborative learning. Journal of Computer Assisted Learning 23(1), 1–13 (2007)

[5] Hernández-Leo, D., Asensio-Pérez, J.I., Dimitriadis, Y.: Computational Representation of Collaborative Learning Flow Patterns using IMS Learning Design. Journal of Educational Technology and Society 8(4), 75 (2005)

[6] Hernández, D., Asensio, J.I., Dimitriadis, Y., Villasclaras, E.D.: Pattern languages for generating CSCL scripts: from a conceptual model to the design of a real situation. In: Goodyear, P., Retalis, S. (eds.) E-learning, Design Patterns and Pattern Languages, pp. 49–64. Sense Publishers (2010)

[7] Hernández-Leo, D., Villasclaras-Fernández, E.D., Asensio-Pérez, J.I., Dimitriadis, Y.: Generating CSCL scripts: From a conceptual model of pattern languages to the design of real scripts. In: Goodyear, P., Retalis, S. (eds.) E-learning Design Patterns Book (in press)

[8] Hernández-Leo, D., Villasclaras-Fernández, E.D., Jorrín-Abellán, I.M., Asensio-Pérez, J.I., Dimitriadis, Y., Ruiz-Requies, I., Rubia-Avi, B.: Collage, a Collaborative Learning Design Editor Based on Patterns Special Issue on Learning Design. Educational Technology & Society 9(1), 58–71 (2006)

[9] Hwang, G.-J., Yin, P.-Y., Hwang, C.-W., Tsai, C.-C.: An Enhanced Genetic Approach to Composing Cooperative Learning Groups for Multiple Grouping Criteria. Educational TEchnology & Society 11(1), 148–167 (2008)

[10] IMS Learning Design Best Practice and Implementation Guide (November 10, 2003a], http://www.imsglobal.org/learningdesign/ldv1p0/imsld_bestv1p0.html

[11] Koper, R., Olivier, B.: Representing the Learning Design of Units of Learning. Educational Technology & Society 7(3), 97–111 (2004)

[12] Koper, R., Tattersall, C.: Learning Design: A Handbook on Modeling and Delivering Networked Education and Training. Springer, Heidelberg (2005)

[13] Ounnas, A., Davis, H., Millard, D.: Semantic Web-based Group Formation for E-learning. In: ESWC, PhD Symposium, Tenerife, Spain, pp. 51–55 (2008)

[14] Ounnas, A., et al.: A Framework for Semantic Group Formation. In: IEEE International Conference on Advanced Learning Technologies, ICALT 2008, Spain, pp. 34–38 (2008)

[15] Pérez-Sanagustín, M., Hernández-Leo, D., Blat, J.: Conditioning factors for group management in blended learning scenarios. In: 9th IEEE International Conference on Advanced Learning Technologies, ICALT 2009, Riga, pp. 233–235 (2009)

[16] Zarraonandia, T., Dodero, J.M., Fernández, C.: Crosscutting Runtime Adaptations of LD Execution. Educational Technology & Society 9(1), 123–137 (2006)

[17] Recourse (July 2009), http://www.tencompetence.org/ldauthor

A Pedagogical Approach for Collaborative Ontologies Building

G.R. Mangione[1], E. Mazzoni[2], F. Orciuoli[3], and A. Pierri[3]

[1] CEMSAC – University of Salerno
[2] Faculty of Psychology – ALMA MATER STUDIORUM (University of Bologna)
[3] DIIMA – University of Salerno

Abstract. The collaborative ontology is a research domain linked to concepts of "extended cognitive context" and knowledge building in co-participation. The result of a survey about the main ontology engineering methodologies and ontology authoring tools, not only witnesses the fact that collaborative ontology authoring fortifies the process of ontology engineering, but also indicates that the collaborative ontology development and harmonization is not well supported by any of the existing ontology authoring tools or environments. These tools do not use a relevant pedagogical collaborative frame, as a collaborative writing approach for shaping the design features of cooperative building. Also the process of ontology building does not take into account what we can call "rich tagging", that is the extraction of ontologies maturing through text produced and shared at a networking layer. In this proposition we present a cscl driven "ontology design model". In this model, the ontology building process is maintained and validated by the encounter of 1) top-down level, where the collaborative writing scripts directs the development of authoring tools for the collaborative ontologies design and 2) bottom-up level, where the collective learning spaces such as forums and wikies, revisited by a semantic structure, are functional to the ontology extraction and validation in the learning experience.

1 Introduction to Collaborative Learning Ontology

The pervasive use of ontologies in learning and knowledge management fields, requires groups of research for efficient and effective approaches to the ontology development.

The Educational Ontologies are conceived as an instrument for modeling, sharing and reusing the knowledge in complex learning experiences. The research has allowed us to exhaustively look at the ontological solutions for the learning and to manage domains more easily. The ontologies for learning are developed in order to:

- **Share common understanding of the learning information structure among humans or software agents.** For example, if several different learning solutions share and publish the same underlying ontology of the terms

T. Daradoumis et al. (Eds.): Technology-Enhanced Systems and Tools, SCI 350, pp. 135–166.
springerlink.com

they all use, then software agents can extract and aggregate the information from these environments. The agents can use this aggregated information to answer to user queries or as an input to other educational applications;

- **Enable reuse of domain knowledge.** For example, if one group of researchers has developed common learning ontologies, others can simply reuse it for their domains. Libraries of reusable ontologies are provided on the web (e.g. the Ontolingua ontology library: http://www/ksl.stanford.edu/software/ontolingua/, or the DAML ontology library: http://www.daml.org/ontologies/)
- **Make domain assumptions explicit**. Explicit specifications of domain knowledge are useful for new users who must learn what some terms in the domain mean.
- **Separate the domain knowledge from the operational knowledge.** We can describe a task of configuring a learning material from its components according to a required specification and implement a program that does this configuration independently of the learning material and components themselves.
- **Analyse domain knowledge.** Formal analysis of learning terms is valuable when both attempting to reuse existing ontologies and extending them.

Today, the didactic ontologies design process is an innovative and competitive research theme.

The following table outlines five primary approaches to the ontology design [27]:

Table 1 Approaches to the ontology design

Approaches	Basic for design	Process
Inspirational	Individual's viewpoint about the domain	The concepts were created to accommodate the representation and processing of knowledge within any system devised for supporting decision-making
Inductive	Specific case within the domain	The Ontology is developed by observing, examining, and analyzing a specific case(s) in the domain of interest. The resulting ontological characterization, for a specific case, is applied to other cases in the same domain.
Deductive	General principles about the domain	This involves the filtering and distilling of general notions so that they are customized to a particular domain subset. It can also involve filling in details, effectively yielding an ontology that is an instantiation of the general notions.

Table 1 (*continued*)

Synthetic	Set of existing ontologies, each of which provides a partial characterization of the domain.	A developer identifies a base set of ontologies, no one of which subsumes any other. The traits of these base ontologies are synthesized to develop a unified ontology. Because it embraces multiple ontologies, the result may be prone to adoption by its adherents and an opportunity presented for them to interact in a coherent fashion.
Collaborative	Multiple individuals' viewpoints about the domain, possibly coupled with an initial ontology as an anchor.	The development is a joint effort reflecting experiences and viewpoints of persons who intentionally cooperate to produce it. The process itself could range from being strongly anchored, with a proposed ontology as a starting point for iterative improvements, to comparatively unstructured serendipitous discussion.

Choosing a collaborative approach to the ontologies design is linked to a strong awareness of how the dynamic and participatory review may impact on the good maintenance of a domain.

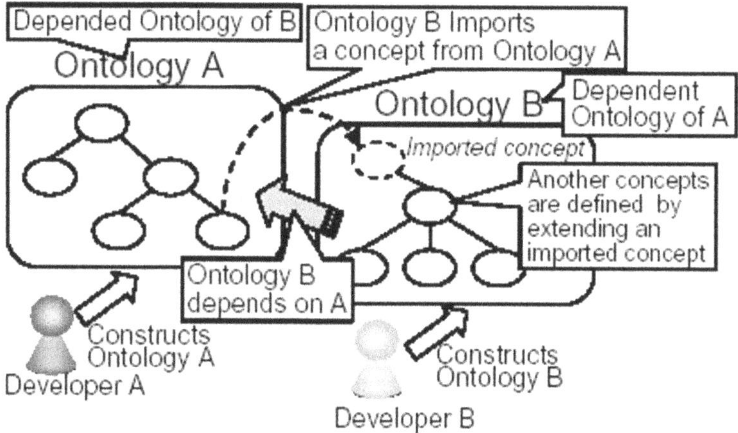

Fig. 1 A simple scratch about collaborative ontology development [25]

The ontologies designed by a knowledge community are more efficient than the ones individually designed and managed by knowledge engineers. So the collaborative ontology editors can strengthen community participation in the ontology development and maintenance process because users are enabled to autonomously change knowledge and look at changes that are triggered by their actions.

The resulting collaborative ontology is a joint effort reflecting multiple individuals' viewpoints instead of a single's viewpoint as it happens for others approaches [3]. The research about collaborative ontology design and maturing is linked to the idea of collective intelligence and its concepts of *extended cognitive context* [13] and *knowledge building in co-participation.* In this context, the concept of collective intelligence is referred to the manipulation of both connection - communication and negotiation – comparisons [8].

A collaborative approach to the construction of didactic ontologies answers to the need for a process of value creation of incremental type.

The research on the *Ontology Development Methodologies* indeed, seems to indicate that higher value can be realized by means of a continuous and incremental deployment model. When this model is applied to the process of ontology development, basic implementation increments appear as follows:

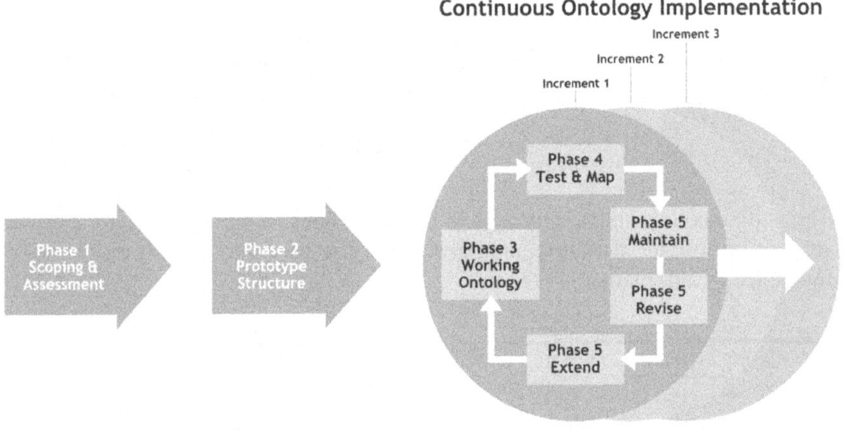

Fig. 2 Ontology development process[1]

Then, the phases of creating a working ontology, testing, maintaining, and finally revising and extending it, are repeated over multiple increments. In this manner the deployment proceeds incrementally and only as the learning process occurs.

The incremental value of a knowledge representation through the ontologies finds in the collaborative approach a driving and facilitating element.

In Figure 3, the development process is reviewed according to the collaborative component. The figure depicts the collaborative ontology lifecycle into four main phases: *creation*, *versioning*, *evaluation* and *negotiation*. The community experts (or dedicated ontology engineers) develop the domain ontology (the *Community* part of the *Creation* component). They use tools for the continuous ontology

[1] Retrieved online http://www.mkbergman.com/908/a-new-methodology-for-building-lightweight-domain-ontologies/ - October 2010.

evaluation and *versioning* to maintain high quality and to manage changes during the development process. If the amount of data suitable for the knowledge extraction is too large to be managed by the community, the *ontology learning* takes its place. Its results are *evaluated* and partially integrated into the more precise reference community ontology. The integration is based on alignment and merging covered by the *negotiation* component. The *negotiation* component takes its place also when interchanging or sharing the knowledge with other independent actors in the field.

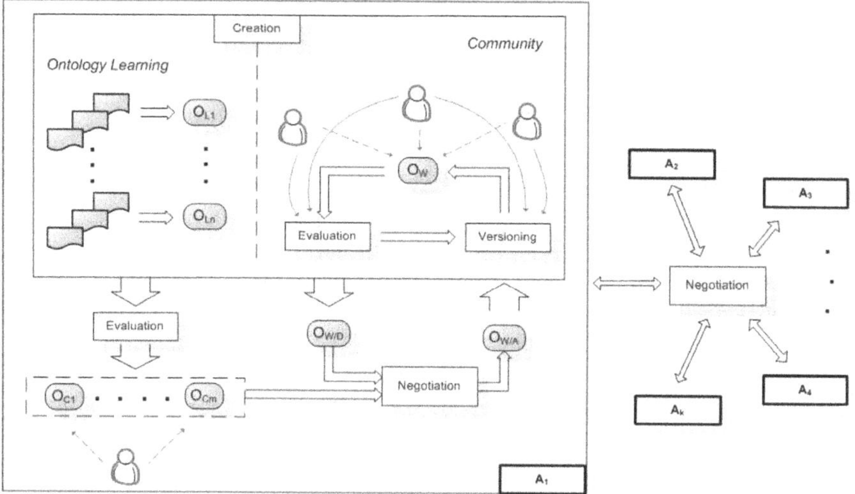

Fig. 3 Collaborative ontology lifecycle [43]

Despite these advantages and conceptualized approaches, the most of the established methodologies for ontology engineering, focus on the knowledge development in a centralized manner [26].

That is, related to the consideration that the collaborative approach to ontology building and maturing [32] may require more time and effort to deployment as opposed to other approaches (inspirational, inductive, deductive, synthetic) due to the iterative cycles of the consensus building mechanism, but the payback is deemed to be better in the long term [9].

In a socio-constructivist perspective [7], the reasoning on knowledge community and collaborative work represents one of the most interesting research challenges on learning ontologies [14]. The openness to constructivist paradigms and the synergy between web ontologies and knowledge communities required to look at innovative solutions for the collaborative building of formalized knowledge in order to enable possible processes encouraging maintenance and updating.

In order to improve the process of design and development of didactic ontologies, our work suggests an approach that, making use of CW (Collaborative Writing) principles and models, integrates creation processes structured from the top, with reviewing processes from the bottom.

2 About Collaborative Building of Ontologies

Building ontologies in a collaborative way has been an ongoing research topic and requires an organized process and a flexible environment that enable the development of communities of knowledge, linking learners and practitioners with experts.

A formalized approach, based on ontology, useful to represent a domain knowledge in distributed and shared learning experiences, needs specific methods and tools able to support a dynamic process of negotiation and renegotiation of lexicons and meanings.

In order to search for innovative ways to complement the collaborative support in the available solutions, it is important to analyze the existing approaches, and to examine weaknesses and strengths that each of them owns when providing a collaborative support.

In the following sections, we have classified the collaborative ontologies authoring solutions in two categories: 1) CW spaces for a top level authoring by teachers and expert domain and 2) social semantic space for a bottom level amelioration of the design previously done by the teacher.

The two above categories will be analysed by reporting and comparing solutions now available and possible weaknesses on which to intervene in order to innovate the instructional design sector.

2.1 Collaborative Ontology Authoring Tools

The building of collaborative ontologies has become an active area of innovative research and practice solution.

A number of ontology authoring tools have got features that allow to facilitate concurrent development of the ontology by a group of experts-users[6]. These tools that can support collaborative functions, mainly provide locking mechanisms for each ontology resource (e.g. class/concept, individual/instance, property/relation) to ensure safe development conditions [28].

We can examine the main ontology engineering methodologies being developed and currently practiced, and survey widely used ontology authoring tools [33].

Usually, two main collaboration modes are adopted by ontology editing tools to support the ontology development:

- The first one is a model where every user accesses the same version of an ontology and where changes are immediately visible to everyone (*synchronous mode*).
- In the second model, users can have their personal sandbox space and can integrate their changes with the master version at a later time (*asynchronous mode*).

The majority of the existing ontology editors belong to the second category and are standalone desktop applications that consider ontology construction as an isolated and detached task, where ontology engineering experts explicitly sit together with knowledge workers to model the ontologies. They do not provide a collaborative environment support to collaboration but in a restricted way, that is proposing limited work modes that cannot cover all the specific requirements of a collaborative writing.

Several variants and mixed approaches are also used by the current ontology management tools, built on a client-server architecture supporting different collaborative approaches.

An overview of the more relevant authoring tools currently developed and used is given below,

The **OntoLingua**[2] tool provides a distributed collaborative environment, that supports World Wide Web services in order to browse, create, edit, modify and use ontologies, like the Ontology Editor which allows these actions through the web browser. Among other features it worths mentioning the multi-user support via write-only locking and user access levels. This tool has an outstanding user community around the world.

WebODE[3] is based on and created to provide technological support to methodology and is an advanced ontological engineering workbench for all the main ontology related activities, for easy development of ontology-based applications and finally to integrate ontologies in information systems. It supports the multi-user activity by synchronization, authentication and access restrictions per user groups and allows work shifts on the same collaborative product.

The **WebOnto** tool developed by the Knowledge Media Institute of the Open University in England and by using a graphical interface, aims at being an easy-to-use yet powerful tool for creating and maintaining ontologies. It has collaborative capabilities and supports multi-users through a global write-only locking with change notification. [27]

OntoEdit is a methodology-based ontology construction tool that supports an iterative development process thanks to three phases: a requirement specification phase, a refinement phase, and an evaluation phase. The professional version has two plug-ins to support collaboration (OntoKick and Mind2Onto). OntoEdit uses MindManager to facilitate brainstorming and discussion sessions about building ontology structures. Individual developers and workgroups on the peer-to-peer network can easily participate in a joint session in creating a mind map.

The peer-to-peer communication of the mind map tool provides the necessary workgroup functionalities. The OntoEdit ontology server employs a very fine-grained locking mechanism and transaction management system to coordinate the concurrent accesses and modifications to the shared ontology, to ensure a safe and consistent ontology development environment.

Hozo [25] is an ontology development tool with ontologies visualization. It supports distributed construction of ontologies in which each ontology is revised asynchronously by different developers.

---

[2] Ontololingua guided tour
http://www-ksl-svc.stanford.edu:5915/doc/frame-editor/guided-tour/index.html

[3] WebODE Ontology Engineering Platform
http://webode.dia.fi.upm.es/WebODEWeb/index.html

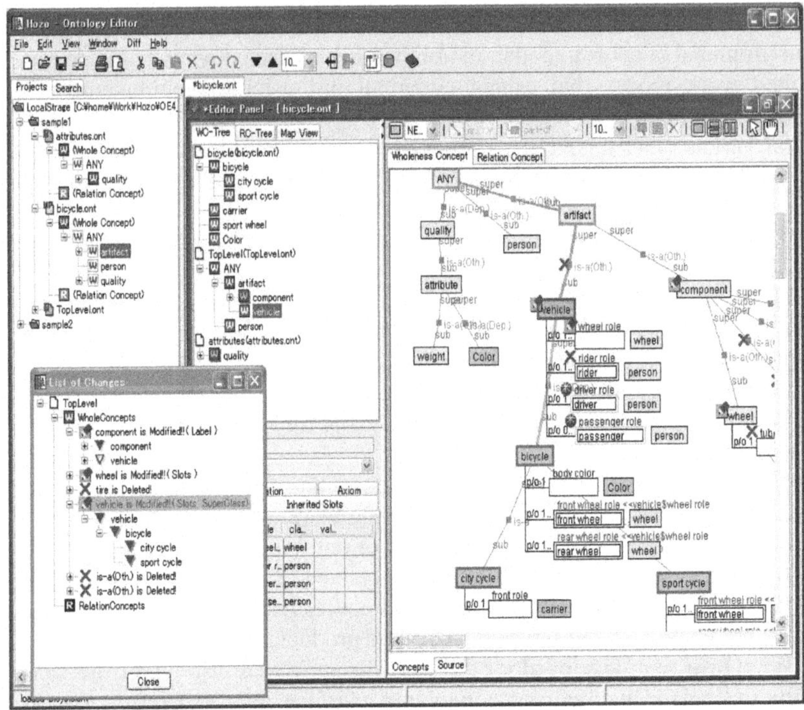

Fig. 4 Snap shot of ontology editor in Hozo[4]

In order to implement a consistent distributed development of ontologies, Hozo[5] provides two functionalities: management of the dependencies between ontologies (to import concepts from other ontologies) and keeping and restoring consistencies of ontologies when they have changed (showing the list of changes and selecting "accept" or "reject" the change option). It also supports access control (lock/unlock) and version management of ontologies on the server.

PROTEGE' [16] is a platform-independent tool widely used in the clinical and medical domain for creating and editing ontologies. Protégé tool[6] adopts a client-server architecture allowing users to concurrently apply changes on a single ontology. Collaborative Protégé [34] is an extension of the existing Protégé system that supports collaborative ontology editing.

Co-Protégé is a process oriented groupware application which supports the collaborative development of an ontology in a dedicated workspace. It is able to offer the required functionalities for externalizing and publishing; the divergence management component, which is in charge of making explicit the contribution by reactions (divergence occurrences and discussion threads) and lastly the awareness

---

[4] Retrieved online http://km.aifb.kit.edu/ws/ckc2007/Tools.htm - October 2010

[5] Hozo Ontology Editor http://www.hozo.jp/

[6] Protégé Ontology Editor http://protege.stanford.edu /

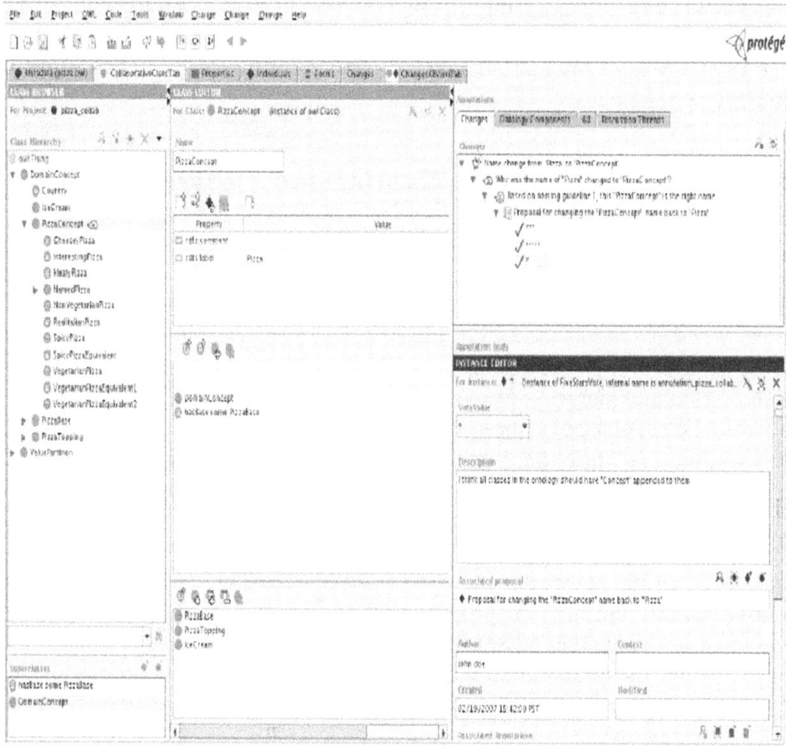

Fig. 5 The collaborative GUI of Protegé [44]

component which facilitates internalization. In addition to the common ontology editing operations, it enables annotation of both ontology components and ontology changes. It also supports the searching and filtering of user annotations based on different criteria. Some specific voting mechanisms have been implemented which can be used to express personal comments or judgements on the changes proposed for parts of an ontology. The users may view the changes proposed by other co-authors and may indicate their preferences.

In order to support the collaborative analysis and definition of domain ontologies, consensus building in the development process, interoperability, and reusability are the primary requirements which are common to all tools. Each of these ontology development tools has its strength in one or two of these aspects, and its weakness in others. Part of the reason for this diversity is that each tool stemmed from a different domain and each of them is tailored to a particular ontology design methodology, or a specific development process model.

The deficiency in these tools is fundamentally linked to the coordination of different work groups involved in collaborative editing, discussion and annotation and is mainly due to the absence of a mechanism able to give back, in real time, information on the other users working on specific tasks and activities.

The result not only witnesses the fact that the collaborative ontology authoring is the inherent nature of the ontology engineering, but also indicates that the collaborative ontology development should rely on authoring tools or environments based on the CW schemes.

2.2 Social Space for Ontology Maturing and Nurturing

The methods for creating an ontology suffer from an engineering-oriented direction where the domain is circumscribed and formalized through an ontology collaborative building process executed by an experts group.

This top-down paradigm inevitably tends to leave the potential user base out, not only in the creation process but also in a process as important as that of revision, modification and inclusion of new concepts[4][5]. This strengthens a bureaucratic, centralistic or, at least, static vision of knowledge.

Also, we cannot see how the literature to-date does not abound solutions and prototypes capable of enhancing the cooperation from the bottom (between learners), to update and improve the ontology created top down by teachers using authoring tools.

The whole of terms, keywords and notes used by learning community within a collaborative space, represents a proper shared repertoire, so it is necessary to stress on the one hand, the possibility to activate evaluation processes of the knowledge co-built in collaboration moments, and on the other hand, the instantiation of processes and methods monitoring meaning sharing, creation and update of the knowledge ontology.

For a long time, the formal education has taken advantage of a restricted vision essentially characterized by a "dominant model" [11], tied to an instructional approach that has contributed to increase the research misalignment with respect to the transformative processes supported by the Computer Supported Collaborative Learning (CSCL) [15]. Consequently, it is necessary to rethink of solutions such as ontology learning tools, that may integrate the knowledge engineering life-cycle, through the acceptance of a contribution derived not only from expert groups but also from the interaction of community learners in instructional spaces. The formalizing approach based on ontologies for the knowledge representation in distributed and shared learning experiences, should provide methods and tools able to support a dynamic process, through which lexicons and meanings are evaluated and used as an added value for the ontology nurturing.

Adopting a collaborative approach for the ontology maintenance is a challenging research topic from the point of view of the benefits it can bring to conventional approaches [41].

The use of social components for the knowledge systematization can drive to significant results in bridging the deep conceptual gap between how to represent the knowledge of educational environments (considering educational theories) and how to maturing it adequately.

The vision, that we propose, is a community of autonomous and networked users, like students, who cooperate in a dynamic and open environment. Each participant will organize some piece of knowledge according to a self-established vocabulary, and will create connections and negotiate meaning with other users within the community. Augmenting the involvement of users, by enabling community members to actively participate into the ontology evolution process is a key factor to achieve a community common ground.

Starting from an existing ontology and allowing users to freely edit ontology classes, according to their personal vocabulary, can significantly improve the ontology maintenance process, complying with the knowledge drift.

Particularly, the attention is focused on what is called "rich tagging" for the ontology learning, a building process that makes use of a bottom up method, which receives feedbacks from the lower part (the networking layer) and extracts more knowledge from the text produced and shared by user groups.

The ontology learning, mainly based, until today, on the knowledge extraction from textual and structured collaborative spaces, can find in the virtual learning community a variety of concept structured according to hierarchies, dictionaries and unique and shared taxonomies at a formal level. The relations between peers in learning semantic spaces (such as wikies or forums) can offer numerous advantages for the ontology design and could become a significant input for the bottom up improvement, development and maturing process [1].

In literature a few studies still in progress share this trend towards the social ontology design.

IkeWiki is one of the most representative Semantic Wikies. It uses annotated relations among pages. IkeWiki allows to annotate links, to type pages, and context dependent. The page names will be class names and the relations will be class relations. Therefore, each wiki page can be formalized and reused for the semantic searching and navigation. It not only provides with searching results but also with relational information between searching keywords and results.

Another related Semantic Wiki is **COW**. It provides an explicit ontology editor outside the Wiki text. In the ontology editor, users can apply ontological functions by using web based forms such as class management, property management, instance management, and ontology version management.

Some recent works describe the *Template-based Semantic Wiki* to satisfy collaborative ontology management requirements for semantic web applications. Domain experts can participate in domain knowledge population, evaluation, refinements, and ontology extraction by means of our approach.

The ontologies which are improved and used as a community, reflect the knowledge of users more effectively than the ontologies maintained by knowledge engineers, who struggle to capture all the variety taking place within a lively community.

The interaction of a social component within an ontology validation and revisioning process requires a system for the analysis of content and relations emerging from the semantic spaces. Also the visualization and analysis of social networks has become an active topic of research. We have focused on how to use the knowledge structure that is currently used by learners in the collaborative spaces, in order to extract meaningful conceptual relations at the ontology level. Moreover, the extraction of these relations will be adopted to further improve the

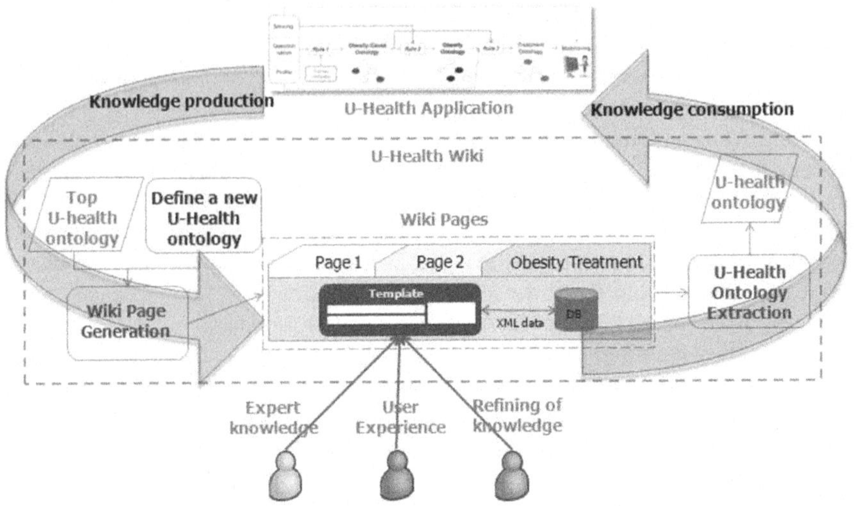

Fig. 6 Template based wiki for collaborative Ontology management [45]

collaborative ontology nurturing and exploitation of the knowledge domains. The huge amounts of social activities and generated contents have to be properly organized and filtered to preserve the benefits from an online collaboration.

While reporting on the experiments of SNA on online social networks, the research has shown a lack of techniques in the SNA application on content representations of social networks. These data cannot be represented only using raw graphs as the classical SNA algorithms do, without some loss of knowledge.

We can observe in literature how Semantic Web frameworks are being used to answer to the problem of representing and exchanging the knowledge. These Semantic web frameworks provide a graph model (RDF1), a query language (SPARQL) and schema definition frameworks (RDFS and OWL) to represent and exchange the knowledge online [35].

They provide a whole new way of exploiting the key features of social networks in order to manage their way of capturing social networks in far richer structures than raw graphs. Recently, social interactions through web 2.0 social platforms have raised lots of attention in the semantic web community. Several ontologies have been used to represent social networks. In all these experiments, researchers have reduced the expressivity of the social network representations to simple graphs, highlighting the lack of tools that can be directly applied on rich typed representation of social networks [36].

To annotate the social network content representations with SNA indices, the SemSNA's [37] ontology has been recently designed, the starting point for giving an answer to such a problem. SemSNA is an ontology able to model concepts that are used in SNA, such as degree or centrality.

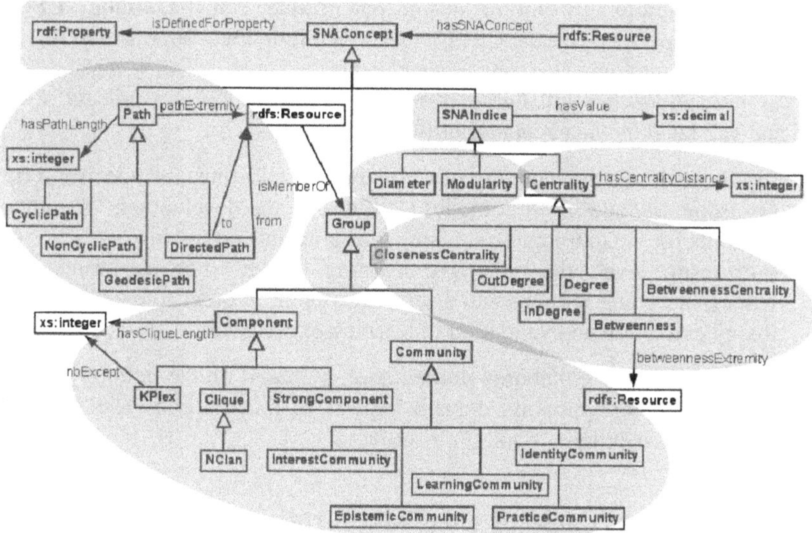

Fig. 7 SEMSNA

But at the current state, a specific methodology is missing, that might take into account the lower concept layer, represented by collaborative learning semantic spaces [1][2], to enrich a specific learning ontology (used in formal learning experience) with social texts, for which SNA tools help find relevant concepts.

2.3 A CSCL Driven Model for Collaborative Ontology Construction and Maturing

After the considerations expounded in the previous sections, we will explain in this paragraph our point of view by collecting some important observations about the ontologies, which base on research field outputs enabling to develope a more appropriate model for the collaborative ontologies construction and maturing.

Our work merges together the collaborative learning with the semantic web and provides a framework for the formalization of a new model of ontology construction and nurturing.

Indeed, from the point of view of the authoring systems there is a lack in terms of management flows for the collaborative writing; while from the point of view of social maturing there are experimentations and solutions suitable for updating and validating the ontologies built in the previous phase.

In our vision the Computer Supported Collaborative learning (CSCL), is used as an approach sustaining a collaborative ontology design and maturing in which both the top down and bottom-up level of the collaborative learning process are synergically interconnected.

Coherently with the challenges expressed in the collaborative learning, we introduce the Collaborative Ontology Construction & Maturing (COCM) model that takes into consideration an ontologydesign perspective, called **Ontology CSCL *driven***. In such a perspective, the CSCL is seen as a functional approach to review the collaborative ontology authoring environments.

In this proposition the collaborative ontology construction is a process maintained and validated by the encounter of two action levels:

- **top down level:** the modality of a collaborative writing, and specific script (as *sequential* and *parallel* mode) [18], directs the development of authoring tools for the ontologies matching, merging and versioning;
- **bottom-up level:** the collective learning spaces, such as forums and wikies, are being rethought in a semantics optic in order to enable the ontology extraction and validation in learning experience.

This chapter describes a collaborative methodology based on the ontology development framework that supports domain experts to reach a consensus through iterative ontologies evaluations and improvements.

3 Top Down Approach for the Collaborative Ontology Construction: To Increase the Value of Teachers' Contribution

The ontology construction takes advantage of a collaborative process which involves the direct cooperation among individuals or groups of domain experts, knowledge engineers or/and software agents, or the indirect cooperation through reuse, merging or adaptation of the previously published, and autonomously developed ontologies. Typically, in such settings, different participants only have partial knowledge of the domain, and hence can only partially contribute to the domain.ontologies. A common task could be to involve a mechanism of refinement of a predefined ontology, or the integration of several partial ontologies in order to obtain a coherent ontology that covers a much larger portion of that domain.

Thus, there is an urgent need for principled approaches and flexible tools allowing individuals to collaboratively build, refine, and integrate the existing ontologies, as needed in specific contexts. Therefore, the domain experts would need tools that support the collaborative ontology development and include the collaboration as an integral part of the ontology development itself.

For this purpose, this section intends in a first part to provide some specific scenarios related to the collaborative building of ontologies and, in a second part, a tool for a top down approach, cscl script based, able to implement these scenarios.

3.1 Collaborative Ontology Construction Scenarios

This section presents some modes adopted for the ontology construction in a collaborative way.

Currently, several are the methodologies about collaborative ontology construction which have been proposed. The Formal Concept Analysis (FCA) [29] for instance, constructs the ontology automatically basing upon the information analysis theory which is data sensitive.

Therefore, wrong results may be generated if noisy data have been entered. Moreover, PROMPT constructs the ontology by means of the metadata editing and concepts similarity computation. However, the metadata editing is a difficult process and the flexibility of similarity computation is limited. The online ontology editors have been provided. In the Collaborative Ontology Building (COB) [30] and OntoWiki [31], although users can collaboratively edit the ontology online, the administrator also has to manage the ontology manually. Furthermore, the lack of a convergence methodology results in that the constructed ontology may be apt to be subjective. In summary, several technical issues are emerging in the collaborative ontology construction. Accordingly, how to assist users to contribute their knowledge easily, how to integrate and coordinate different opinions of concepts associated with relations contributed from various users, and how to evaluate and maintain an acceptable ontology are our concerns in this section.

The collaborative approach to the ontology design requires the development of reflecting experiences and viewpoints of persons who intentionally cooperate to produce it. Chances for relatively wide acceptance are enhanced if these persons are diverse in the contributions they make. This helps reduce blind spots in the ontology and enrich its content. On the other hand, coordination of the design process may suffer if too many persons are directly involved in.

The process itself could range from being strongly anchored, with a proposed ontology as a starting point for iterative improvements, to comparatively unstructured serendipitous discussion. In order to execute a collaborative approach, a consensus-building mechanism needs to be employed. The famous model of the cognitive processes involved in a writing task of Flower and Hayes [38] is a good starting point for constructing a model by which to support collaborative writing. It identifies three main phases in the process:

- *Planning*, including the generation of information relevant to the task, organizing information, and setting goals.
- *Drafting or translation*, the turning of plans and ideas into text to meet the goals.
- *Reviewing*, this combines evaluating the text and editing either the text itself or the ideas and goals.

There are different modes of collaborative building of ontologies:

- Parallel Mode
- Sequential Mode
- Mutual Mode

The **Parallel Mode,** based on the use of Definition and Execution Processes, is considered for coordinating the collaborative activities. The parallel editing mode foresees a start-up activity, a set of intermediate editing activities, parallelly carried out by the community Domain Experts, and a final activity after which a final ontology is released. During the setting of the collaborative environment, associated

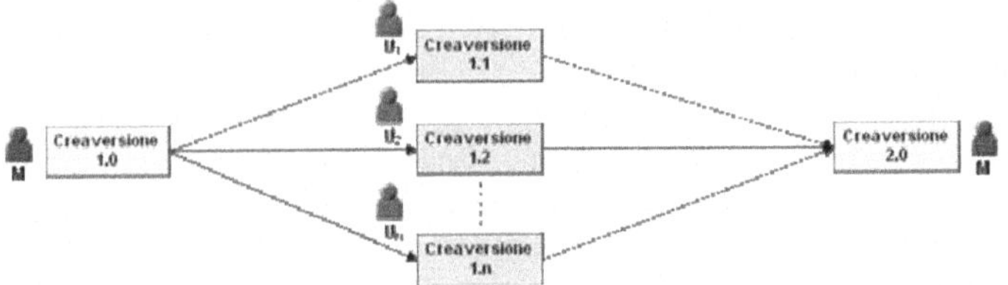

Fig. 8 Parallel Mode

to an editing process in Parallel Mode, the coordinator defines the number of editing activities that build up the process and assigns them to a specific community of Domain Experts.

In **Sequential Mode** the final ontology is obtained through several refinement iterations. Each iteration is performed by only one domain expert that accesses to a previous version of the ontology (released by the domain expert who triggered the previous iteration), refines it and releases a new version of the ontology. It will become the starting ontology for the domain expert responsible for the next iteration. A new iteration can start only when the previous iteration ends.

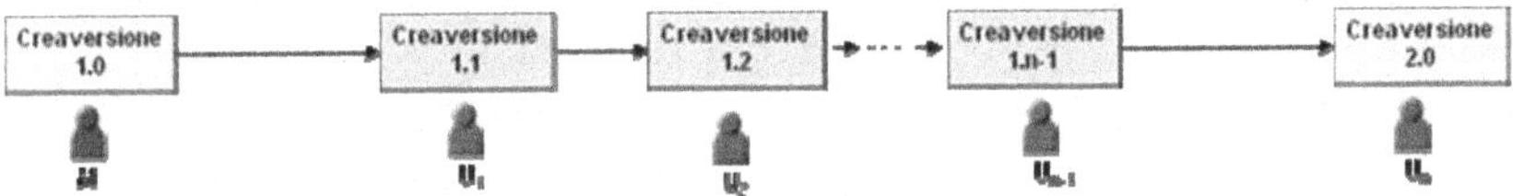

Fig. 9 Sequential Mode

The **Mutual Mode** is based on Wikies, a typical collaboration tool of Web 2.0[39][40]. In Mutual Mode domain experts define ontology concepts creating Wiki pages and ontology relations establishing links between Wiki pages.

We will focus our attention on the definition of specific scenarios related to both sequential and parallel modes.

A Scenario for the parallel mode

Description

The teacher of the course "Trust management for the quality of the e-learning" at the Firenze University, Mrs. Pettenati, wants to define an ontology to describe a specific domain. She asks three collaborators (A, B, C) dislocated in different geographical areas for help.

The better solution seems to be the use of a community of work among the four people cooperating. The teacher defines all the start-up operations (definition of learning objectives, interactions' modalities with the environment). After that, the teacher defines three macro-concepts through a specific ontology that will be refined by the three collaborators. Each of them will take into consideration one of the macro-concept to be exploded and deliver the work according to a fixed time. In the final phase the teacher will harmonize the three ontologies obtained in order to have a final result.

This is not a simple operation since the teacher must avoid overlapping concepts and must test that the final domain has been opportunely represented. Once the teacher has obtained the final version, she can store it within a repository in order to share it with other users, after having set some appropriate permissions.

In the following figure, we show the process for the collaborative ontology building in the Parallel Mode:

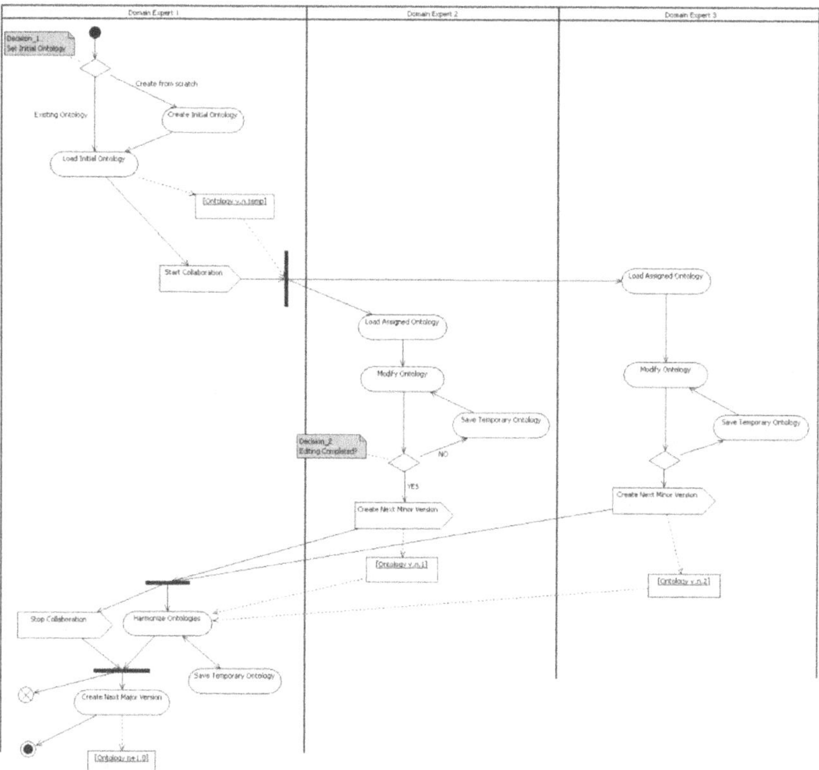

Fig. 10 Activity Diagram for the collaborative building of an ontology in a Parallel Mode.

A Scenario for the sequential mode

Description

The teacher of the course "Infobrokering for the didactic quality" at the University of Firenze wants to create an ontology for modeling the knowledge of the course. To achieve this result, he collaborates with a colleague A, that works for the University of Salerno and is an expert in the modeling of a domain through specific ontologies.

The teacher creates a little community of work composed of two experts. He sets the learning objectives, identifies the participants and defines other information necessary to arrange for the collaboration environment.

After that, his colleague builds a first version of an ontology on "Strategies of retrieval within the Web 2.0". This ontology does not include aspects related to the social networking that will be included in a second version. The first version is shared among the components of the community to be notified through an alert. In such a way the teacher can refine the ontology as for what concerns the "navigation's styles" and create a new version of the ontology.

His colleague receives this new version and complete the work by adding the ontology's components related to social networking, asking for a help to another colleague B.

The colleague B enters the work community, the teacher has added him after having been informed about the participation of a new component. The colleague B logs-in to the environment and works to the second draft of the ontology, issuing, in a second time, a third version of it.

In the final phase the teacher, after having made a few minor reviews, delivers the final version of the ontology. He is able now to organize and publish the course for the students of the master to use it.

In the following figure, there is a diagram representing the collaborative building of ontologies in a sequential mode:

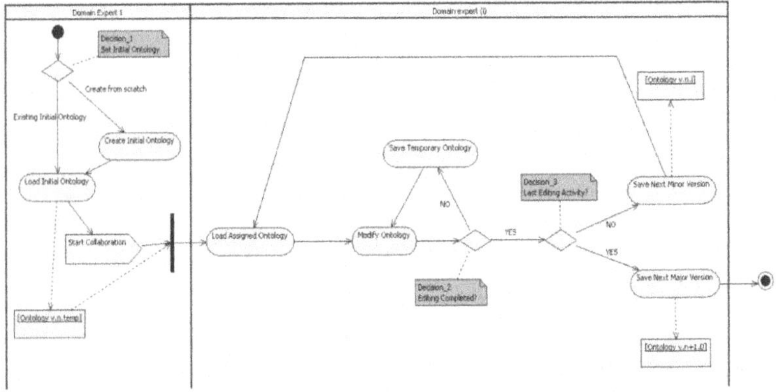

Fig. 11 Activity Diagram for the collaborative building of an ontology in a in Sequential Mode

3.2 A Tool for the Collaborative Building of Ontologies

In the top down level collaborative building of ontologies, an important step is to have the consensus of a community teacher that takes part into this work [16]. The research suggests that the focus be moved from the product to the process, where the construction of collaborative ontologies is supported by specific tools that give the chance to interact with the knowledge domain and to modify it through a user-friendly process.

The adoption of this approach, requires the construction of an ontology that will include different participants' experiences, competencies and points of view.

Karapiperis and Apostolou [9] presented a consistent methodology for the Collaborative Ontology Construction, articulated in four different phases: preparation, anchoring, iterative improvement and application. Furthermore, in [10] a report with detailed interviews to members of ten different ontology engineering projects is illustrated. The interviews were conducted either on the phone or in person. A lack and also a request for user-friendly tools able to support the collaborative ontology construction orchestrating single user asynchronous tasks, clearly emerge after analyzing the report.

Usually, two main collaboration modes are adopted by ontology editing tools to support the collaborative ontology development. The first one is a model where everyone is allowed to access the same version of an ontology and changes are immediately visible to everyone (synchronous mode). In the second model, users can have their personal *sandbox* space and can integrate their changes with the master version (asynchronous mode). Several variants and mixed approaches are also used by the existing ontology management tools.

Our idea consists in defining a tool for the collaborative ontology construction, based on the principles of the methodology presented in [9], that takes into account lacks and needs emphasized by the interviews [10], that also aims at overcoming the knowledge engineering expertise, needed by the existing tools and that finally exploits typical human–computer interactions provided by the Web 2.0 applications.

The defined tool provides a validation phase based on a semantic forum wiki engine, which allows the participants to reach a consensus for the final ontology.

The collaboration processes executed through coordination and cooperation tool functionalities (CCT) [11], can be classified in *Sequential* and *Parallel*. The Sequential processes imply that work is completed and then handed off from one participant to another.

Thus, each participant is guided by what has been done before.

This approach gives more benefits if the process starts from a complete draft of the result to achieve and needs an artifact exploiting the collective intelligence expresses by a group of persons. The Parallel processes foresee the presence of a coordinator in order to stimulate the participants and put together all the contributions coming from each member of the collaboration group. Each participant yields separately its own artifact, meeting the guidelines stated by the coordinator. The coordinator harmonizes all the received artifacts into a schema commonly approved.

This approach is particularly advantageous when the coordinator divides the whole work into parts with a few overlapping points, assigns each part to a single participant and, lastly, collects and assembles all the partial results in a complete result.

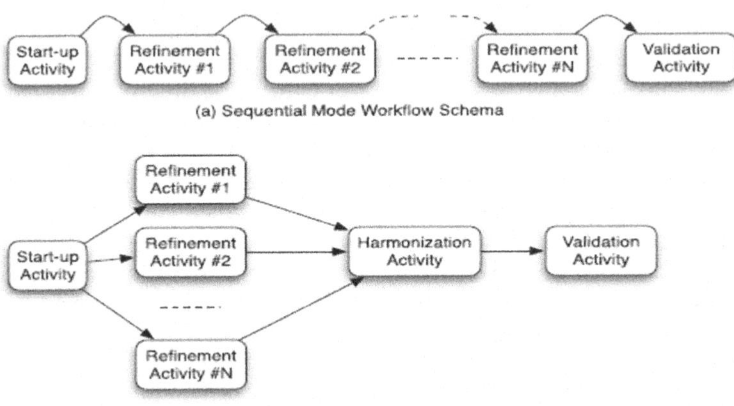

Fig. 12 Workflow schemas for sequential (a) and parallel (b) processes.

In the following section, we provide details on the activity types defined in Figure 11

- **Start-up Activity:** this activity is due in order to edit a first draft of the ontology used as an initial version to guide the collaboration process.
- **Refinement Activity:** this activity is used to modify a previous version of an ontology in order to obtain a more refined version to be stored into the Artifacts Management Module. The activity defines a new version of the ontology.
- **Harmonization Activity:** this activity is executed when the harmonization of two or more ontologies representing frames of the same domain is required. The harmonization task is performed by a semi-automatic Ontology Merging Tool that supports the user in the merging process. The result of this process can be refined by users using a Visual Ontology Editor.

Basing on the work presented in [11] that allows the collaborative construction of an ontology making recourse to three approaches (sequential, modal and parallel), also the bottom up collaborative construction has been analysed, in order to increase the value of contributions given by the users. In this way we might have a more complete collaborative ontology construction than the one introduced in [11], since it takes into account different viewpoints from the main actors involved in the teaching/learning process.

In particular, the **Validation Activity** that in [11] is executed at the top down level, in this new solution is executed by involving all the participants engaged in collaborating in the semantic spaces (forums and Wikies) and in creating the last concurrent refinements.

4 Bottom Up Collaborative Ontologies Building: To Increase the Value of Students' Contribution

A collaborative e-learning experience could be very important when used for assessing the knowledge acquired by each individual learner during a learning session and offering, in case of negative results, personalized remedial works able to fill identified knowledge gaps with an adaptive learning path. In an attempt to re-define the learning process as a direct function of social construction and sharing of knowledge, the research community played an important role, whose results have conveyed into the development of semantic social spaces [17].

These spaces, opportunely reviewed and structured, permit to generate bottom level ontologies in a dialogical and collaborative modality and by SNA methodology to analyze the composing Semantics that revolves around to target concept.

4.1 Ontology Extraction and Validation from a Semantic Space

Collaborative spaces such as Knowledge Forum and Wiki with a semantic component (associated to Collaborative Learning Activities) pursue the constructive approach proposed in AOMS and utilized to validate or integrate the original ontology. The wiki and forum semantic annotation will be designed at two levels:

- As for the first level, we will provide end-user ontologies, taxonomies, controlled vocabularies and concept map, created with OWL[7] (Web Ontology Language) and SKOS[8] (Simple Knowledge Organization System) to support them in the process of annotation of user-generated contents.
- As for the second level, we will use semantic schemas like SIOC[9] (Semantically-Interlinked Online Communities) in order to model data generated during online Wiki and Forum sessions. The advantage offered by the second level is represented by the ability to create a semantic layer accessible through standard query languages like SPARQL[10].

The two levels are integrated by an architecture, commonly used in Semantic Web applications, that provides with the use of Upper Ontologies (second level) and Domain Ontologies (first level). The Semantic Wiki Engine (SWE) is implemented by extending an open source wiki engine ScrewTurn[11] that allows the management of semantic Wiki Spaces (SWS) [2]. These areas may represent a learning ontology where the concepts are linked together. The Semantic Knowledge Forum (SKF) [1] is implemented by extending an open source forum YAT and becomes a space capable of supporting the learning ontology through the ability to trace and extract ontologies from the organization of discourse structures.

[7] http://www.w3.org/TR/owl-features/
[8] http://www.w3.org/2004/02/skos/
[9] http://sioc-project.org/
[10] http://www.w3.org/TR/rdf-sparql-query/
[11] ScrewTurn Software, ScrewTurn Wiki, http://www.screwturn.eu/

A Semantic Link Network (SLN) is a model to intuitively represent the semantic relationships between scaffolds.

In this paper, we use the SLN to organize the conversational script with semantic associations. The semantic associations represent the binary relationships between different scaffolds. Each scaffold is bound by default to the concept father (title of the forum) with a *has part* report while the user can choose which report to tie to the next scaffold. Specific thinking type related to the scaffold concept (i.e. Question, Opinion, Suggestion, Recommendation, Request) are used to develop the concept objective so granular and oriented to the script. Once validated the ontology from collaborative activities, the SWE and SF are converted into an e-learning ontology (AOMS XML-BASED INTERNAL RAPRESENTATION), annotated with metadata, indexed and stored for users of intelligent learning system. The comparison with the starting ontology facilitates the validation process and, possibly, by the same device exploded in top-down, a process of matching and versioning starts.

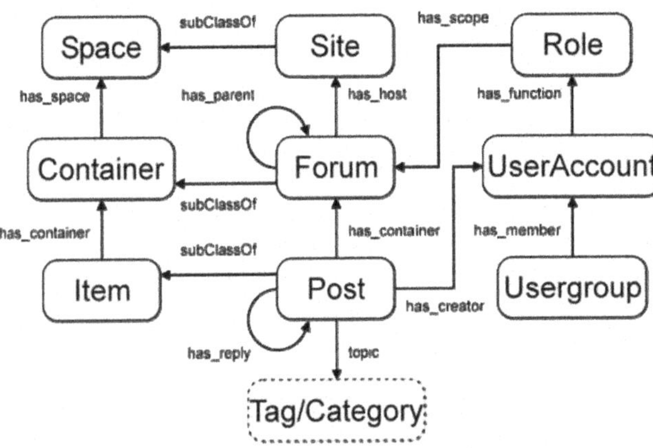

Fig. 13 SIOC Overview (from http://rdfs.org/sioc/spec/)

Figure 13 shows the SIOC schema able, in our vision, to model information coming from Web 2.0 collaboration tools like blogs, wikies and forums. SIOC provides a semantic organization of user-generated content that we can easily link to thesauri, controlled vocabularies, concept maps, lightweight ontologies and so on. The following code lines show a piece of forum data modeled in SIOC:

my:PsychologyForum **rdf:type** *sioc:Forum;*
my:PerceptionPost **rdf:type** *sioc:Post;*
  sioc:has_container *my:PsychologyForum;*
  sioc:topic *my:Perception.*
my:CognitionPost **rdf:type** *sioc:Post;*
  sioc:has_container *my:PsychologyForum;*
  sioc:topic *my:Cognition.*

More in details, the above code define a discussion forum called *my:PsychologyForum* (an individual of *sioc:Forum* class) and two posts called *my:PerceptionPost* and *my:CognitionPost* (individuals of *sioc:Post* class). The prefix *my* represents a sample namespace. The posts definitions also include the topic (*sioc:topic* property) *my:PerceptionPost* and *my:CognitionPost* are about. In particular the used topics (*my:Perception* and *my:Cognition*) come from a controlled vocabulary defined by using the SKOS schema. The following code lines illustrate how the controlled vocabulary is constructed:

```
my:Attention rdf:type skos:Concept.
my:Cognition rdf:type skos:Concept.
my:Perception rdf:type skos:Concept.
my:Emotion rdf:type skos:Concept.
my:Motivation rdf:type skos:Concept.

my:PsychologyTopic rdf:type skos:Concept;
        skos:narrower my:Attention;
        skos:narrower my:Cognition;
        skos:narrower my:Perception;
        skos:narrower my:Emotion;
        skos:narrower my:Motivation.
```

More in details, the above code lines state that *my:Attention, my:Cognition, my:Perception, my:Emotion* and *my:Motivation* are all individuals of *skos:Concept* class and can be use to "tag" instances of *sioc:Post* class in our semantic knowledge forum. Furthermore, we use the *skos:narrower* property in order to assert that *my:Attention, my:Cognition, my:Perception, my:Emotion* and *my:Motivation* are narrower than *my:PsychologyTopic* concept.

The aforementioned semantic representation allows the definition of simple SPARQL queries used, for instance, to find all posts related to a specific "tag" and consequently enables the application of SNA methods that we will show below.

4.2 Assessment of the Contribution Derived by Group Activities over Terms and Concepts Defined in Educational Ontology

One of the most important (and innovative) aspect of our proposition is the application of Network Text Analysis (NTA) to textual data arising from the collaboration between students in CSCL environments. This application is intended for making a comparison between the terms that characterize concepts of an ontology of a specific domain of knowledge (predefined by the teacher in the definition of the course contents, and those terms arising from collaborative semantic spaces (forums and wikis) in which students discuss and construct knowledge on the domain proposed.

Social Network Analysis (SNA) is an established technique of inquiry [7][8][18][19]for studying interactions in e-learning context and analyzing the relation between learning processes and the relational structures (SNA indicators) of groups and individuals (members). While this represents a classic type of application of SNA (simply it is applied to interaction in CSCL environments), many researcher [9][20][21][22][23] have proposed to apply SNA not only to interactions but also to textual data (Network Text Analysis – NTA), e.g. for deriving the structure of knowledge (the connections between lemmas) of the discourses/discussions (in a web forum or in a chat) within a group or the structure of the subjects treated by an individual or a group in texts created by a blog or a wiki. Thus in NTA nodes represent words and not people like in most classic SNA data. This means that first of all we have to perform some analysis that permits to derive the lemmas that will constitute the adjacency matrix (the typical SNA matrix representing the presence and if needed frequency of relations between nodes) for carrying out the analysis (Table 2).

Table 2 Adjacency matrix of textual data representing the co-occurrence between lemmas (nodes) in a web texts (lemma 3 and lemma 4 are associated 1 time while lemma 1 and lemma 2 are associated 3 times, etc.).

	Lemma 1	**Lemma 2**	**Lemma 3**	**Lemma 4**	**...**
Lemma 1	-	3	1	3	...
Lemma 2	3	-	2	2	...
Lemma 3	1	2	-	1	...
Lemma 4	3	2	1	-	...
...	...	...	...	...	-

Hoser and colleagues [35] have outlined that SNA applied to the Semantic Web is a very interesting field of inquiry but it is emerging slowly. So, they advocate the systematic development of Semantic Network Analysis (SemNA) and in their contribution they provide evidences of the promising set of indices of the SNA for analyzing ontologies and Semantic Web applications (in particular the centrality analysis). In another contribution, Mika [36] formalized a tripartite model of social networks and semantics in which ontologies are characterize by three different classes of nodes: actors, concepts and instances. A key aspect of this contribution is that in one of the two case studies proposed by the author for showing the usefulness of the model, he analyzed a large scale folksonomy system by using the same data we previously proposed (tab. 2) i.e. the "co-occurrence network of ontology learning". Similarly Dietrich, Jones and Wright [37] proposed a mixed approach in which social networking and semantic web technology are integrated to share knowledge within the software engineering community. Even though all these studies are very interesting and propose some creative and innovative solutions, they do not allow us to compare the textual production of a given community to a predefined ontology representing a specific field of knowledge.

So, in our proposition, the Network Text Analysis is the crucial technique of inquiry that permits:

a) to analyze the textual production of individuals and groups in specific collaborative e-learning contexts;

b) to match the textual production with the ontology pre-defined for the same field.

The basic idea is that NTA allows the teacher to compare the relevance and the structure of the contents created by students during their textual production with those of the Ontology predefined for a specific field of knowledge. The following paragraph is devoted to explain the steps characterizing this procedure.

4.3 Text Analysis, Network Text Analysis and Ontology Comparison

The tool performing the analysis we have described above, acts in two consequential steps, the first devoted to the text analysis and the second to the NTA.

The first step is characterized by a text analysis similar to those carried out by text analysis software (like T-Lab, TaLTaC, Atlas-T, Alceste, etc.). For instance, in a study carried out at the Faculty of Psychology of the University of Bologna, Mazzoni, Guidi and Musacchio [24], have analyzed the structure of a subject discussed by students during a web forum about whether Psychology is completely a science discipline. For the text analysis of the web forum discussion, authors have used T-Lab for deriving the matrix of similarity i.e. the matrix that shows the co-occurrences of terms, namely the strength of their association 'inside a given text". The matrix of similarity is analogous to the adjacency matrix (tab. 2) and can be easily converted into the last one, in which the terms became the nodes and the strength of their associations represents the strength of their relations. In a similar way, the tool we propose in this contribution applies a text analysis to some selected texts.

Before turning on the analysis described above, the teacher has the possibility to select the texts on which to apply the analysis (Tab. 3).

After the teacher has defined the analysis context, the tool runs a text analysis to give the teacher a further possibility to control lemmas derived from the analysis. Thanks to this passage, the teacher has the possibility to delete those lemmas that are considered as "non relevant" and/or redefine those lemmas that can be integrated into a same term (e.g. psychologist-psychologists; teacher-teachers; etc.). To help the teacher during this step, the tool shows some possible redefinition so that he/she simply has to decide whether to accept or not the integration or the deletion.

After this step, the tool builds the matrix of similarity that shows the co-occurrences between the lemmas, namely the strength of their association 'inside a given text", and if the teacher decides to proceed without viewing it (normally the matrix of similarity is more interesting for research interests, for this reason the tool does not show it unless expressly asked in the setting box), the tool directly converts it into an adjacency matrix.

Table 3 The possibilities the tool offers to a teacher to perform text analysis.

First he can decide about to operate an analysis on:	After having decided the level of the analysis (class, group, student), the teacher has also the possibility to select the tools on which to carry out the analysis:
o The entire class group's textual productions.	o All the semantic spaces (e.g. web forum or chat messages, blog or wiki contents construction) which a particular group or a single student has participated in.
o The textual productions of specific subgroups which he/she has created (by selecting one or more subgroups);	o One or more semantic spaces which he/she can select.
o Finally, the textual productions of specific students (by selecting one or more students).	

During the second step of analysis, the tool applies the Network Text Analysis (NTA) to the adjacency matrix data derived during the first step. Normally the teacher is not aware of all these processes in the sense that the tool directly shows the tables and the graphs resulting from the NTA.

The NTA carried out by the tool we are proposing is based on two main types of indices.

- The first is the degree centrality (and degree centralization) of lemmas i.e. their relevance by considering the quantity of the associations with each other. In other words, lemmas that have more associations (co-occurrences) with others are more central while those with less connections are more peripheral.
- The second index is the eigenvector centrality of lemmas that measures their relevance on the ground of the associations with other lemmas that are themselves relevant. In other terms, lemmas that are associated with many relevant lemmas are more central than those that are associated

with many peripheral lemmas. In contrast with the previous index, this second considers not only the quantity of connections but also their relevance.

The results of this analysis is a three columns table (tab. 4) showing in the first two columns the list of lemmas (from the most central to the most peripheral) coming from the degree and eigenvector centrality analysis and in the third column the ontology predefined for the specific fields of knowledge.

Table 4 An example of output after the NTA coming from the study carried out at the University of Bologna [30]

Degree Centrality		Eigenvector Centrality		ONTOLOGY
Lemmas	**Centrality index**	**Lemmas**	**Centrality Index**	**Lemmas of the specific knowledge domain**
clinico	0,767045	medico	0,310375	Sperimentazione
capire	0,767045	clinico	0,301637	Laboratorio
medico	0,744318	capire	0,224247	Ricerca
persone	0,721591	persone	0,197850	Comprendere/capire
professione	0,704545	professione	0,163278	Studiare
scientifico	0,653409	trovare	0,161212	Formazione/università
pensare	0,636364	lavoro	0,157355	Metodi di ricerca
trovare	0,625000	scientifico	0,157133	Analisi dei dati
parlare	0,602273	pensare	0,152533	Selezione del campione
lavoro	0,590909	scuola	0,148556	Professionalità
Corresponding lemmas: 2				

The Table 4 shows the output for the teacher. In this case, before carrying out the analysis in the setting box the teacher has selected to evaluate only the 10 most relevant lemmas resulting from the Centrality Indicators and those ones more relevant defined for the ontology. The tool automatically highlights how the two left column entries (analysis of the students' text production) correspond to those in the right column (terms of ontology predefined by the teacher). This representation allows the teacher to be aware of, on the one hand, the similarity/difference between knowledge and concepts covered by the students (during their collaboration) and those which were supposed to be learnt (ontology), and on the other hand the correspondence between the relevance of the concepts discussed by the students and the structure of concepts and sub-concepts in the ontology. In the example (Table 4), only two lemmas ("professione" and "capire", which mean "profession" and "to understand") of the students' discussions correspond to the terms defined in the ontology about "Psychology as science". At the same time, one of the two lemmas ("professione") is not as central in the ontology as it is for the students' discussion.

Thus, by considering the centrality indices, the tool we propose allows the teacher to:

- compare the knowledge construction (by considering lemmas of textual productions) to those requested by the predefined ontology of a specific domain of knowledge;
- compare students or groups to each other about their performance according to the correspondence between the knowledge construction in their texts production and the ontology of a specific domain of knowledge.

Finally, by means of two other analyses of the NTA (i.e. neighborhood and cohesion) the tool also allows the teacher to view the structure (Figure 14) of the connection between lemmas characterizing the students' textual productions (entries of the left columns of the output).

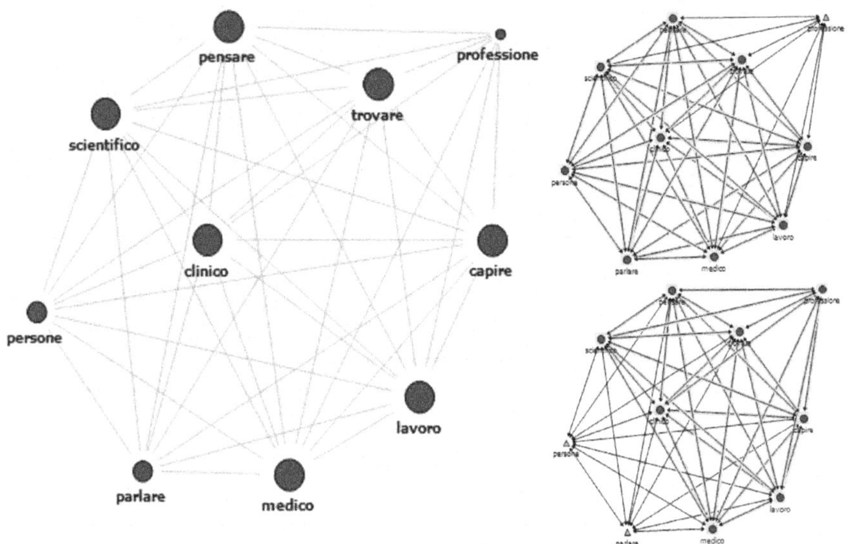

Fig. 14 connections between the lemmas

The Figure 14 shows the structure of the connections between the 10 lemmas of Table 4. On the left there is the graph derived from the neighborhood analysis in which the connections structure between lemmas is shown. On the right, the two little graphs derived from the cohesion analysis highlight two principal substructures of the students' text production. This further possibility allows the teacher not only to compare the quantitative correspondence between lemmas of the students' knowledge construction and lemmas of the ontology, but also to analyze how these lemmas are structured, which means having a representation of what type of representation the students are constructing about a specific domain. For instance, the graph represented on the right of Figure 14 (at the bottom), shows a structure composed of lemmas like "profession", "to think", "to find", "to understand", "work", "scientific", "clinical" and "medical". These associations are clearly connected to the very critical situation Psychology in Italy is facing, (finding a job in this domain is a difficult challenge), by considering the three main contexts in which Italians imagine the work of a psychologist be: research (scientific context) and clinical contexts.

5 Conclusions

In this chapter we have proposed an innovative approach to the design and maturing of an ontology based knowledge domain. In this perspective the CW is seen as a functional approach to the review of collaborative ontology process environments.

In this model the development of an ontology is a collaborative process that is maintained and validated by the encounter of two action levels. We have developed (and here presented) a tool that, at a **top down level,** allows the teachers to build collaborative learning ontologies by combining specific collaborative writing scripts and that, **at the bottom-up level,** allows the community of learners to take part in this maturing ontology experience by learning activities and text production in semantic collaborative spaces.

The innovative aspect of such an ontology building and maturing method, permits at the top down level to execute collaboration processes through coordination and cooperation tool functionalities (CCT) and the classification in Sequential and Parallel scripts. With respect to a networking learning context, it consists on the one hand in comparing the knowledge construction (by considering lemmas of texts production) to those requested by the predefined ontology of a specific domain of knowledge, and on the other hand, in the possibility to evaluate the knowledge acquired in the different phases of the collaborative activities, allowing the teacher to compare the quantitative correspondence between lemmas of the students' knowledge construction and lemmas of the ontology.

These first results encourage us to continue our work and to improve the methodology and the tools prototype, integrating them with the innovative e-Learning Platform IWT. In the future we will develop the proposed Web-based environment, evaluating it in ARISTOTELE (a co-funded project under the R&D 2007-2013, EC FP7) focusing on the Corporate context. The proposed approach is especially suitable for the Corporate Learning and Knowledge Management, where the necessity to fill the gap between the knowledge modeled by the organization and the specialized and integrative visions of the employees is pressing to maturing and update an ontology domain.

After that a more rigorous and quantitative experimentation will be made whose results will be presented in a future paper.

References

[1] Li, Y., Dong, M., Huang, R.: Toward a Semantic Forum for Active Collaborative Learning. Educational Technology & Society 12(4), 71–86 (2009)

[2] Schaffert, A., et al.: A Semantic Wiki for Collaborative Content Creation. In: Reich, S., et al. (eds.) Semantic Content Engineering, pp. 188–202. Trauner, Linz (2005)

[3] Dutra, M., Ghodous, P., Gonc]alves, R.: Resolving. Collaborative Conflicts through an Ontology-based Approach. In: Proceeding of the CE 2008, London, Dublin (August 2008), http://dx.doi.org/10.1007/978-1-84800-972-1_2
[4] Euzenat, et al.: Ontology Matching. Springer, Heidelberg (2007)
[5] Flouris, G., Manakanatas, D., Kondylakis, H., Plexousakis, D., Antoniou, G.: Ontology change: classification and survey. The Knowledge Engineering Review 23(2), 117–152 (2008)
[6] Jerome, et al.: Ontology Matching. Springer, Heidelberg (2007)
[7] Matteucci, M.C., Carugati, F., Selleri, S., Tomasetto, C., Mazzoni, E.: Teachers' Judgment from a European Psychosocial Perspective. In: Ollington, G.F. (ed.) Teachers and Teaching: Strategies, Innovations and Problem Solving, pp. 31–55. Nova Science Pulishers, Inc., NY (2008)
[8] Mazzoni, E., Gaffuri, P.: Monitoring Activity in e-Learning: a quantitative model based on web tracking and Social Network Analysis. In: Juan, A.A., Daradoumis, T., Xhafa, F., Caballe, S., Faulin, J. (eds.) Monitoring and Assessment in Online Collaborative Environments: Emergent Computational Technologies for E-learning Support, pp. 111–130. IGI Global (2009)
[9] Karapiperis, S., Apostolou, D.: Consensus building in collaborative ontology engineering processes. Journal of Universal Knowledge Management 1(3), 199–216 (2006)
[10] Seidenberg, J., Rector, A.L.: A methodology for asynchronous multi-user editing of semantic web ontologies. In: Sleeman, D.H., Barker, K. (eds.) K-CAP, pp. 127–134. ACM, New York (2007)
[11] Gaeta, M., Orciuoli, F., Ritrovato, P.: Advanced ontology management system for personalised e-Learning. Knowledge-Based Systems 22, 292–301 (2009)
[12] Wilson, S., et al.: Personal Learning Environment e sistemi educativi: una sfida al modello dominante. Journal of e-Learning and Knowledge Society (2) (2007)
[13] Bonaiuti, G.: E-learning 2.0, I Quaderni di Form@re, vol. (6). Erickson (2006)
[14] Aroyo, L., Dicheva, D.: The New Challenges for E-learning: The Educational Semantic Web. Educational Technology & Society 7(4), 59–69 (2004)
[15] Orvis, K.L., Lassiter, A.R.L.: Computer-supported collaborative learning: best practices and principles for instructors. The Idea Group, Hershey (2007)
[16] Petrucco, C.: Le Prospettive Didattiche del Semantic Web. In: Atti Didamatica 2003. TED 27-28, pp.168–176 (Febbraio 2003)
[17] Domingue, J.: Tadzebao and WebOnto: Discussing, Browsing, and Editing Ontologies on the Web. In: Proceedings of the 11th Knowledge Acquisition, Modelling and Management Workshop, KAW 1998, Banff, Canada (April 1998)
[18] Trentin, G.: Apprendimento in rete e condivisione delle conoscenze. Franco Angeli, Milano (2004)
[19] Aviv, R., Zippy, E., Ravid, G., Geva, A.: Network Analysis of Knowledge Construction in Asynchronous Learning Networks. Journal of Asynchronous Learning Networks (JALN) 7(3), 1–23 (2003)
[20] Popping, R.: Knowledge Graphs and network text analysis. Social Science Information 42(1), 91–106 (2003)
[21] Popping, R., Roberts, C.W.: Network Approaches in Text Analysis. In: Klar, R., Opitz, O. (eds.) Classification and Knowledge Organization: Proceedings of the 20th Annual Conference of the Gesellschaft für Klassifikation, pp. 381–898. Springer, University of Freiburg (1997)

[22] Batagelj, V., Mrvar, A., Zaver_snik, M.: Network analysis of texts. Language Technologies, Ljubljana, 143–148 (2002)
[23] Diesner, J., Carley, K.M.: Using Network Text Analysis to Detect the Organizational Structure of Covert Networks. In: Proceedings of the North American Association for Computational Social and Organizational Science (NAACSOS) Conference, Pittsburgh, PA (July 2004)
[24] Mazzoni, E., Guidi, L., Musacchio, G.: Network Text Analysis (NTA) applied to web interactions: from networks to social representations. working paper (2009)
[25] Kozaki, K., Sunagawa, E., Kitamura, Y., Mizoguchi, R.: Distributed and Collaborative Construction of Ontologies Using Hozo. In: Proc. WWW 2007 Workshop on Social and Collaborative Construction of Structured Knowledge, Banff, Canada, CKC (May 2007)
[26] Tempich, C., Pinto, H.S., Sure, Y., Staab, S.: An Argumentation Ontology for Distributed, Loosely-controlled and evolving Engineering processes of ontologies (DILIGENT). In: Gómez-Pérez, A., Euzenat, J. (eds.) ESWC 2005. LNCS, vol. 3532, pp. 241–256. Springer, Heidelberg (2005)
[27] Holsapple, C.W., Joshi, K.D.: Collaborative Approach in Ontology Design. Communications of the ACM 45(2) (2002)
[28] Sure, Y., Erdmann, M., Angele, J., Staab, S., Studer, R., Wenke, D.: OntoEdit: Collaborative Ontology Development for the Semantic Web. In: Horrocks, I., Hendler, J. (eds.) ISWC 2002. LNCS, vol. 2342, pp. 221–235. Springer, Heidelberg (2002)
[29] Ganter, B., Stummer, G., Wille, R.: Formal Concept Analysis: Mathematical Foundations. Springer, Berlin (2005)
[30] Bao, J., Caragea, D., Honavar, V.: Towards Collaborative Environments for Ontology Construction and Sharing. In: International Symposium on Collaborative Technologies and Systems, Las Vegas, USA, May 14-16 (2006)
[31] Auer, S., Dietzold, S., Riechert, T.: OntoWiki – A tool for social, semantic collaboration. In: Cruz, I., Decker, S., Allemang, D., Preist, C., Schwabe, D., Mika, P., Uschold, M., Aroyo, L.M. (eds.) ISWC 2006. LNCS, vol. 4273, pp. 736–749. Springer, Heidelberg (2006)
[32] Braun, S., Schmidt, A., Zacharias, V.: Ontology maturing with lightweight collaborative ontology editing tools. In: Gronau, N. (ed.) 4th Conf. on Professional Knowledge Management - Experiences and Visions, Workshop on Productive Knowledge Work (ProKW 2007) GITO, Berlin, Germany, pp. 217–226 (2007)
[33] Noy, N.F., Chugh, A., Alani, H.: The CKC Challenge: Exploring Tools for Collaborative Knowledge Construction. IEEE Intelligent Systems 23(1), 64–68 (2008), doi:10.1109/MIS.2008.14
[34] Sebastian, A., Fridman Noy, N., Tudorache, T., Musen, M.A.: A Generic Ontology for Collaborative Ontology-Development Workflows. In: Gangemi, A., Euzenat, J. (eds.) EKAW 2008. LNCS (LNAI), vol. 5268, pp. 318–328. Springer, Heidelberg (2008)
[35] Erétéo, G., Gandon, F., Corby, O., Buffa, M.: Semantic Social Network Analysis. In: Web Science (2009), http://journal.webscience.org/141/
[36] Erétéo, G., Buffa, M., Gandon, F., Grohan, P., Leitzelman, M., Sander, P.: A State of the Art on Social Network Analysis and its Applications on a Semantic Web. In: SDoW 2008, Workshop at ISWC 2008 (2008)
[37] Erétéo, G., Gandon, F., Corby, O., Buffa, M.: Semantic Social Network Analysis. In: Web Science 2009 (2009)

[38] Flower, L., Hayes, J.R.: A Cognitive Process Theory of Writing. College Composition and Communication 32(4), 365–387 (1981)

[39] Augar, N., Raitman, R., Zhou, W.: From e-learning to virtual learning community: Bridging the gap. In: Liu, W., Shi, Y., Li, Q. (eds.) ICWL 2004. LNCS, vol. 3143, pp. 301–308. Springer, Heidelberg (2004)

[40] Notari, M.: How to use a wiki in education: Wiki based effective constructive learning. In: Proceedings of the 2006 International Symposium on Wikis, Odense, Denmark, August 21-23, pp. 131–132 (2006), http://www.wikisym.org/ws2006/proceedings/p131.pdf (retrieved November 2006)

[41] Davies, J., Fensel, D., van Harmelen, F.: Towards the Semantic Web: Ontology-driven Knowledge Management. John Wiley & Sons, Inc., New York (2003)

[42] Sure, Y., Erdmann, M., Angele, J., Staab, S., Studer, R., Wenke, D.: OntoEdit: Collaborative ontology development for the semantic web. In: Horrocks, I., Hendler, J. (eds.) ISWC 2002. LNCS, vol. 2342, p. 221. Springer, Heidelberg (2002)

[43] Novacek, V., Laera, L., Handschuh, S., Zemanek, J., Völkel, M., Bendaoud, R., Hacene, M.R., Toussaint, Y., Delecroix, B., Napoli, A.: D2.3.8v2 Report and Prototype of Dynamics in the Ontology Lifecycle (2007)

[44] Tudorache, T., Noy, N.: Collaborative Protégé. In: Proc. WWW 2007 Workshop on Social and Collaborative Construction of Structured Knowledge, Banff, Canada, CKC (May 2007)

[45] Lim, S.-K., Ko, I.-Y.: Collaborative Ontology Construction using Template-based Wiki for Semantic Web Applications. Presented at the International Conference on Computer Engineering and Technology, ICCET (2009), http://ieeexplore.ieee.org/xpls/abs_all.jsp?arnumber=4769581&tag=1 (October 2010)

[46] Erétéo, G., Gandon, F., Corby, O., Buffa, M.: Analysis of a Real Online Social Network Using Semantic Web Frameworks. In: Bernstein, A., Karger, D.R., Heath, T., Feigenbaum, L., Maynard, D., Motta, E., Thirunarayan, K. (eds.) ISWC 2009. LNCS, vol. 5823, pp. 180–195. Springer, Heidelberg (2009)

A Model-Based Approach to Designing Educational Multiplayer Video Games

N. Padilla Zea, N. Medina Medina, F.L. Gutiérrez Vela, and P. Paderewski

Abstract. Several studies support the benefits that collaborative learning offers to students' overall development. It has also been shown that introducing new technologies into the educational field as motivational tools improves learning. Based on the strong evidence that CSCL (Computer-Supported Collaborative Learning) is an effective way of learning, analysis of the quality of collaboration occurring in these kinds of processes has become an important research field. Starting from these two realities and with the aim of assessing the collaboration that occurs during an educational process involving educational video games with group activities, this paper presents a model of a Video Game Supported Collaborative Learning (VGSCL) system. By means of various models related to the learning and game processes, the quality of collaboration occurring during this process can be analyzed, and the game can be adapted to make both the play and learning experiences more enjoyable and effective.

1 Introduction

Nowadays, video games are a preferred activity for many children (and adults). Although one might think that video games are devices only for amusement or entertainment, several studies [1][2] have shown that they can be powerful personal and educational tools for children.

The most important games for purposes of this research are educational video games, which teach certain content related to the academic curriculum. In this context, according to the model presented by Vygotsky [3], the game acts as a *mediator* in the learning process because educational content is implicitly present inside the game. One of the factors that contribute to the success of this type of video game is the fun component, because students, once they become players, face educational challenges without being aware of them [4]. For this reason, the design of educational video games must be realized without losing the basic element of relaxation. To accomplish this, it is necessary to have a good story and to design the tasks of the video game to create a situation in which accomplishing an enjoyable task can be equivalent to learning the educational content inside it.

Specifically, the interest in this research is focused on the use of collaborative activities in educational video games. These create additional difficulties in the design process because new relationships and premises in group tasks become apparent. To make this process easier, a set of design guidelines [5] has been defined to favor collaborative processes between group members and to retain the many

T. Daradoumis et al. (Eds.): Technology-Enhanced Systems and Tools, SCI 350, pp. 167–191.
springerlink.com

advantages that this type of learning offers [6][7]. To ensure that students obtain all possible benefits from collaborative learning, it is necessary for collaboration to be effective. A group of people sitting around the same table and working on the same task can nonetheless be working in a non-collaborative way [8]. To avoid this situation, the collaboration that occurs in the course of a task must be analyzed to make it possible to introduce changes to the process to improve collaboration and to observe the students' evolution over time. To achieve this goal, it is necessary to record and organize all useful information for later analysis. The only way of getting this information is to model the system properly and exhaustively, recording interesting data about players, groups, and the game process.

To address this situation, the authors present a model-based proposal, which defines a set of models that are thought to be necessary to analyze collaboration quality during the learning process. Thanks to this analytical process, available information can be used to adapt both the learning and game-playing processes to obtain better results.

The paper is organized as follows: Section 2 presents the main characteristics and elements of VGSCL systems. Section 3 explains in detail the models proposed, and Section 4 describes how information is recorded in these models. Finally, Section 5 outlines conclusions and suggestions for future work.

2 The VGSCL System

Modeling a VGSCL process has two main objectives: to give teachers a tool with which students can achieve educational goals in an attractive way, and to monitor how students work to enable adaptation of the game (learning) process to improve the results. Both of these objectives, as mentioned earlier, need to model players' activities during the game from three perspectives: learning, entertainment, and interaction with other players. Then multiple conclusions can be deduced from the models and an exhaustive analysis performed of collaboration and other factors that influence learning. Practically speaking, to implement these models, three main elements of collaborative learning must be taken into account: 1) educational and recreational goals, 2) educational and recreational tasks, and 3) interaction between players/students.

2.1 Goals

The goals that a student/player must achieve during the learning process can be educational or recreational, and both of these are related because recreational goals contribute to achieving educational goals, as defined in the General Tasks and Goals model (Section 3.1). Both sets of goals can change dynamically during the learning process.

According to the number of people that address a goal, two types of goals can be defined: individual and group goals. *Individual goals* must be achieved by each of the players independently; *group goals* must be achieved by the group in a common way. Group goals favor interdependence among group members.

In addition, group goals can be classified according to the number of players that can achieve a goal at one time. This classification yields competitive and noncompetitive goals. *Competitive goals* are those whose tasks can be achieved by only one player at a time. In this case, when a player wins, the rest of the players usually lose. This kind of goal is not customarily presented in an educational context, but it is especially important at a recreational level because it includes an additional motivation. *Noncompetitive goals* can be achieved by all players independently of whether some players achieve them before others.

Individual and group goals can be assigned by the teacher or by players (students) themselves. In the first case, the teacher assigns certain goals to the player (or group) that has to achieve them. This option is used when the teacher thinks that a player (or group) must learn, practice, or reinforce a specific part of the curriculum. As a result, the teacher wants learners to create recreational goals related to the educational goals that the teacher needs them to achieve. In the second case, each player addresses the goals that he wants to during the game, alone or in a group. This option is used when the group is homogeneous with regards to the educational content and students can decide which aspects they prefer to learn. This second kind of play favors the development of planning and decision-making skills in students.

2.2 Tasks

To achieve their goals, players (students) must carry out a set of tasks. The way in which tasks are related to goals is shown in the Tasks and Goals Model (Section 3.1). In some cases, to achieve a goal, students can carry out different subsets of tasks. Similarly to goals, the set of tasks can be assigned by the teacher or by the students themselves. To complete a task, a student must carry out a set of activities, which will include a set of resources to perform them. Tasks can be individual or group tasks, and three types of group activities can be further distinguished:

- *Simultaneous*: All group members working on the task must address each activity at the same time. This kind of task requires synchronization and favors collaboration inside the group and individual learning for each group member.
- *Ordered task*: To achieve the task, some group members must take part. It is not necessary to do so at the same time, but activities must be done in a specific order. This means that before an activity can be performed, all previous activities must have been performed before. Ordering constraints may affect some of the activities in a task, but not others. In any case, this way of performing tasks favors interdependence between group members because they are aware of the fact that they need their partners to complete the task.
- *Unordered task*: Each player can carry out his own subset of activities when he wants, without influencing the work of the rest of the group members. To complete a task, a student simply needs to perform its activities. This kind of task also improves cooperative attitudes because group members confront the task as a team, but each member has responsibilities only for part of the task.

With regard to how players can achieve a goal, three game modes are permitted:

- *Free*: A player (student) addresses the tasks that he wants to, and which goal is addressed by these activities does not matter.
- *Goal-oriented*: A player (student) can address only tasks that help him to achieve his goals. There are two possibilities:
- *All at the same time*: A player has a set of available tasks related to a set of goals, and he can carry out these tasks in any order. In this way, the player can progress in parallel on all the goals pursued.
- *One by one*: A player must achieve one goal before confronting another. For this reason, he must carry out tasks related to one goal until that goal is achieved.

Similarly to goal assignment, the game mode can be assigned by the teacher, or the group can decide it if they are capable of doing so.

2.3 Interaction

During a collaborative video-game-based learning process, three types of interactions can occur: communication, collaboration, and coordination (the 3 C's model [9]). The authors believe that detecting and classifying messages between group members while they are interacting is a basic element of the analytical process. To make this classification easier, the authors propose a strong message categorization [10] scheme with three categories:

- *Communication Messages*: These are messages that intend to inform the recipient about something. This category includes, for example, question-and-answer exchanges or social messages which provides information about relationships existing inside the group.
- *Collaboration Messages*: These messages are generated during a collaborative situation, for example generating a proposal or asking for resources.
- *Coordination Messages*: This kind of messages serves to organize methods or strategies to achieve a goal. Decision-making or planning tasks are included in this category.

3 Modeling the VGSCL System

The current proposal is a model-based approach focused on the analysis of collaboration in video games with the intent of performing future adaptations which improve the learning process in general and the collaborative process in particular. To this end, the set of models can be divided into four main subsets: 1) models to define and monitor educational content; 2) models to specify and monitor recreational content; 3) models to relate educational and recreational content, which define how recreational content satisfies educational requirements; and 4) user modeling to monitor students' educational evolution while they are playing the game.

Because each of these models is designed by a different teacher or by users in general, it is necessary to reconcile how they refer to educational content and to areas of knowledge. For example, consider a video game with the educational goal of teaching addition. This objective can be called *Teach addition* or *Learn additions* or *Additions* or *Adding…* The main problem is that the same objective can exist with different names in each of the video games. This entails excessive effort and missing data.

To avoid this problem, a set of dictionaries is proposed for groups of users with similar characteristics, which contains all the general names used by the proposed system. For this reason, when a user needs to incorporate a concept as a goal, task, or other entity, it is necessary to choose that concept from the dictionary. If this concept does not exist in the dictionary, then it can be added. In addition, if a concept has relations with other concepts, these other concepts are stored with this dictionary key, thus making the definition of educational and game processes easier. In this paper, terms starting with capital letters are elements that the authors have included in the dictionary.

3.1 Models of Educational Content

This set of models must be defined by the teacher or by the school's pedagogical team. In these models, they must describe which subjects are going to be taught and how this content can be represented in terms of goals and tasks. In addition, the sets of goals and tasks must be related to define which tasks enable the students to achieve which goal. For this purpose, the authors have defined three models: 1) Educational Model, to define the general educational project, 2) Educational Goal Model, to define each of the objectives that students must achieve, 3) Educational Task Model, to describe the set of tasks included in the learning process. Each of these models is described below.

3.1.1 Educational Model

The *Educational Model* enables the teacher to specify what content he wants to teach by means of a video game. In particular, it makes it possible to establish an overall learning strategy to cover a particular subject or a set of subjects.

Because the learning process throughout the system should be based on the design of educational content to be taught or reinforced by means of the game, it seems clear enough that the values for all the attributes in the model must be known and specified *a priori*. Furthermore, because the educational project should not change during this process, the initial content of the model should not be changed. Table 1 shows the attributes included in this model.

Table 1 Attributes of the Educational Model.

Attribute	Description	Domain
Identification	Internal identification	x: x ∈ [EM0000, EM9999]
Knowledge Areas	Knowledge areas to be taught in the educational process	{x: x is a Knowledge Area}
Educational Age	Ages for which the learning process is designed	(x, y): x ∈ [0, 99], y ∈ [0, 99], x≤y
Previous Knowledge	Knowledge that that user must have before starting the learning process described in this educational process	{x: x is an Educational Goal or an Educational Task}
Educational Goals Model	Set of educational goals to be taught in this educational process, with prerequisite relationships between them	{x: x is an Educational Goal}

The *Identification* attribute is assigned automatically when the model is introduced into the system. All the models in the system are identified using a similar identifier, in which the two first characters are letters identifying the type of model and the rest of the identifier consists of numbers.

Because a learning process can cover one or more subjects, and to extend the applicability of the model, a concept broader than the academic subject is needed to classify the learning process. To meet this need, the *Knowledge Areas* attribute is used in the educational model. A knowledge area can include multiple subjects which must deal with the same concepts. The value of this attribute limits the set of educational goals to be included in the learning process, because a goal without this knowledge-area attribute cannot be included in the system. The set of educational goals of an educational model is expressed as an *Educational Goals Model* to specify the prerequisite relationships among goals. When the teacher selects a goal from the dictionary, there can be several sets of subobjectives associated with it. To include a goal in the educational model, the teacher must choose the specific subset of goals to achieve from among those related to the main goal.

In *Previous Knowledge*, the teacher can specify which goals and tasks the learners must have mastered before starting this educational process. It is possible to check whether a student has accomplished a goal or task by referring to his Student-Player Model (Section 3.4.1).

Educational Age is used to specify the proper range of student ages to learn this content. By using this attribute, the inclusion of an attribute like *course* is avoided to make the application more flexible. For example, different countries can vary the age of the children in the same course, the number of courses that exists in each grade, and so on.

3.1.2 Educational Goal Model

In the system proposed here, educational content is organized according to *goals* and *tasks*. A goal is a general content or skill that is mastered by completing a set of tasks. Furthermore, each task can contribute to achieving one or more Educational Goals. The Educational Goal model includes educational information about the goal itself and its relations with educational tasks. The attributes included in this model are shown in Table 2.

Table 2. Attributes of the Educational Goal Model.

Attribute	Description	Domain
Identification	Internal identification	x: x ∈ [EG0000, EG9999]
General Name	General name describing the main content to be learned	x : x is a General Name
Knowledge Area	Knowledge area into which the goal falls	x: x is a Knowledge Area
Transverse Areas	Auxiliary knowledge areas addressed in this educational goal	{x: x is a Knowledge Area}
Educational Age	Corresponding educational ages for this educational goal	(x, y): x ∈ [0, 99], y∈ [0, 99], x≤y
Educational Content	Educational content included in this educational goal	Natural Language
Goal Model	Goal hierarchy	{x: x is an Educational Goal}
Tasks and Activities Model	Set of educational tasks and activity paths, with prerequisites among them, and the formula to calculate the final mark based on the tasks performed	{(x, y): x ⊆ Educational Tasks, y is a formula}

Once an Educational Goal has been defined, its attributes will not change during the process. Therefore, these values will usually not be modified.

Besides the internal identifier, a *General Name* (from the dictionary) is assigned to the goal to describe it in a general form; this name represents the content to be learned by achieving this goal. Attributes related to knowledge areas are intended to establish which sets of knowledge are associated with each area of educational content. These attributes are also included in the Educational Task and Activities Model to distinguish between a main area, into which the educational content falls specifically (*Knowledge Area*), and other areas related to the goal or task, but in which it is addressed peripherally or less deeply (*Transverse Areas*).

In addition, an attribute to indicate the proper *Educational Age* associated with this educational goal is also needed so that goals can be chosen in relation to the corresponding age range in the educational model. This range must fall in the range of the educational model to be included.

In principle, the *Educational Content* for the goal should be explained in natural language to clarify what content students are going to learn by achieving this educational goal, but standards and specifications for educational content, such as those adopted by the ASPECT Project[1], may also be included.

[1] http://www.aspect-project.org/, accessed on March 22, 2010.

Because educational content can be organized hierarchically, goals and subgoals can be defined. Only leaf goals can have tasks associated with them. To model this situation, the educational goal model has the *Goal Models* attribute. If this goal is not in a leaf, then it has an associated goal model; otherwise, it has a *Tasks and Activities Model.*

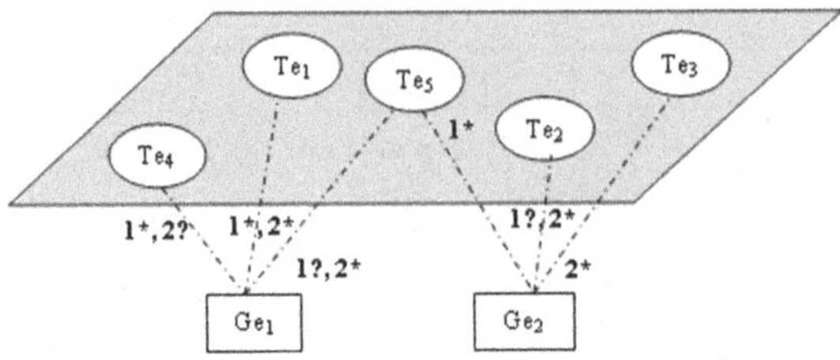

Fig. 1 Paths that a player can use to achieve goals Ge_1 and Ge_2.

The *Tasks and Activities Model* contains one or more sets of educational tasks which contribute to achieving the goal; these tasks form *educational task paths.* Within this set of available educational tasks and activities, some are obligatory and others optional. This means that students must perform all the tasks and activities which are required to achieve a goal, but not necessarily the optional tasks, which will serve to provide additional dimensions to the learning process. To identify each of the options available to achieve a goal, *paths* are defined, each identified by a number. Therefore, *a path is a set of tasks and activities associated with an educational leaf goal that makes it possible to achieve that goal.*

In this model, the symbol * is used to represent obligatory tasks and the symbol ? to indicate voluntary tasks. Because a task can be associated with one or more goals, the obligatory/optional symbol must be attached to each association, not to the task. This means that a task can be obligatory for one goal, but optional for another. For example, in Fig. 1, to achieve goal Ge_1, there are two possible paths: 1) the first, labeled as **1**, in which the player has to perform $\mathbf{Te_1}$ and $\mathbf{Te_4}$ obligatorily and $\mathbf{Te_5}$ optionally; and 2) the path labeled as **2**, in which there are two mandatory tasks ($\mathbf{Te_1}$ and $\mathbf{Te_5}$) and one optional task ($\mathbf{Te_4}$). Similarly, to achieve goal Ge_2, players have two alternative paths: 1) path **1**, where $\mathbf{Te_5}$ is mandatory and $\mathbf{Te_2}$ is optional; 2) and path **2**, where $\mathbf{Te_2}$ and $\mathbf{Te_3}$ are mandatory. In the model, the set of valid paths to achieve a goal can be specified as a list of pairs, where:

- The first element of the pair is the set of tasks in the path, expressed by using some formal or semiformal technique to specify the restrictions between tasks and activities, such as that proposed by Paternò [11].
- The second element is an algebraic expression to calculate the score for the goal in terms of the scores obtained on each of its tasks. This provides a mechanism for giving prominence to one task over others by assigning weights.

3.1.3 Educational Task and Activities Model

To achieve an educational goal, students must perform a set of educational tasks and activities. As previously stated, each of these goals can be achieved by means of different sets of tasks, and these tasks must be defined in the system. Similarly to goals, tasks can be divided into subtasks. The information included in the system with regard to educational goals is shown in Table 3.

Table 3 Attributes of the Educational Task and Activities Model.

Attribute	Description	Domain
Identifier	Internal identification	x: x ∈ [ET0000, ET9999]
General Name	A general name describing the main task to be performed	x: x is a General Name
Knowledge Area	Knowledge area into which the task falls	x: x is a Knowledge Area
Transverse Areas	Transverse knowledge areas encountered in this educational task or activity	{x: x is a Knowledge Area}
Educational Content	Educational content to be mastered in this task	Natural Language
Educational Tasks and Activities Model	Set of educational tasks and activities with prerequisites among them, and a formula to calculate the final score according to the tasks completed	{(x, y): x ⊆ Educational Tasks and Activities, y is a formula}

The attributes in this model are similar to those in the previous model: an internal identifier, a general name from the dictionary, the main knowledge area and transverse areas, the educational content included in this task or activity, and a tasks and activities model.

As was seen in Fig. 1, an educational task or activity can contribute to achieving one or more educational goals, and to achieve a goal, it may be necessary to perform several tasks or activities. When the teacher chooses a General Name for an educational goal or task, related tasks and activities are included in the dictionary key for this educational goal. However, the teacher can configure which of these task and activity paths should be proposed and the order in which tasks and activities must be done. This information is specified by means of the *Educational Tasks and Activities Model*. For each of the options included in the model, it is necessary to specify how to calculate the final score by means of a formula.

3.2 Models for Recreational Content

The previous subsection explained the set of models to specify the educational part of the system proposed here. Because the system is being designed to enable implicit learning by means of video games, the next task is to define a set of models to specify the video game to be used to teach the educational content. Because the system is intended to monitor both the learning and playing processes, these models include information for the teacher, who will know general information about the video game, but not the details of its implementation.

To maintain a direct relationship between these two aspects of the system, the recreational content has been defined based on the same elements used for the educational content. Thus, the general aspects of the video game are defined in the Game Model, the challenges are separated out into the Video Game Challenges Model, and a set of Video Games Stages and Levels Models is provided to enable the student to overcome the challenges posed by the game.

3.2.1 Game Model

The attributes of this model describe the main characteristics of the game. To select a video game, the teacher will use the knowledge areas in which the student will work; these will be specified in the Educational Model. With this information, the teacher can choose any one of the video games which includes one of these areas, although the more knowledge areas are included in both the Game Model and the Educational Model, the more useful the game will be.

Table 4 gives details of the attributes of the Game Model.

Table 4 Attributes of the Game Model.

Attribute	Description	Domain
Identifier	Internal identifier	x: x ∈ [GM0000, GM9999]
Name	Name of the video game	Natural language
Knowledge Areas	Set of knowledge areas addressed in this video game	{x: x is a Knowledge Area}
Age	Recommended age range	(x, y): x ∈ [0, 99], y ∈ [0, 99], x<y
Difficulty	General difficulty of this video game	x: x ∈ {High, Normal, Low}
Interaction	Collaboration, coordination, and competition features	Natural language
Mode	Game model selected by the teacher	x: x ∈ {Free, By goals all at the same time, by goals one by one} [5]
Type	Describes the game type	x: x ⊆ {Action, Adventure, RPG, Strategy, Simulation, Race, Fight, Puzzle, Musical, Sport, Platform, Gun} [12]
Device	Game device on which the video game runs	x: x ∈ {PC, Nintendo DS, Nintendo Wii, XBOX 360, PSP, iPhone, iPod Touch}
Story	Thread of the game	Natural language
Multimedia	Set of multimedia effects included into the video game: graphics, sounds, or animations	Natural language

This model includes attributes which describe the game and help the teacher to select the proper video game for the educational content to be taught and to configure some of its characteristics.

First of all, similarly to the previous models, there is an internal *Identifier* and the *Name* of the game. Then, because the video game has been developed to help the student learn educational content, the model specifies the *Knowledge Areas* that this video game covers, the recommended *Age* for the video game, and its general *Difficulty*, to guide teachers in selecting a video game for a particular group of students. In addition, other aspects of the model, such as the *Interaction* characteristics that the game incorporates and the default *Mode* of the game can be of interest for the teacher when selecting the proper video game.

Next, there is a set of attributes which specifically describe the game characteristics. The *Type* and the main features of the *Story* of the game enable the teacher to choose one game or another with regard to the preferences of the students or to other parallel activities being conducted at school. The *Device* attribute indicates the technology for which the game has been developed and includes the main devices on the market. In addition, a set of *Multimedia* effects is specified. Among other possibilities, the teacher may use the information contained in this attribute to select a game more or less appropriate for a group of students with disabilities. For example, a game that does not include animations might be more appropriate than another for students with attention deficit disorder.

3.2.2 Video Game Challenge Model

As previously stated, to maintain uniformity between the educational process and the recreational process, educational games for this system are also modeled including elements in two categories. These categories are parallel to those used in previous models: *goals* in educational models are *challenges* in video game models, and *tasks and activities* in educational models are *stages and levels* in video game models.

Video Game Challenges are milestones that players must achieve. To overcome a challenge, the player must complete a series of Video Game Stages and Levels, through which, besides overcoming the challenge posed by the game, he obtains knowledge. Table 5 shows the attributes of the Video Game Challenge Model.

Table 5 Attributes of the Video Game Challenge Model.

Attribute	Description	Domain
Identifier	Internal identifier	x: x ∈ [VC0000, VC9999]
General Name	General name to describe the main challenge in this goal	x: x is a General Name
Description	Natural-language description of this goal	Natural language
Video Games	Set of video games in which this challenge is included	{x: x is a Game}
Video Game Challenges Model	Set of Video Game Challenges with prerequisites between them and formulas to calculate a score	{(x, y): x ⊆ Video Game Challenges, y is a formula}
Video Game Stages and Levels Model	Set of task paths to complete the Video Game Challenges with prerequisites between them and formulae to calculate the score for each of the paths.	{(x, y): x ⊆ Video Game Stages and Levels, y is a formula}

The model includes an internal *Identifier*, a general *Name* to describe the challenge, and a natural-language *Description* to explain what the player must do. Although each of the video games has a specific set of challenges, there may be general challenges that can be used in several video games. To accommodate this possibility, the model includes a set of *Video Games* to which a particular goal can be related. In any case, this attribute is automatically filled in when attributes are included in video games to avoid inconsistencies from manual data entry.

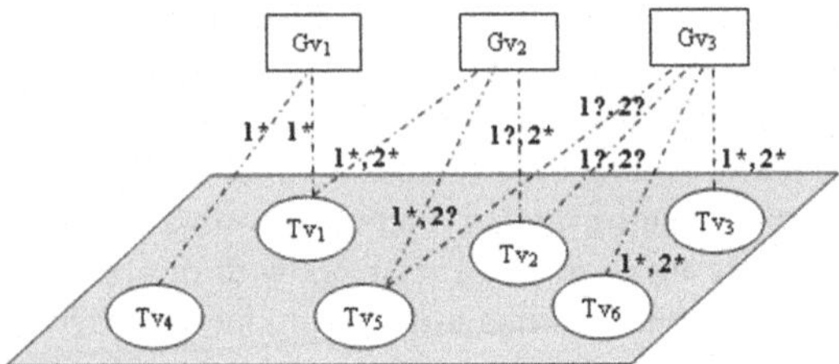

Fig. 2 Paths that a player can use to overcome challenges Gv_1, Gv_2 and Gv_3.

Similarly to the educational models, challenges in video games can also be hierarchical. These relationships are described, as in the educational models, by including a *Video Game Challenges Model.* If this challenge is a leaf challenge, it includes a *Video Game Stages and Levels Model*, as in Educational Goals, to specify the set of stages and levels paths to be completed to overcome the challenge (Fig. 2).

3.2.3 Video Game Stages and Levels Model

The Video Game Stages and Levels Model enables the player to overcome the different challenges that he faces during the game. Some of these attributes have meanings similar to those in the previous models and will not be explained again. These attributes are: *Identifier*, *General Name,* and *Description.* Table 6 shows the set of attributes that define a Video Game Stage or Level.

First of all, this discussion will focus on the *Category* attribute, which indicates whether the task is mostly a goal, a route map, a puzzle, a riddle, a dialogue between characters, or another kind of task. This small set of possibilities is proposed here, but teachers can include others in the repertoire if needed. Next, the number of *Players* needed to solve this task is specified, and, if more than one, how players must address the activities (*Type*): all at the same time, in a specific order, or in any order. If the activity is only for one player, then the value of this attribute is Null.

Desirable Features describes the characteristics that could be useful to complete this stage or level more easily. A similar attribute is included in the user model (Section 3.4.1) for comparison. In this way, it could be better to choose the group member who is the best communicator, strategist... to perform a stage or level, depending on its characteristics.

Table 6 Attributes of the Video Game Stages and Levels Model.

Attribute	Description	Domain
Identifier	Internal identifier	x: x ∈ [VS0000, VS9999]
General Name	General name describing the main task to be performed	x: x is a General Name
Description	Natural-language explanation of what the player must to do to complete the task	Natural language
Category	Category to which this task belongs, according to the challenges proposed	x: x ∈ {Goal, Map, Puzzle, Dialogue, Strategy, Riddle…}
Players	Number of players that must participate in this task	x: x ∈ N
Type	When this task is for a group, specifies the way in which the players must address it. Otherwise Null.	x: x ∈ {Null, Simultaneous, Ordered, Unordered}
Desirable Features	Set of desirable features of players performing this task to obtain better results	{x: x ⊆ {Diversifier, Organizer, Orderly, Ambitious, Conformist, Goal-oriented, Explorer, Chatty, Strategist, Investigator, Prestigious, Influential, Communicator, Collaborator, Coordinator}}
Difficulty	General difficulty of this Video Game Task	x: x ∈ {High, Normal, Low}
User Control	Specifies whether this task is performed with or without user control	x ∈ {yes, no}
Resources	Set of resources or tools needed to complete the stage or level. These are specific to each game.	Natural language
Stages and Levels Model	Set of Video Game Stages and Levels with pre-requisites between them and formulae to calculate the score.	{(x, y): x ⊆ Video Game Stages and Levels, y is a formula}

The next attribute is *Difficulty*, which is related to the general difficulty of this stage or level. This attribute enables the teacher to choose one path or another for a specific student to avoid the situation in which excessively difficult video-game challenges have a negative impact on the learning process. It is also possible to include stages or levels which do not require the user to solve them, but which rather form part of a transition or are intended to give information. For such tasks, the model includes the *User Control* attribute, with a default value of "yes."

The next attribute is *Resources,* which describes which resources among those offered during the game are needed to complete this stage or level. The player or some of his partners must have this resource to be able to complete the stage or level.

Finally, and again similarly to the educational process, a "task" (stage) can have "subtasks" (levels). Therefore, in the case of a stage, a *Stages and Levels Model* will be included to describe these relations as well as their prerequisites and a formula to calculate the final score in this path.

3.3 Model to Relate Educational and Recreational Contents: The General Tasks and Goals Model

This model constitutes the key of the system proposed here because it makes it possible to connect the educational and recreational contents in such a way that completing a stage or level in the video game is equivalent to completing the related educational tasks. Similarly, overcoming a challenge in the video game is equivalent to achieving the related educational goals. The relationship between educational and recreational contents is described in the General Tasks and Goals Model.

A video game can be represented as a set of challenges that players must overcome. Because of the special characteristics of the video games discussed here, two sets of goals exist, a set of educational goals and a set of goals related to game mechanics. This means that two goal-trees can be constructed: a tree of didactic goals mapped onto a tree of video-game challenges. As previously stated, to achieve an educational goal, students must complete a set of educational tasks, and to meet a recreational challenge, players must complete a set of stages and levels.

To use video games as CSCL tools, the educational content must be hidden inside the recreational content. Because educational video games try to achieve implicit learning (without the users' being aware of their learning process), it was judged necessary to define the didactic goals and the recreational goals separately. However, both types of goals must be connected, so that when the user overcomes a challenge in the video game, he also achieves the didactic goal or goals associated with that satisfied recreational goal. It is therefore proposed to establish two levels in the General Tasks and Goals Model: *Educative Level* – L_E (lower layer in Fig. 3) and *Video Game Level* – L_V (upper layer in Fig. 3).

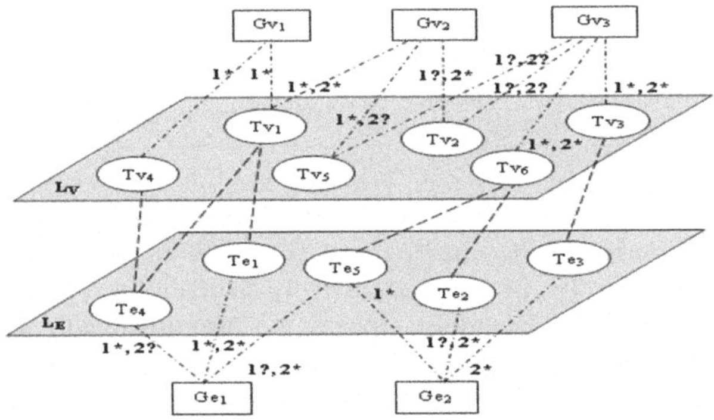

Fig. 3 General Tasks and Goals Model.

Let Tv_i be the set of stages and levels included in L_V (*Video Game Stages and Levels*) and Te_i the tasks and activities included in L_E (*Educational Tasks and Activities*). The establishment of relations among the tasks at both levels is called *implementation* (*Implements* attribute in Table 7). The stage or level, Tv_B, implements the educative task or activity, Te_A, whether or not Tv_B in the video game is useful for teaching the educational content of Te_A. This relation is represented by a broken line connecting the contents in both levels (for example, Tv_4 and Tv_1 implement Te_4 in Fig. 3). In this way, for each stage or level in L_V, one or more task(s) in L_E must implement it.

Nevertheless, as can be observed in Fig. 3, not all stages and levels in L_V have to be associated with tasks or activities in L_E, because some stages and levels in the video game can be simply entertaining, without any educational content.

In line with the presentation of other models in this paper, Table 7 specifies the attributes included in this model. Despite the internal *Identifier*, in this model, an educational model and a video game model are related. To give the teacher more flexibility, the model includes the possibility of starting from a specific educational model and introducing changes if needed. This feature is represented in the model by means of three attributes: 1) the *Educational Model,* which points to a previously defined educational model that is already included in the system; 2) if a value is indicated in the *Educational Goals*, it means that the teacher has changed the paths or prerequisites between educational goals that were included in the educational model; and 3) *Educational Tasks and Activities* may also indicate changes with regard to the initial educational model. If no educational goals or tasks are indicated, then the teacher has not changed the initial educational model, and the system uses the model described in the system.

Table 7 Attributes of the General Tasks and Goals Model.

Attribute	Description	Domain
Identifier	Internal identifier	x: x ∈ [TG00000, TG99999]
Educational Model	Educational Model to which Educational Tasks included in the lower level refer	x : x is an Educational Model
Educational Goals	Educational goals included in this video game	{x: x is a Educational Goal}
Educational Tasks and Activities	Educational tasks and activities included in this video game	{x: x is a Educational Task or Activity}
Video Game Model	Video Game Model to which Video Game Tasks included in the upper level refer	x : x is a Video Game Model
Implements	List of Educational Tasks that implement each of the Video Game Tasks	{(x, y, z): is a Video Game Task, y ⊆ {Educational Tasks}, z is a formula}

Assume a game related to the field of language, which has three educational objectives: spelling, synonyms, and verbs. This game is inspired by the story of Snow White and the Seven Dwarves. There are seven characters, the dwarves. The video-game challenges are three: 1) catch butterflies in the wood, 2) come into the witch's house, and 3) wake Snow White up.

Challenge 1) is individual. To overcome this challenge, each dwarf has to catch several butterflies. Each of these butterflies has a letter, and he has to build a word using these letters. The difficulty level modifies the difficulty of the selected words or the number of letters in the word or introduces letters which do not contribute to building the word. Entering the witch's house (challenge 2) is a group challenge: players have to pair off and create pairs of synonyms to be able to enter. If their selected words are not synonyms, they cannot open the door. The last challenge (number 3) is also a group one. When they are in the house, one of the dwarves takes a verb out of a trunk. According to the difficulty level, this can be an infinitive or a specific form of the verb. Using this verb and some of the words that have been built, the players have to make a sentence. When they have made the sentence, they put it in the potion book to obtain the potion that will wake Snow White up.

Overcoming challenge 1 implies achieving an educational goal related to spelling, because the player has been able to build a correct word. Thus, *catching butterflies* implements *spelling*. Similarly, when dwarves have been paired by joining synonyms, the players have learned how these words are related, so *pairing dwarves* implements *synonyms*. Finally, and similarly to the two previous examples, completing *wake Snow White up* implements *verbs* because to obtain the potion, players have to build a correct sentence with the correct form of the proposed verb and using the words constructed in the previous challenges.

3.4 User Modeling

To help the teacher monitor the learning process, the system includes both individual and group models to record educational marks and game preferences for each student and group. In this way, the system can analyze information and provide reports about how well students are learning.

3.4.1 Student-Player Model

An important part of this system is to know how much the student has learned by means of the video games by recording and organizing all relevant information. The authors consider that the best way of doing this is to model the user properly and exhaustively.

The proposed Student-Player Model is composed of four perspectives related to the aspects to be studied: Personal, Educational, Video Game, and Interaction.

The *Personal Perspective* contains general information about the user. The specific design of this perspective is shown in Table 8, but a few elements of it will be discussed further. First of all, it contains some considerations related to disabilities (the *Visual Problems*, *Auditory Problems*, *Motor Problems*, and *Cognitive Problems* attributes). This information is included with a list for each type of disability. Some of these are included in the table, but the teacher can include more terms in the dictionary if needed.

With regards to candidate roles, three different attributes have been included, each related to one of the perspectives. The *Candidate Educational Role* attribute refers to the type of learning the candidate is accustomed to do. For example, if the student prefers to engage in free learning (Section 2.2), he is a diversifier; if he performs learning by pursuing goals all at the same time, an organizer; if he learns by addressing goals one by one, an orderly person. In addition, if he is accustomed to performing all optional tasks and activities (section 3.1.2), he is ambitious; if not, a conformist. Here a certain number of possible roles are included, but the teacher can include others in the dictionary.

The *Candidate Video Game Role* and *Candidate Interaction Role* attributes work in a similar way: a set of roles is proposed, but the teacher can include others. In particular, roles proposed for the video-game perspective are related to the attribute categories in the Video Game Stages and Levels Model (section 3.2.3), and interaction roles are related to Social Network Analysis (SNA) indices calculated from group interactions (Section 3.4.2).

Table 8 Attributes of the Personal Perspective of the Student-Player Model.

Attribute	Description	Domain
Identifier	Internal identification	x: x ∈ [SP00000, SP99999]
Name	Student name and surname	Natural language
Age	Student current age	x: x ∈ [0, 99]
Educational Age	Recommended age in the current grade	x: x ∈ [0,99]
Sex	Student gender	{Male, Female}
Nationality	Student nationality	x: x is a nationality
Visual Problems	Student visual problems	{(x[2], y): x is a Visual Problem, y is a percentage}
Auditory Problems	Student auditory problems	{(x[3], y): x is a Auditory Problem, y is a percentage}
Motor Problems	Student motor problems	{(x[4], y): x is a Motor Problem, y is a percentage}
Cognitive Problems	Student cognitive problems	{(x[5], y): x is a Cognitive Problem, y is a percentage}
Candidate Educational Role	Ordered list of roles that student is accustomed to play in educational tasks. The first element is the most usual.	(x, y, z): x, y, z ∈ {Diversifier, Organizer, Orderly, Ambitious, Conformist…}
Candidate Video Game Role	Ordered list of roles that student is accustomed to adopt when playing a video game. The first element is the most usual.	(x, y, z): x, y, z ∈ {Goal-oriented, Explorer, Skillful, Conversationalist, Strategist, Investigator…}
Candidate Interaction Role	Ordered list of roles that student is accustomed to play in group activities. The first element is the most usual.	(x, y, z): x, y, z ∈ {Prestigious, Influential, Communicator, Collaborator, Coordinator, Isolated}

[2] For example, color-blindness, visual deficiency…

[3] For example, dystrophy.

[4] For example, deafness, hypoacusis…

[5] For example, epilepsy, ADDH (Attention-Deficit Disorder with Hyperactivity).

The next perspective is related to student educational achievements and contains information about the educational goals and tasks that a particular student must address, has addressed, or has finished. The teacher checks the educational results in this perspective so that if the student has achieved the goal, the mark obtained is available. Table 9 shows all the attributes of *Educational Perspective*.

Attributes in this perspective are initially empty, and they are updated when a teacher plans new goals or tasks or while the student is playing. The list of *Proposed Educational Goals* can be updated at the beginning of the game or while the game is running, depending on the game mode (Subsection 3.2.1). If the game mode is free (students address different goals according to their preferences), then a new goal will be added to the list each time a goal is addressed for the first time. If the game mode is governed by goals (all at the same time or one by one), the list of goals must be specified before starting the game, and the student is allowed to address only goals that are on the list.

The list of *Faced Educational Goals* is composed of tuples of four elements: the first is an educational goal in the *Proposed Educational Goals*, the second is the proportion of the goal achieved, the third the date on which the student started this goal, and the last the set of video games that the player has used to address this goal. The last attribute is necessary because the same educational goal can be worked on using different video games. In each game, the educational tasks undertaken to achieve the objective may be different (and, of course, the video-game tasks that support the educational tasks will be different). Similarly, the list of *Achieved Educational Goals* also consists of a set of tuples, but each of them with five elements. The first is an educational goal obtained from the list of *Faced Educational Goals*, the second is the mark obtained for this goal, the third the date on which the student started the goal and the date when he finished the goal, and the last the set of video games used to achieve the goal.

When a *Faced Educational Goal* reaches a value of one in the proportion of a task completed, then the *Achieved Educational Goals* list is updated because all tasks associated with this goal have been completed. At this time, this tuple is deleted from the *Faced Educational Goals* list and information is updated in the *Achieved Educational Goals* list by the following transformations:

- The Educational Goal is copied into a new tuple. Table 9 indicates that the first element in the tuple for the list of Achieved Educational Goals is obtained from the list of Proposed Goals. This is because this goal will be deleted from the list of Faced Educational Goals.
- The mark obtained is calculated using formulae associated with each of the educational tasks. If this goal already has another previous mark, this mark is updated, unless the teacher has locked out this possibility. In that case, the mark does not change even if the student achieves the goal again.
- The third element is also copied from the third element in the faced-goal tuple.
- The fourth element is the current date, to indicate when this goal was achieved.
- The list of *Video Game Identifiers* is copied into the fifth element.

Once these transformations have been made, the tuple is added to the Achieved Educational Goals list. Because this goal has already been achieved, it is also deleted from the Proposed Educational Goals list.

Attributes related to Educational Tasks work similarly to those just explained, but with regard to tasks instead of goals.

Table 9 Attributes of the Educational Perspective of the Student-Player Model.

Attribute	Description	Domain
Proposed Educational Goals	List of Educational Goals to be learned by the student	{x: x is an Educational Goal}
Faced Educational Goals	List of Educational Goals that the student has started, proportion of tasks completed, date when the goal was started, and the video games used to achieve the goal	{(x, y, z, t): x ∈ Proposed Educational Goals, y ∈ [0, 1], z is a date, t is a set of Video Game Identifiers}
Achieved Educational Goals	List of Educational Goals that the student has achieved, marks obtained, date at which goals were achieved, and video games used to achieve goals	{(x, y, z, w, t): x ∈ Proposed Educational Goals, y ∈ [0, 10], z is a date, w is a date, t is a set of Video Game Identifiers}
Proposed Educational Tasks	List of Educational Tasks that the student must perform	{x: x is an Educational Task}
Faced Educational Tasks	List of Educational Tasks which the student has started and proportion of activities that has already been performed	{(x, y, z): x ∈ Proposed Educational Tasks, y ∈ [0, 1], z is a date}
Passed Educational Tasks	List of Educational Tasks that the student has passed and marks obtained	{(x, y, z, t): x ∈ Faced Educational Tasks, y ∈ [0, 10], z is a date, t is a set of Video Game Identifiers}

The third perspective of the Student-Player Model, related to video-game preferences, is called the *Video Game Perspective*. This perspective is intended to adapt the game to avoid the situation in which difficulties in the game cause educational difficulties. Attributes in this perspective can be seen in Table 10.

This set of attributes makes it possible to adapt the video-game challenges to each user. The *Device* on which this player usually plays and previous *Experience* in video games can be filled in when the model is introduced into the system. This attribute is scaled according to Card et al. [13], with 1 meaning beginner, 2 novice, 3 intermediate, 4 expert, and 5 master, and can be updated as the user gains experience with video games. The rest of the attributes are closely related to the video games played in the system and are initialized with a default value until data have been analyzed and updated in the model. In particular, lists of tasks will be initialized with an empty list, *Length* with the value medium, and *Quantity* with Null.

Table 10 Attributes of the Video Game Perspective of the Student-Player Model.

Attribute	Description	Domain
Devices	Devices that this player uses to play video games	{x: x is a Video Game device}
Experience	Experience that this player has with video games	x: x ∈ [0, 5]
Best Task	Ordered list of Category Tasks in which this player obtains the best results	{x, y, z: x, y, z ∈ { Goal, Map, Puzzle, Dialogue, Strategy, Riddle}}
Worst Task	Ordered list of Category Tasks in which this player obtains the worst results	{x, y, z: x, y, z ∈ {Goal, Map, Puzzle, Dialogue, Strategy, Riddle}}
Preferred Task	Ordered list of Category Tasks which this player prefers to perform	{x, y, z: x, y, z ∈ {Goal, Map, Puzzle, Dialogue, Strategy, Riddle}}
Rejected Task	Ordered list of Category Tasks which this player prefers not to perform	{x, y, z: x, y, z ∈ {Goal, Map, Puzzle, Dialogue, Strategy, Riddle}}
Length	Preferred length of tasks performed	x: x ∈ {short, medium, long}
Quantity	Preferred quantity of tasks included in a single stage of the video game	

Attributes related to tasks (*Best/Worst Task* and *Preferred/Rejected Task*) are expressed as ordered lists to classify several categories according to the preferences of each player. Table 10 contains the same values that were indicated in the *Category* attribute in the *Stages and Levels Model* (Section 3.2.3), but all values included for that attribute can be used here. Each of these lists is composed of three elements in such a way that the first element is the one that best fits the description of the attribute. In this way, the system has more information to choose tasks if there are no tasks in a specific category.

The two last attributes are the preferred *Length* of a task and the *Quantity* of tasks that each player prefers to face. Length is measured by the number of activities included in a task. The second attribute is related to the number of tasks in a stage. Considering that each stage can include one or more goals, this attribute is defined in terms of the number of tasks in paths for each of the proposed goals and varies for each particular game.

The last perspective of this model is called the *Interaction Perspective* and is related to how the user interacts with the other group partners. In particular, the values of the attributes in this perspective are related to group tasks. Thus, for each group task, the messages exchanged between the members of the group are recorded in the model, which differentiates whether the message is for communication, collaboration, or coordination. This perspective uses a message categorization scheme previously defined by the authors [14] and based on the 3 C's model [9]. The main elements of this categorization are schematically outlined below:

- Communication
    - Questions / Answers
    - Sharing information
    - Checking information
    - Social exchanges
- Collaboration
    - Proposal: Statement, negotiation, counteroffer
    - Help: Asking for, negotiation, solution
    - Resources: Asking for, user identification, negotiation, solution
- Coordination
    - Making decisions: Identification, negotiation, vote, agreement
    - Group identification
    - Planning tasks: Identification, negotiation, work distribution

This form of data analysis for the *Interaction Perspective* is based on SNA, and therefore the attributes of this perspective have been designed taking into consideration how SNA is carried out.

3.4.2 Group Model

Similarly to the way the student-player was modeled, it is necessary to model the group as a whole. Group modeling could be accomplished as an addition of individuals to the group, but this is not enough in this context because implementing CSCL techniques can lead to groups which are more that the sum of their parts.

Nevertheless, a direct analogy between individual and group models exists. For this reason, the Group Model is also composed of four perspectives: the first, *Identifying Perspective*, includes meta-information about the group; the second, *Educational Perspective*, deals with learning achieved by the group; the third perspective, the V*ideo Game Perspective*, includes preferences of group with respect to the game; and finally, the *Interaction Perspective* analyzes how the group interacts as a whole. The information in this last perspective makes it possible to measure the quality of interaction and to determine if quality of collaboration and learning achievements are related.

The *Identifying Perspective* contains general information about the group, including an *Identifier*, the *Date of Creation* of the group, and a list of *Members* of the group. In addition, if the group has come into existence as a result of a change in the membership of a previously existing group, this previous group is considered as the *Father*, and the new group is designated as its *Child*. Thus, a tree-shaped structure of groups defines the evolution of groups over time.

Perspectives on educational and video games are very similar in both models, but in this case they contain information about the features and achievements of the whole group (group tasks and goals). However, the *Interaction Perspective* in the group model (Table 11) is quite different because it contains information about the general characteristics of the group. In particular, four attributes are included for each of the categories in the message categorization scheme. The measurements considered are [15]:

- *Most prestigious member*, the one who receives the most messages.
- *Most influential member*, the one who sends the most messages.
- *Network density*: the ratio of contacts actually made and total possible contacts in the network.
- *Adjacency matrix:* stores all the messages exchanged by the group members. The number in position M[i][j] represents the number of messages that member i has sent to member j.

Table 11 Attributes of the Interaction Perspective for the Group Model.

Attribute	Description	Domain
Most Prestigious Member for Communication (there are also attributes "Most Prestigious Member for Coordination" and "Most Prestigious Member for Collaboration")	Group Member who receives the most communication messages (or coordination or collaboration, when corresponding)	x: x is a Group Member
Most Influential Member for Communication (Coordination and Collaboration)	Group Member who sends the most communication messages (or coordination or collaboration, when corresponding)	x: x is a Group Member
Network Density for Communication (Coordination and Collaboration)	Ratio of number of contacts made by group members and the total possible number of contacts for communication messages (or coordination or collaboration, when corresponding)	x: x=Effective_Relations / Possible_Relation, x ∈ [0, 1]
Adjacency Matrix for Communication (Coordination and Collaboration)	Matrix containing all the communication messages (or coordination or collaboration, when corresponding) exchanged by the group members	MM_{ixi}: M is an squared matrix about communication with dimension i (MR_{ixi} for coordination, ML_{ixi} for collaboration)
Most Prestigious Member for Communication (Coordination and Collaboration)	Group Member who receives the most messages (of any type: communication, coordination, or collaboration)	x: MM[i][x] > MM[i][j] ∀i, j (using MR for coordination and ML for collaboration)
Most Influential Member for Communication (Coordination and Collaboration)	Group Member who sends the most messages	x: MM[x][j] > MM[i][j] ∀i, j (using MR for coordination and ML for collaboration)
Network Density for Communication (Coordination and Collaboration)	Ratio of the number of contacts made by group members and the total number of possible contacts	x: x ∈ [0, 1]

4 Recording Information in the Models

The user models record significant information at three levels: educational, game, and interaction, which enables a full analysis of individual students and the groups to which they belong. However, a model is also needed to store details of all the events that happen during the execution of the game, for example to study sequences of play or the evolution of the players. Thus, the user model enables an evaluation of learning and interaction as a basis for proposing improvements and adaptations in interaction, group formation, planning objectives and tasks, and educational game design. Periodically, relevant information is detected in the *State of the Game Model* and updated in the corresponding models. This can be configured to be done at the end of a game session or at specific intervals.

The information maintained by this model is an event log. The log records events that have been defined as interesting, that is, all the significant events in the game. These events will be analyzed later to identify relevant information about the players, the group, and the context in which tasks are executed.

Two parts can be distinguished in this model: the individual player area and the group area. The group area records information concerning tasks that the group performs together. The individual player area records information concerning tasks performed by each player alone. In both areas, the following set of elements for each task is recorded:

- The ID of the task, obtained from the Tasks and Goals Model. When the task is a group task, the type of task is also recorded.
- The point in the game at which the task is being carried out: the stage or level of the video game in which the players are at this moment.
- The goal being pursued: The goal of the video game that the player or group desires to achieve by performing that task at that point in the video game.
- Beginning of the task: Date and time at which the task started.
- End of the task: Date and time at which the task will finish. Along with the beginning of the task, it possible to know the duration of the task.
- Set of activities: For each activity developed during the task, the following data are recorded: beginning of the activity, end of the activity, number of failures, and score obtained.
- Other interesting events: Type of the event (check status, lend a resource, etc.), beginning of the event, end of the event, and other data needed depending on the type of event.

In addition to the information just mentioned, in the group area, the following information is maintained for each task:

- The members of the group who are working on the task. Because all members do not take part in all the activities of a group task, the subset of members that executes each particular activity is stored. This information enables the teacher to determine the degree of participation of each student during the educational process implemented in the video game.
- The group members who are exchanging messages. For each activity, the messages sent are recorded. Associated with each message the following data are stored: date, time, communication tool, sender, recipient, and message content. Through the message log, the teacher can track the coordination and planning processes performed by the students.

The information recorded in this model must be analyzed and summarized to be updated in the corresponding models. To perform these transformations, the authors are designing a platform to include all these models and to process the information recorded here and to generate, in addition, general improvement reports to be used by the teacher.

5 Conclusions and Further Work

This paper has proposed a set of models to analyze collaborative learning in educational video games. Four groups of models, each with a specific aim, have been presented. The first group of models is intended to describe the educational project, which content will be taught, and how various parts of the content are related. The second group describes the characteristics of the video games included in the system, specifying recreational challenges, stages, and levels in a similar way as for the educational models, which enables a relationship between levels in both groups of models to be established in the third group. This third set of models has been defined with two levels, one containing educational tasks and goals and the other containing video-game stages and challenges. Graphically, tasks in both levels can be linked to specify which video-game tasks provide each of the elements of educational content. The elements of the fourth group of models are the student-player model and the group model. These models are especially important for the teacher because they contain information about what students have learned and how they have collaborated in their groups.

The authors are in the process of defining a platform to integrate these models with a set of modules to analyze and summarize the information in them, to offer the teacher general improvement reports and to carry out automatic and semiautomatic adaptations of the video game. In this manner, the authors believe that this collaborative work can be carried out in the best possible way.

Acknowledgments. This study has been financed by the Ministry of Science and Innovation, Spain, as part of the DESACO Project (TIN2008-06596-C02-2) and the F.P.U Programme.

References

[1] Nussbaum, M., Rosas, R., Rodríguez, P., Sun, Y., Valdivia, V.: Diseño, desarrollo y evaluación de videojuegos portátiles educativos y autorregulados. Ciencia al Día 3(2), 1–20 (1999)

[2] Lacasa, P., Martínez-Borda, R.: Aprendiendo con los videojuegos. Universidad de Alcalá de Henares in collaboration with Electronic Arts España (2007)

[3] Mooney, C.: Theories of Childhood: An Introduction to Dewey, Montessori, Erikson, Piaget, & Vygotsky. Redleaf Press Minnesota (2000)

[4] McFarlane, A., Sparrowhawk, A., Heald, Y.: Report on the Educational use of Games. Teem Publications, Cambridgeshire (2002)

[5] Padilla Zea, N., González Sánchez, J.L., Gutiérrez, F.L., Cabrera, M.J., Paderewski, P.: Diseño de videojuegos colaborativos y educativos centrados en la jugabilidad. In: Proceedings, 10th Simposio Internacional de Informática Educativa, Salamanca (Spain), pp. 1–6 (2008)

[6] Jong, B., Chan, T., Wu, Y., Lin, T.: Applying the adaptive learning material producing strategy to group learning. In: Pan, Z., Aylett, R.S., Diener, H., Jin, X., Göbel, S., Li, L. (eds.) Edutainment 2006. LNCS, vol. 3942, pp. 39–49. Springer, Heidelberg (2006)

[7] Mendoza, P., Galvis, A.: Juegos Multiplayer: Juegos colaborativos para la educación. Master's Thesis, Universidad de los Andes, Bogotá, Colombia (1998)

[8] Collazos, C.A., Ochoa, S.F., Mendoza, J.: La evaluación colaborativa como mecanismo de mejora de los procesos de evaluación del aprendizaje en un aula de clase. Revista Ingeniería e Investigación 27(2), 72–76 (2007)

[9] Ellis, C.A., Gibbs, S.J., Rein, G.L.: Groupware: some issues and experiences. Communications of the ACM 34(1), 39–58 (1991)

[10] Padilla Zea, N., González Sánchez, J.L., Gutiérrez, F.L., Cabrera, M.J., Paderewski, P.: From CSCL to VGSCL: a new approximation to collaborative learning. In: Proceedings of the 1st International Conference on Computer-Supported Education, pp. 329–334 (2009)

[11] Paternò, F.: Formal reasoning about dialogue properties with automatic support. Interacting with Computers 9, 173–196 (1997)

[12] González Sánchez, J.L.: Diseño de Videojuegos Adaptados a las Educación Especial. Master's Dissertation, University of Granada (2007)

[13] Card, S.K., Moran, T.P., Newell, A.: The Psychology of Human-Computer Interaction. Lawrence Erlbaum, Hillsdale (1985)

[14] Padilla Zea, N., González Sánchez, J.L., Gutiérrez, F.L.: Collaborative learning by means of video games: an entertainment system in the learning processes. In: Proceedings, 9th IEEE International Conference on Advanced Learning Technologies (ICALT), pp. 215–217 (2009)

[15] Hanneman, R.A., Riddle, M.: Introduction to Social Network Methods. Free online textbook (2005), `http://www.faculty.ucr.edu/~hanneman/nettext/` (accessed March 11, 2010).

A Framework to Foster Collaboration between Students through a Computer Supported Collaborative Learning Environment

A. Bayón, O.C. Santos*, J. Couchet, and J.G. Boticario

aDeNu Research Group. Artificial Intelligence Department,
Computer Science School, UNED,
C/Juan del Rosal, 16. Madrid 28040, Spain
abayon1@alumno.uned.es, {ocsantos,jcouchet,jgb}@dia.uned.es
http://adenu.ia.uned.es

Abstract. Computer Supported Collaborative Learning (CSCL) environments facilitate the management of collaborative tasks. However, these systems do not usually provide the personalization features required to adapt the learning experience to the student needs, a drawback that can affect the collaboration objective and ultimately the learning process. Nevertheless, there have been several research approaches that have progressed on providing intelligent features to support management, tracking and evaluation tasks in collaborative settings. In particular, we propose a framework that provides adaptive collaboration support for a CSCL environment framed in an open and standards-based learning management system. Our proposal combines adaptation rules defined in IMS Learning Design specification and dynamic support through recommendations via an accessible and adaptive guidance system. A partial prototype of this approach has been implemented and a formative evaluation was carried out to guide the on-going work. The implementation offers CSCL courses following a methodology called Collaborative Logical Framework and has been run in a real world scenario at the Madrid Science Week 2009.

1 Introduction

Whenever a Computer Supported Collaborative Learning (CSCL) [1] course is running, tutors and administrations expect that learners do their contributions collaboratively. Moreover, learners expect that tutors or the system could guide them while they are working. In addition, the tutors would like to i) pre-empt students' needs and offer them personalized information before their doubts appear and ii) promote the collaborative work for those students who are reluctant to collaborate.

Building a system to support these kind of courses should consider these issues during the design and development. To this, the system should adapt their facilities at every moment to the users' needs. The intention is therefore to offer an

* Corresponding author.

T. Daradoumis et al. (Eds.): Technology-Enhanced Systems and Tools, SCI 350, pp. 193–219.
springerlink.com

adaptive system to help both tutors and learners in managing the collaboration process over time.

Intelligent support is intended to facilitate the management during the design, development and analysis of the collaborative learning experience. From that support students may improve their engagement in the collaboration process and instructors alike are to be benefited in terms of different management tasks, such as grouping students, selecting a moderator, assessment of the collaboration or detecting emergent roles and undesired interactions. In particular, a critical issue is the interaction analysis, which is usually not an easy task due to the large number of students and, therefore, the high number of interactions. Thus, when designing a collaborative learning environment, an intelligent support has to be provided to analyze the student collaboration regularly and frequently with little intervention by the instructor. If possible, this support should be done in a domain independent way [2].

From that perspective, at the aDeNu research group we have proposed the Collaborative Logical Framework (CLF) as an extension of the Logical Framework Approach [3] to foster collaboration between students and the tutor through a CSCL environment. Collaboration is one of the main strategies in e-learning [1] and there are clear indicators to support its success [4]. However, many issues depend on the management of the required adaptations to the users' needs. It is out of the scope of this paper to discuss on the selected collaborative indicators on student's performance, which is a hot research issue in itself [4, 5, 6]. In turn, we take as a starting point a tentative list of indicators selected in [7] and focus on how these (or other) indicators could be obtained and used within a framework to support collaborative learning.

Some authors have also studied the power of IMS Learning Design (IMS-LD) to model collaborative scripts [8], showing how typical CSCL interaction patterns can be captured with IMS-LD (with a detailed knowledge of the specification required) [9]. The originality of our approach resides in providing the hooks for the dynamic guidance at runtime [10]. The IMS-LD is a generic and flexible language to enable a wide range of pedagogies in online learning to be expressed, including those that depend on adaptive features. Nevertheless, practice has demonstrated that IMS-LD has limitations in supporting some particular collaborative scenarios [11]. Still, it is the most powerful language to describe adaptations at design time in a standard and interoperable way.

Before introducing our approach, in the following section we review existing approaches that support the development of collaborative frameworks. The third section introduces the component-based framework designed to provide adaptive collaboration support for a CSCL environment. The required adaptive capabilities are achieved building a model of each student, which is central in the framework. Next, the user model together with the student's interactions gathered through a tracking and auditing module, are used to cope with foreseen and unforeseen situations. Foreseen situations can be managed through rules in the IMS-LD specification [12]. In turn, unforeseen situations can be managed in terms of dynamic adaptations via a semantic educational recommender system that provides accessible and adaptive guidance [13].

After that, the fourth section introduces the mapping of the theoretical framework into an open standard-based Learning Management System (LMS) named OpenACS/dotLRN [14]. It presents the current prototype which integrates the CLF into the OpenACS/dotLRN infrastructure to provide the collaboration support by offering CSCL courses through the CLF methodology. Section 5 describes a formative evaluation of this prototype which took place at the Science Week 2009, an event promoted by the Madrid city hall in conjunction with the Madrid public universities. Finally, we draw some conclusions and comment on the future works.

2 On Adapting Collaborative Frameworks

Collaborative frameworks should depend strongly on adaptive features. Communication plays such a central role in managing collaboration that there is not collaboration without communication [4]. Considering the unmanageable number of communication instances in nowadays' collaborative environments, teachers and students alike have difficulties in taking advantage of the richness involved in those instances. To deal with this issue, data mining techniques can be applied to support an automated process of interaction analysis [15]. In this way, the instructors' workload can be reduced in terms of analyzing the student collaboration. Monitoring results can be displayed using different techniques, such as simple attribute-value sets [16], [17] or via graphic tools [18], [19]. Students themselves can benefit from simple and usable visual tools [6], [20]. Monitoring student collaboration can be a goal on its own and there is ample research on this issue [21], [22], [23].

Beyond monitoring, analyzing students' interactions depends on desired outcomes. The first, and foremost objective is to ascertain whether collaborative learning takes place [4]. The assessment or evaluation of collaboration may have various purposes. The ultimate goal is to evaluate aptitudes and capacities of fellow students to collaborate, i.e. the achievement of teamwork skills and functional know-ledge [24]. Another important analysis goal is to support just the management of collaboration. To that end different tasks may be involved: grouping of students according to their collaboration [2] selecting an administrator or moderator [25], detecting emergent roles and undesired interaction patterns [26].

Research studies can be characterized by the source of data that they use, the inference method, and the process that they apply to their results [27]. Regarding data sources, collaboration interactions are the main origin [28] but collaboration context or circumstances may affect the collaboration performance [29]. Information on activity, initiative or acknowledgment, may come from analyzing student interactions in forums [2], [19], [30], [18]. Further, statistical analysis of forum interactions may reveal student constancy or regularity indicators, which implicitly consider time variables [2]. Redondo et al. [31] proposed a model of collaboration with a set of indicators (e.g. number of system accesses per student, number and mean of contributions made, kind of the contributions, depth of the discussions…). According to the learner's interactions, the indicators were updated and the learner's model was compared through fuzzy logic rules with a priori model of the suitable collaboration, which experts had foresaw.

As for the process and techniques to support the aforementioned tasks, machine learning-based studies have focused on analyzing student collaboration [31], [5]), Students were grouped according to their collaboration using unsupervised classification techniques (clustering) [32], [33] and collaboration metrics were constructed using supervised classification techniques [2].

Regarding the application of the collaboration analysis, there is an increasing interest in finding general and transferable features [34], [35], [36], [2]. The purpose here is to make the approach transferable and reusable in terms of collaboration models that can be supported by ontologies, inferred features that are domain independent and implementation drawing on processes that can be applied to different e-learning systems [2]. In looking for transferability open student models focus on featuring student-modeling issues and as they are to be managed by the students themselves, the models must be meaningful to the students [37]. Consequently, these open student models should be independent of the system or the learning platform. The responsibility for learning decisions lies with the learner [38]. This latter aspect, students that have access to their models, is meant to support metacognitive information on collaboration, which is intended to help learners in improving their control over the learning and collaborative processes [39]. Actually, collaboration itself is a metacognitive feature (it helps self-regulating the collaborative learning process) and meta-cognitive skills can help students and improve their learning [40]. Further, self-regulation is more than the regulation of cognitive activities (metacognition), since it also involves motivational and emotional aspects [41].

Another important issue, not so much explored for the time being, in providing adaptive features in collaboration is to generate recommendations, which are meant to improve student collaboration [42]. Recommendations can be used to guide the student to perform specific actions in order to help her on her task, as well as encourage for participation and improve the team work.

Apart from the aforementioned issues, there is vast research arguing on the advantages of using collaborative learning and significant efforts have been made on characterizing the main aspects involved [43], [44], [4], [1].

From the above analysis on the state of the art follows that in order to support collaboration in current and future frameworks, there is a need for an intelligent support that facilitates its management during the design, development and analysis of the collaborative learning experience and supports both students and instructors.

3 Component-Based Adaptive Framework

This section introduces the framework that we have defined as a component-based model that supports the implementation of adaptation features in a CSCL environment guided by learning design. This approach was first introduced in the aLFanet project (IST-2001-33288) where it was proposed the combination of design and runtime adaptations through the e-learning lifecycle. Details on the proposal to extend the use of IMS-LD to support the CLF are described elsewhere [10].

In this paper we focus on the advances carried out in the last years at aDeNu group. In particular, we present a framework to manage adaptations in collaborative environments supported both by instructional design (foreseen situations) and recommendations (unforeseen situations). The main framework features are shown the Fig. 1.

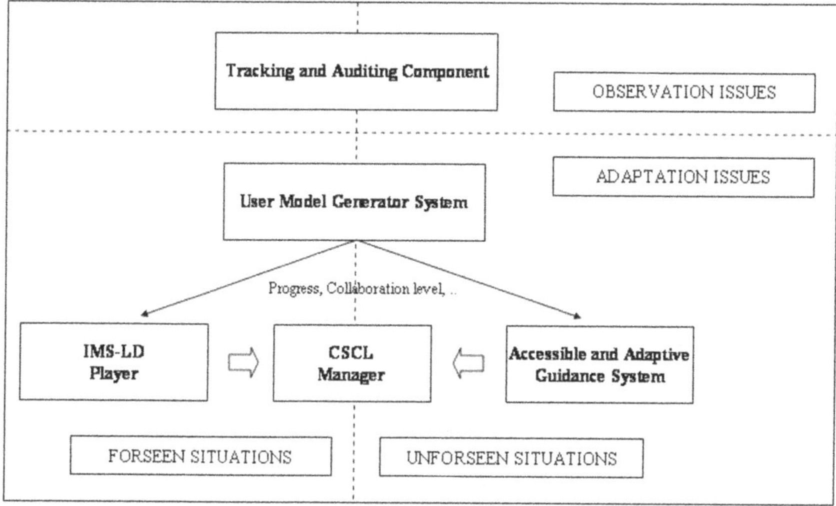

Fig. 1 Component-based framework to support adaptive features in CSCL environments guided by learning design

Five architectural components are identified in Figure 1:

- The CSCL Manager: LMS module in charge of supporting the collaboration among students and tutor.
- The Tracking and Auditing Component: tracks the LMS web usage and fills a repository of interactions with these data.
- The User Model Generator System: generates the user model for each learner involved in the learning scenario. This user model gathers information about an individual user that is essential for an adaptive system to provide the required adaptations (i.e. foreseeable and unforeseen situations).
- The IMS-LD Player: LMS module that runs the design-time adaptation framework (i.e. adaptive features). It is described in terms of an XML schema and can provide support for a meta-language that can be used for describing the CSCL Manager contents and configurations through a scripting language.
- The Accessible and Adaptive Guidance System: provides help and dynamic guidance to the users like a 'virtual tutor', both in their interaction with the LMS and in the effective use of the teaching-learning experience.

The CSCL Manager has to support the CLF methodology as described in [7] (see Fig. 2). In particular, the collaboration process is defined as an initial interaction stage and a set of phases, each one composed of three stages: individual, collaboration and agreement, as is shown in the Fig.2. In the individual stage, each student produces her proposed solution without any contact to her mates. In the collaboration stage, participants are given access to the contributions of their mates, and are asked to rate and comment on them. Due to the discussions taken place, each participant can produce new versions of their own contributions, where the version is the new learner' answer for the same predefined task. The agreement stage requires that one of the participants in the activity is assigned the moderator role. The responsibility of this role entails building the agreement solution based on the contributions from all the participants (following a similar approach as in the collaboration stage, that is, the rest of participants should rate and comment the moderator's version, and she can produce new versions based on this feedback). The moderator selection can be done manually by the teacher (i.e., CLF manager) or automatically by the CLF, taking into account the interactions from previous stages.

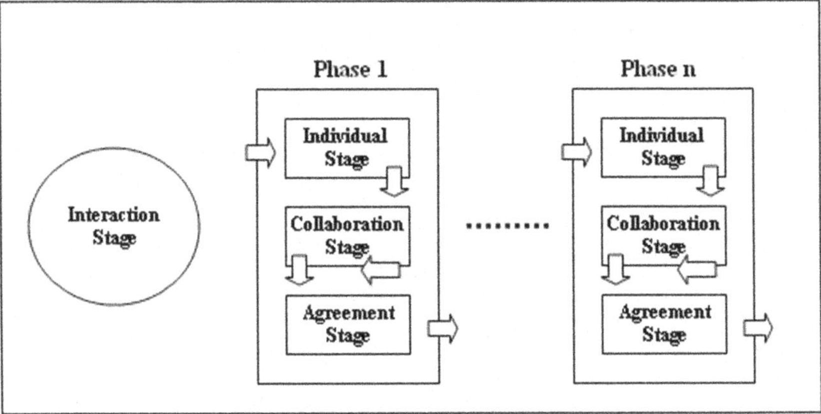

Fig. 2 Collaborative Logical Framework

Two different adaptation situations (see Fig 1.) are to be supported: i) adaptation to foreseen situations and i) adaptation to unforeseen situations.

3.1 Adaptation to Foreseeable Situations

The IMS-LD Player aforementioned is in charge of managing foreseeable situations. In the specific case to manage the collaboration, the IMS-LD language is to be extended to describe the support required by the CSCL Manager. Several information should be gathered through a script to define the content and the course management, such as the following: a) Name of the CLF, b) If active for all users, c) Starting date for each phase, d) Days to work to provide the consensus solution, e) Parent phase for task, f) If automatic change between different stages in the

CLF, g) Number of tasks per phase, h) Task question, i) Description of the CLF, j) Name of each phase, j) Days to work in the user solution, l) Name of the task, m) If agreement solution should consider all the individual solutions, n) Number of phases, o) Type of task, and p) Number of advices for students.

These data involves the description of the elements of the CLF, the permission and roles involved, the length, etc. By having this information mapped into an IMS-LD structure, it could be reused in different LMS.

Those items can provide the clues to start the CSCL activity using the designed framework. However, in order to add adaptation to the system, some other aspects related to collaboration in the script that defines the content of the course can be included. It is important to remark that the adaptation proposed is based in the way the players –CLF and IMS-LD– interact each other during the activity. The information is grouped in five elements:

- **Groups configuration:** Considers the way the students take part in the activity. They will work in groups. Three parameters to manage this concept can be defined:
  - **Maximum number of members in a group:** it should be neither too large nor too small in order to get good metrics during interaction. A good number is four.
  - **Number of groups:** Depends of the participants in the activity and the number of people per group.
  - **Automatic grouping:** The activity may be divided into several sub-activities. The system allows re-grouping using clustering techniques at the end of each sub-activity taking into account the participants' behavior as the criteria to regroup. In this way, it is possible to distribute participants with the same profile in different groups, in order they were homogenous. This attribute may have the values TRUE or FALSE.
- **Metrics:** The participant's performance can be measured through metrics concerned to collaboration (participation in forums, ratings and versions). By default more than 100 metrics have been defined, but the system also allows the creation of new metrics using SQL queries. The metrics are available for tutors and students on line in order they can check their behavior during the activity. To add new metric using the IMS-LD structure two parameters are required:
  - **Metric identification:** Number and description of the metric to consider.
  - **Metric definition:** type (active or passive), scope (phases or course) and the SQL query.
- **Collaborative Indicators:** The participants' profiles are identified by collaborative indicators, and the way to defined them is using the metrics. At the moment 12 indicators have been defined following [7]: Participative, Useful, With-Initiative, Non-Collaborative, Insightful, Communicative Thinker-out, Thorough, Inspirable, Inspirator, Insecure and Gossip. Additionally is possible to define new indicators using the metrics in the IMS-LD structure. The information to provide is:

  - **Indicator identification:** Name, description and type (active or passive).
  - **Indicator definition:** code of the metrics used to defined the indicator.
  - **Machine Learning model:** A way to compute the participants' performance is using machine learning techniques. This attribute is used to indicate the location of the model to use.
- **Recommendations:** Adaptation is mainly provided through recommendations. Depending on the way the participant's acts, then the system may offer them different advices. Recommendations to use can be defined in the IMSL-LD structure by means of two items:
  - **Recommendation identification:** Name and Description.
  - **Recommendation definition:** The collaborative indicators which activates the recommendation.
- **General Information:** Information to manage generic scenes of the activity can be also added in the IMSL-LD structure.
  - **Way of computing the indicators:** Two ways are available i) manual rules and ii) machine learning techniques. The former is based on the use of the statistics term called standard deviation, while the latter uses models created from previous activities. It is important to note that when the course ends, the models can still learn from the data gathered from the course interactions.
  - **Rule to define the moderators:** moderators are participants in the CLF with some specific skills (collaborative indicators). To select the moderator, the sum of the values of the indicators included in the participant's profile is used to check if that added value is above the limit proposed as a threshold to select the participant as moderator. The limit is set up in this rule.
  - **Hints:** the previous element (i.e. Recommendations) allows tutors to include hints for every activity. This attribute can be set up to true or false.

3.2 Adaptation for Unforeseen Situations

Unforeseen situations are those situations that cannot be taken into account at design time, mainly because they depend on runtime actions. To provide a dynamic contextual support, recommendations can be provided to the learners during their execution in the course. These recommendations take into account the learner profile and the current context of the learner in the course. As the CSCL Manger stores in the User Model Generator System the computed values for the collaboration indicators, the Accessible and Adaptive Guidance system can use the collaboration profile of the user to generate different recommendations according to it.

Recommendations have several purposes: i) provide students with the specific information to help them in their tasks, ii) encourage students to work collaboratively, and iii) enhance the participation and communication among them. Their definition is based on the collaborative indicators values, and the actions suggested could be any related to the tools used by the CLF (forums, ratings, version creation, etc.). To describe the recommendations, a semantic model has to be used [13]. Table 1 shows some sample recommendations and how some collaboration indicators from those defined in [7] could be involved in the delivering process. In particular, the example involves the Gossip indicator which describes a learner who reads a lot of information without a clear objective, such as surveys, messages, comments and ratings, but does not produce any contribution related to it.

Table 1 Sample recommendations and collaboration indicators involved

	Sample Recommendations	Collaborative indicators
Rec-1	Read this answer form John as it is the best rated by the moment	Gossip= no
Rec-2	Send a message to Mary through the Forum as you have not take part in this discussion, and you can bring her your understanding gained from reading the other contributions	Gossip= yes

The above recommendations are an example of the dynamic recommendations based on the context that could be described for a particular learner. In the case that this user has a Gossip collaborative indicator with the value “no”, and the learner has not read John's answer (which is the best rated). The Rec-1 is generated on the fly and shown to the learner. In turn, when the learner has a Gossip collaborative indicator with value “yes” (i.e. the learner read the contributions of the others), the recommender system offers the Rec-2 in order to encourage the learner to contribute in a forum where has not participated yet, with the understanding gained from her previous readings. Nevertheless, the appropriate recommendations for a course should be designed taking into account the methodology proposed to design educational recommendations by following a user-centered methodology [45].

4 From the Framework to a Prototype

The above approach is being developed on existing open source components, as shown in Fig. 3.

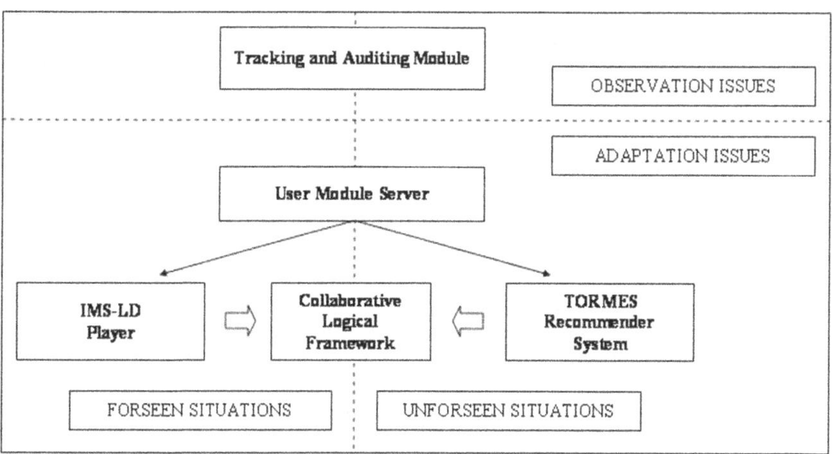

Fig. 3 Instantiation of the framework into open source components.

4.1 Framework Components Mapping

The following subsections describe how each component of the framework is mapped to the OpenACS/dotLRN installation. Some components are internal to the OpenACS/dotLRN environment (i.e. the Tracking and Auditing Module, IMS-LD player and the Collaborative Logical Framework), but the other two (the User Module Server and the TORMES Recommender System) are external services. The communications among OpenACS/dotLRN and these services take place through web services to satisfy the interoperability requirements. The implementation approach presented here can be used as inspiration for other implementations.

4.1.1 Collaborative Logical Framework (CLF)

The CSCL Manager is mapped to the Collaborative Logical Framework (CLF) component. It is implemented as an OpenACS/dotLRN package completely integrated with the rest of OpenACS/dotLRN modules [46]. The CLF extends the Logical Framework Approach, a stepwise methodology that requires individual and collaborative work used by the international cooperation agencies to plan, define and manage their projects [3]. The goal is to facilitate reaching a consensus to deliver a group solution.

The traditional methodology (i.e. the Logical Framework Approach [3]) uses the consensus as the method to solve the different phases in which the projects from the cooperation agencies are structured. The phases are set by agencies members to design and monitor the objective of the project. However, some previous experiences showed that students who participated in the Logical Framework Approach complaint from the lack of collaboration as the only way to collaborate is through messages on a forum. As a result of these findings, we proposed a collaborative extension to the Logical Framework Approach called the Collaborative Logical Framework –CLF [7]. The extension of this methodology to enhance its collaborative features consists in the addition of three stages to every phase: individual, collaboration and agreement, and an initial interaction stage when the CLF process starts, as described in [7].

The CLF is supported by a set of collaborative indicators, which try to define the student's performance in order to facilitate adaptation. The collaborative indicators are defined using different metrics taken from student's interaction over the four CLF stages, focusing in three elements: answers and versions creation, forum participation and ratings. There are two kinds of metrics, those related to the creation of objects (active metrics) and those concerned to visits done on them (passive metrics). For instance, active metrics are the number of versions created, the number of threads where the student is taking part, the number of versions not rated yet. In turn, passive metrics are the time spent in rating a version, what the student visited before creating a new version, or what the student visited after commenting a colleague's answer.

Twelve different collaborative indicators have been proposed trying to detect different behaviors in the course [7]. Half of them are active and half are passive according to the metrics used for their definition. The collaborative indicator definitions consist in a set of rules built with the computed metrics. For instance, the Gossip indicator computes i) what the student does before sending a new message, iii) what she does after sending the message, iii) what she does before rating a new colleague's answer and iv) the number of times she reads others answers. If metrics i, ii and iii say that she usually reads other answers and metric iv is quite above the group average, the system considers the student a Gossip. Other indicators are more focused on the participation level as they consider the contributions done.

The system can use two different methods to compute if the student's behavior meets any of the collaborative indicators. On the one hand, a given set of rules, which define the indicators can be used. On the other hand, models can be built from interactions taken from previous courses or even from the initial interaction stage of the CLF. Machine-learning techniques can be applied to the latter in order to support an intelligent process.

In particular, to automate students' behavior monitoring three processes are defined: i) inferring the collaboration indicators, ii) grouping students and iii) selecting a moderator. The objective is to help the teacher in the process of monitoring the users' participation (i.e. learn the collaboration indicators) and take this information as an entry point to help the management of the collaboration (i.e. grouping users and selecting a moderator). Moreover, this information can also be used to support the collaboration process in a personalized way through dynamic contextual recommendations by the TORMES Recommender System introduced in the next subsection. This recommender takes into account the system context. For instance, it considers a forum's item that is being read by the learner and the collaborative indicators values for this learner, to generate dynamic recommendations adapted to her [13].

To support this intelligent analysis, Weka data mining suite [47] has been integrated into OpenACS/dotLRN to compute the indicators required by the CLF from the interactions following a similar approach as in [48]. The approach consists in applying clustering algorithms to group students according to their collaboration profile, so heterogeneous groups can be built by selecting users from the different clusters. In turn, classification algorithms (in particular, classification trees) are used to classify students regarding their collaboration class and their adequacy to become the moderator. In this way, applying the classification algorithms on those models it is possible to find out if the learner's behavior corresponds to any of the collaborative indicators).

4.1.2 Components to Support the CLF

Several components are necessary to support the functionality of the CLF: the Tracking and Auditing Module, the User Model Server, the Grail IMS-LD Player and the TORMES Recommender System.

First, in order to gather the interactions of the users in the system, a Tracking and Auditing Module has been implemented in the LMS. In particular, as it is very much related to the LMS functionality, it has been implemented as an

OpenACS/dotLRN package [49]. The Tracking and Auditing Module provides useful information to the user model related to the interactions on any OpenACS/dotLRN module. Everything can be tracked, both active actions (based on creation of objects) and passive actions (based on visits to objects) done by the users. However, in order to be efficient, the administrator can properly configure the tracking component to focus on those elements that are relevant for the analysis. The component provides also its own scripting mechanism to configure the information to be gathered.

The User Model Management System is in turn mapped to the User Model Server as is specified in [50]. It builds the user model with data gathered from the users through the LMS (explicitly through questionnaires such as the Felder Learning Styles Inventory [51]) or implicitly from the data tracked by the Tracking and Auditing Module, such as the collaboration indicators computed by the CLF or imported from external portfolios in IMS Learner Information Package specification [52]. In this way, the User Model Server stores the information required for the adaptation purposes, both for foreseen or unforeseen situations. It is independent from the LMS infrastructure, and communicates with it via web services. In this way, the different components involved in the adaptation processes can easily communicate with it, both to store information and to retrieve it.

The IMS-LD Player can control the adaptation support of the foreseen situations. Therefore, it is much related to the LMS functionality and thus, it is also implemented as an OpenACS/dotLRN package called Grail (Gradient RTE for Adaptive LD in dotLRN). In this way, the tutor at design time can specify different adaptation paths, and how the collaboration is to be managed in each of them as described in Section 3.1.

In order to provide an accessible and adaptive guidance, the TORMES Recommender System is designed [53]. The recommender uses the information from the users' interactions (both LMS database and the tracking component) and the user model stored in the User Model Server to generate recommendations that meet the specific user needs. These needs (e.g. collaboration support) have been obtained from educational experts following a scenario-based approach [54]. A formal model has been defined to create the recommendations at design time in terms of applicability conditions and restrictions, which are later used for selecting the appropriate recommendations for the user in the context at hand [13].

4.2 Implementation Details for the Prototype

A first prototype of the framework has been implemented to provide the collaboration support. It consists of the CLF integrated with the Tracking and Auditing Module in an instance of OpenACS/dotLRN 2.4.1 version, using several OpenACS/dotLRN core features (user management, file storage, forums, surveys, workflow, cronjob) and OpenACS/dotLRN extras (ratings package). The user management allows to control the roles and access to the system, the course and the CLF. The file storage, the forums and the surveys offer basic functionality to allow learners produce their contributions in the system. The workflow facilitates the delivery of each of the CLF phases in the order defined. The cronjob is used to activate periodic tasks, such as the computing of the indicators. Finally, the ratings package is very useful to gather feedback from learners on the contributions done by their classmates.

Next we explain how to interact with the CLF in order to start up a collaborative activity base on the phases described in Fig. 2.

Firstly, the activity to develop should be configured (see Figure 4). This action will be in charge of the tutors who better know the contents of the exercise. The CLF component is prepared to manage a course that consists on a sequence of steps to be considered that actually are called phases. For each phase there will be a series of tasks (questions, problems, essays …) to carry out which have to be defined in this site. The stages of the Collaborative Logical Framework (individual, collaboration and agreement) will be applied to the phases of the CLF component. Therefore, the participants will have to work with all the tasks of every phase in the three ways (individually, collaboratively and trying to reach an agreement).

Configuration of the activity also allows the tutors to manage the groups of people taking part, the metrics that will be gathering information related to the interaction in the exercise and the collaborative indicators which will be depicting the user's performance.

The period the activity takes and the time each phase is in the different stages of the collaborative logical framework will be managed automatically by the software from the dates entered during configuration. Tutors or administrations can also modify these dates or the state of the phases manually.

Figure 4 shows the configuration done for the formative evaluation carried out and described in Section 5. In particular, a phase was defined (Tale Creation) with a single task (Writing Text).

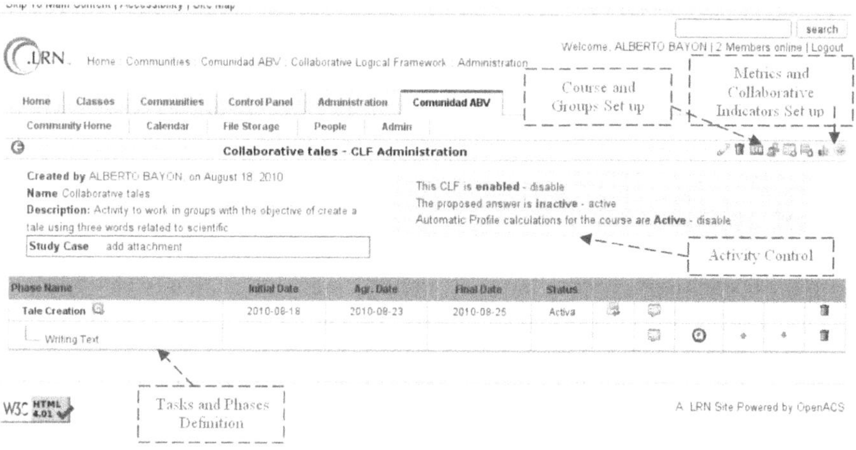

Fig. 4 CLF Administration and Configuration

When learners enter in the community to take part of the activity, they will find the active subjects (phases) of the as presented in Figure 5).

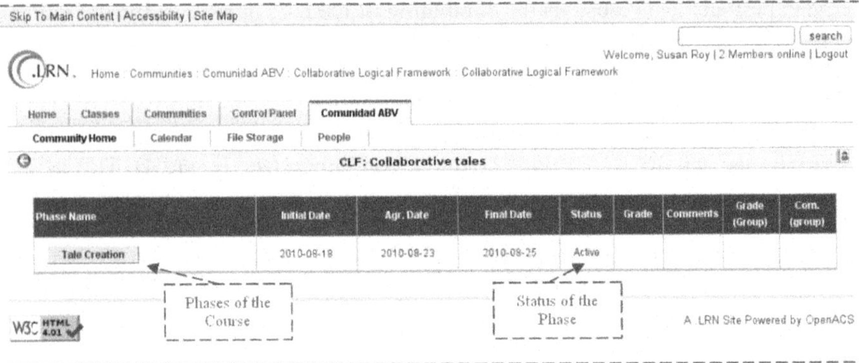

Fig. 5 Entrance to the Course

Inside the course, the learner initially works individually, without communication with their colleagues (see Figure 6).

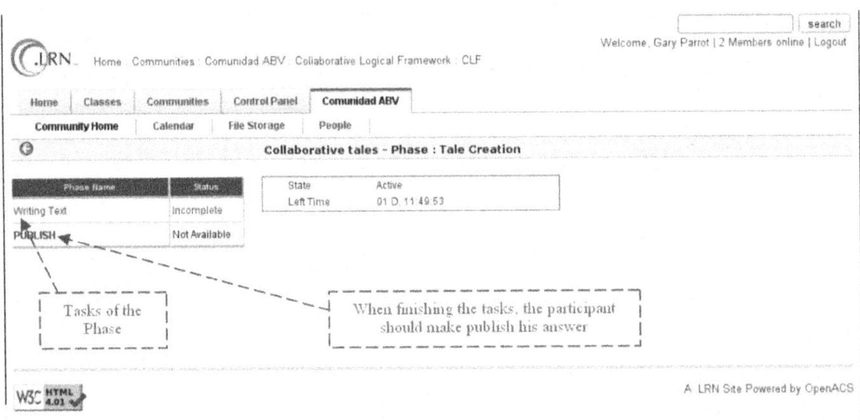

Fig. 6 Working Individually

The student has to answer the tasks defined in the phase, and when finished, she has to make it public in order to their colleagues in the same group could access to it. It has to be taken into account that the objective is to work collaboratively and create a joint answer by all the members of the group. As an example, during the activity proposed for the evaluation of the CLF (see Section 5), the task to answer was to create a short tale using three words (Fig. 7).

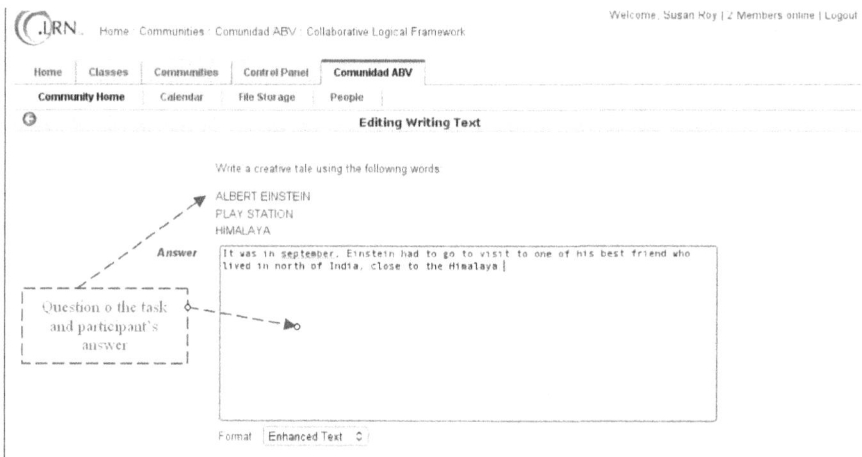

Fig. 7 Answering the question of the task

When the student publishes her answer, then she goes to the collaboration stage (following the stages in Fig. 2). In that moment, she could access her colleagues' answers, give them ratings, enter in the forum associated to those answers or even create new versions of her solution to increase the score gave by her colleagues (Fig. 8). The length of the time that the learners work collaboratively depends of the period defined during the configuration of the course.

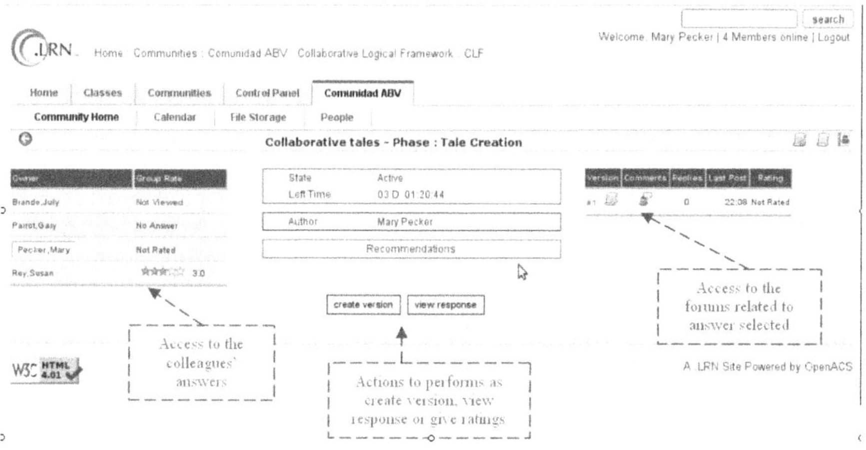

Fig. 8 Working Collaboratively

While the learners work collaboratively, the system is gathering information of their interactions (messages sent to the forums, access to the solutions, ratings ...) and several metrics are computed. Those metrics take part of the collaborative indicators definition, which certainly provides the participants behaviour in the course. Both, the metrics and the indicators are being calculated during the whole activity. In this moment, the twelve collaborative indicators used are those defined in [7].

- Active: Participative (PAR), Useful (USE), With-Initiative (WIT), Non-Collaborative (NOC), Insightful (INS) and Communicative (COM).
- Passive: Thinker-out (THI), Thorough (THO), Inspirable (INP), Inspirator (INR), Insecure (UNS) and Gossip (GOS).

When the collaboration stage ends, the system chooses a moderator for each group. This decision is made using the computed collaborative indicators and some configurable rules defined for the course. For instance, the moderator could be a learner who is Participative, Collaborative, Inspirator and With-Initiative.
The participants and the tutors can view in real time the indicators the system is calculating (Fig. 9).

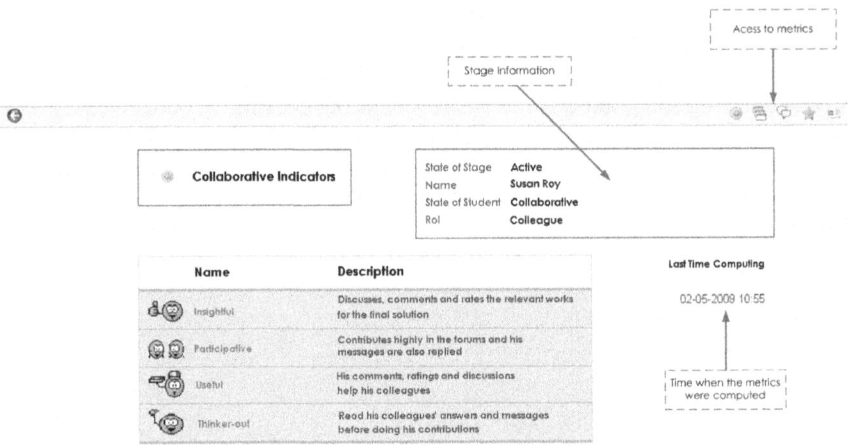

Fig. 9 Collaborative Indicators computed for one participant

The moderator selection marks the start of the agreement stage. The moderator has the responsibility to create a solution with the best ideas from her colleagues' answers, while the others have to help her with suggestions through messages in the forum and giving ratings. The aim is to reach an agreement that will represent the final solution of the group. The interface in this stage is the same that when they were working collaboratively (Fig. 8), but some actions are only available for the moderator (create the answer and create new versions), some only for the other participants in the group (give ratings and view the answers) and all can access the forums to communicate.

5 Formative Evaluation of the Prototype

A formative evaluation was carried out on May 23-24, 2009 during the Madrid Science Week, at the National Museum of Science and Technology (MUNCYT), where the aDeNu research group –as part of the UNED- participated in a public stand exposed to an audience composed mainly by young people between 14 and 17 years old. The activity was organized as a competitive context, in the way that the group who scored better in the CLF and had a highly rated story was given a prize. In particular, a skill-based contest –usually known as gymkhana- was organized. The jury was made up by the aDeNu researchers who were in charge of the experience. Moreover, they were also observing the participants during the experience, and taking notes on the collaboration indicators that they would consider for each participant. This information was very useful to evaluate the system.

5.1 Experience Description

The activity proposed coped with the time restrictions in this type of events. Teams composed of four members were required to participate in a narrative gymkhana consisting in writing a mini-story collaboratively from three words given in advance: the name of a scientist, an object and a place, as for example: Edison, clock and Everest. The total time for the activity was half an hour. To engage the participants in the task and foster collaboration, a pen drive was given to the participant selected as moderator in each of the teams. The best story of the day was rewarded with an iPod for each member of the team that created it. 56 participants took part in the experience.

Because of the nature of the event (science fair) participants were invited to take part in the contest as they came to the stand. So the election of them was completely random. They were mainly secondary school students with high experience in collaborative tools as emails, forums, chats or social networks . This skill was important for the performance of the gymkhana because it made easier that the participants could be involved in the event activities.

The gymkhana was handled by the CLF, so that the participants used the application to write their tales and communicate each other in order to create group story. For this purpose a course was created in the application with only a phase and a task (*Tale Creation*) using the CLF setup (Fig. 4). The activity was developed following the Collaborative Logical Framework Methodology (Fig 3 - Interaction, Individual, Collaboration and Agreement stages). For that reason the participants had to be trained to know how the methodology works. This handicap was solved considering the training as a part of the gymkhana.

Previous to the activity, the participants were introduced to the CLF methodology (interaction stage). The rest of the CLF was structured in a sequence of three stages. In the first (individual stage), the team members worked individually to create their individual version of the mini-story based on the three words previously given. Next, at the second stage (collaboration stage), each one had to publish her version, read the version from their peers, rate them (from 1 to 5), and optionally send comments to the author through the communication forums. As a result of this communication, participants could create new versions of their contributions. Finally, after a pre-established time, the system chose a moderator and the third stage started (agreement stage), where the moderator had the responsibility to create a new version based on the best valued team versions. Other team members rated and commented this version, and the moderator could use those comments in order to create new versions of her proposal. When the time expired (a pre-fixed time), the latest valid version of the moderator became the mini-story result of the team collaboration.

The goal of this formative evaluation was twofold. On the one hand validate the performance of the framework developed from the collaboration point of view and obtain data to produce the adaptive behavior in the coming development phase. On the other hand, to analyze the collaborative indicators definitions, the way the system computes them from different metrics and the scope of them with the objective to decide if the students' performances are properly identified.

To cope with those goals three different sources were settled to gather data:

- The information provided by the CLF application itself: metrics and collaborative indicators.
- A questionnaire filled by the participants (learners) at the end of the gymkhana with several questions about the activity comprehension, bugs detected, CLF methodology impression, relationship between the methodology and collaboration, understanding of the collaborative indicators and disagreements about the collaborative indicators computed for each participant. In particular, the following information was collected: i) information about the user interface and the application functionality, ii) information on the package performance in a multi-user environments, iii) interest of the participants in the proposed activity, iv) user's ability to assimilate the CLF methodology , v) validity of the methodology to manage collaborative activities, vi) agreement of the users with the collaborative indicators that the system has calculated for them, vii) analysis if the metric composition of each collaborative indicator is appropriate according to the interaction between participants and vii) assessment of the CLF management and monitoring facilities offered to the tutors and administrators.
- A questionnaire filled by aDeNu researchers who were watching the participants during the development of the gymkhana in order to detect their behavior during the whole activity from the point of view of the "human eye". The behavior means to measure their performance in the same terms than CLF does with the collaborative indicators (Participative, Useful, With-Initiative …). In this way, at the end of the activity it could be possible to check both, the collaborative indicators computed by the CLF and the others detected by the aDeNu monitors. This analysis is really important in order to decide if indicators are well declared and to get a better definition.

The number of participants was fifty-six, divided into fourteen teams of four members each one. Ten members of the aDeNu group (researchers) were involved in the activity. Regarding the three questionnaires, two were for the participants and the remaining one for the researchers. Before starting the activity, the participants completed a questionnaire about their experience in the use of the Web and the tools associated with it (Q1). At the end of the activity, they completed a questionnaire about their feeling about the CLF and the gymkhana (Q2). The researchers filled the last questionnaire (Q3) during the execution of every group as explained above, taking notes of the behavior of the participants.

5.2 *Information Obtained from the Questionnaires*

The questionnaire Q1 was merely informative and its purpose was to get some feedback about the participant's knowledge of the new technologies. Considering the age of the majority of the participants, the results of those questions were the ones expected: more than 80% had been using Internet since more than one year and the most common tools they used were email, social networks and chats.

The set of questions that made up the questionnaire Q2 had more importance because they focused on the methodology introduced during the gymkhana. The aDeNu members noticed that participants learnt quickly the CLF concepts and were able to work collaboratively with their colleagues. This feeling was certified by the user's answers. Figure 10 shows that the grade of understanding of the methodology was high (i.e. Much or Very Much) for 76% of the participants. Moreover, the answer to the question "*do you think this methodology helps the collaboration?*" was answered affirmative by more than 70% of the participants. This analysis suggests that the CLF can be used as a good way of introduce collaboration in CSCL environments.

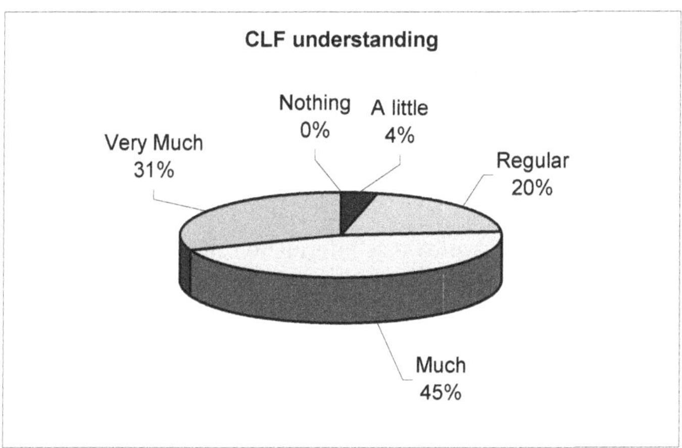

Fig. 10 CLF Understanding

The most important element to analyze from these questions was the influence of the collaborative indicators in the event. The participants were asked if they agreed the collaborative indicators that the application computed for each one. The participants mostly agreed with the results, except for those indicators with negative meaning (no-collaborative - 100% disagree - , gossip – 66%, insecure - 100%), probably because they do not like to be characterized with adverse skills.

On the contrary most of the other indicators where accepted. Figure 11, shows the result of this analysis. For each indicator there are two columns: the first is the number of students computed by CLF during the activity, and the second column depicts the amount of students disappointed with the result. In order to address this issue, the definition of the indicators should be done in a more positive way.

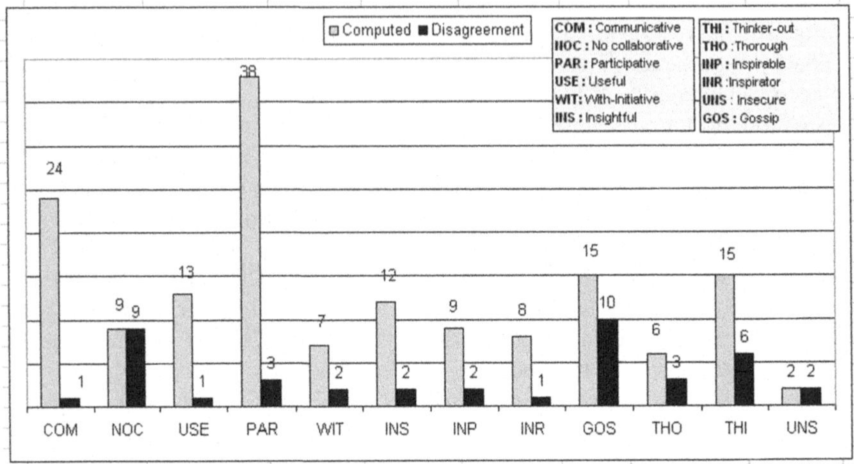

Fig. 11 Agreement with computed indicators

In relation to the collaborative indicators, it is important to comment that the moderator of each group for the agreement stage was selected by the system itself applying some rules to the individual indicators. By asking the participants their level of agreement with the moderator selection, 76% of them agreed with the system decision. Taking into account that the moderator had an extra prize, this result can be considered quite meaningful.

The last group of questions (Q3) provided information about the correspondence between the indicators calculated by the system and the participant's performance observed during the activity by aDeNu members in order to determine: i) if the twelve indicators were properly defined, ii) if the twelve indicators covered all the student's behaviors, iii) if the indicators could identify different performances without overlapping, and iv) if the indicators calculation was similar to tutor's observation. Since this was a formative evaluation, the goal was to gather information to feed the on-going development works to implement the whole approach. No general conclusions about the approach were expected to be achieved at this stage.

Having said that, we comment on the results obtained. When we checked if the indicators computed by the application were similar to the indicators wrote down by the aDeNu researchers who were observing the experience, we noticed some differences between both the observed and computed ones.

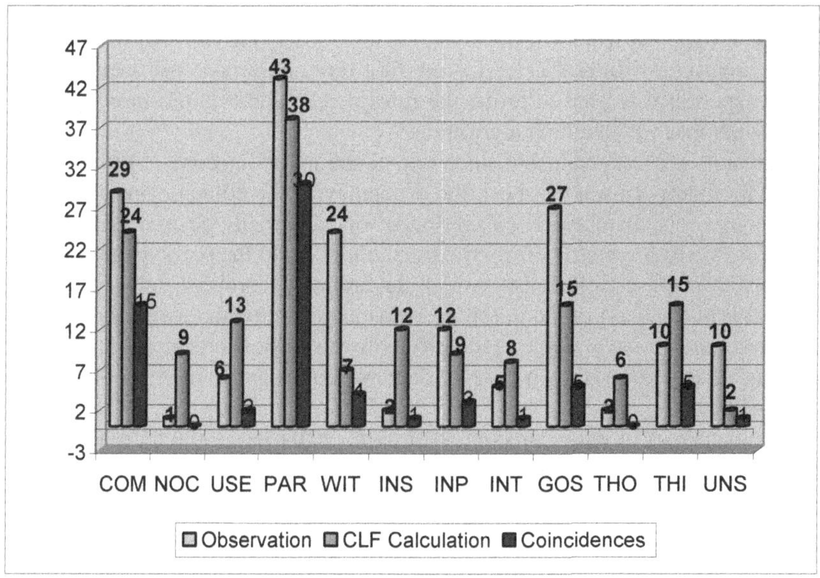

Fig. 12 Computed vs. Observed Indicators

Figure 12 depicts the results of those questions. The first column is the number of students observed by the aDeNu researchers, the second shows those computed by CLF and the third is the number of coincidences of the two first columns. For instance, in case of the Communicative indicator (COM): the aDeNu people considered that 29 of the 54 participants acted in this way just by observation; On the other hand, according to the rules defined to identify the communicative indicator, the CLF classified 24 participants as communicative. 15 of them matched up with the 29 detected by aDeNu researchers. Therefore there were 9 (24 – 15) who were identified as communicative by CLF but not by the tutors, and 14 (29 – 15) who were identified by the tutors, but not by CLF.

We found differences in some indicators with a behavior highly recognizable from the point of view of observation as *with-initiative, gossip or insecure*. In these cases the number of participants identified by the researches was higher than the one provided by the system. This result must be taken into account in order to reconfigure the metrics defining those indicators.

Differences were found on those elusive indicators with not so clear definition as Insightful, Thorough or Useful. The system was able to find a lot of samples (12, 6 and 13 respectively) while the observers merely found a few of cases (2, 2 and 6).

In relation to the differences found, we have to consider that collaborative indicators have been computed by the CLF using some manual rules defined from the metrics. The CLF is able to compute the collaborative indicators using manual rules or classification algorithms from data models (learning machine techniques), but for this later case it is necessary to have a reliable model to work with (from that moment in time the available models were built but their results underperformed those obtained from manual rules). This is due to the fact that the we were using the prototype with real users for the first time. However, the data gathered from this experience is used to refine the rules design and to build more accurate models for the machine learning algorithms.

These results are very valuable since one of the main objectives of this formative evaluation activity was to check the efficiency of the collaborative indicators definitions in a real scenario. The experience was also really positive because the system was able to get metrics from the interaction, build the participants' profiles and establish the base to generate recommendations according to those profiles. Actually, this is the heart of the unforeseen adaptation provided by the system: the recommendations. If we are able to identify how the participants work (and we did even though some adjusts are needed), then we could guide their work through recommendations.

From the analysis of the results obtained in this experience, the focus of the ongoing development has to be put on improving the collaborative indicators definition and the data taken from this activity:

- The definition of the twelve indicators was done using the first approach depicted in [7]. After the experience, we noticed that this approach can be improved because some indicators are not easily recognizable, not even for the tutor's observation, as Thorough (only 2 participants identified by observation) or Insightful (2 participants). They should also be redefined in a more positive way.
- The computing of indicators. One point to review is the metrics taking part of the indicators definitions to know if they really are representing that profile.
- To know if a participant's behavior suits a specific indicator, the system used manual rules based on statistics rates (standard deviation). The rule consists on consider if certain amount of metrics taking part of the indicator definition reach a positive value. The criteria to decide when the metric reach a positive value can also be adjusted.
- Experiments with classification algorithms instead of manual rules to compute the indicators. The data obtained in this experience can be used to improve the models that allow the system to use machine learning.

To sum up results from the answers to the questionnaires it follows that:

- The CLF approach allows the creation of effective scenarios that enable learning through interaction, exploration, discussion and collaborative knowledge construction.
- It is possible to identify the learner's behavior while they are interacting within the activity, through the capture of different metrics that take part in the collaborative indicators definition.
- The CLF component provides appropriate tools to configure the metrics and the indicators definition. This feature provides high flexibility

to the system that can be used to improve its performance with further experiences in order to reduce the differences found.

- The definition of some indicators overlapped the description of others. This result is not surprising as the list of indicators offered was tentative, to be further refined after experiences with users. An alternative approach to improve this issue could be consider less indicators grouping those with a similar meaning (thinker-out and thorough, communicative and participative ...) and consider the label for indicators as a range of values instead of absolute identification. For instance, the system could consider the learners as none, a little, quite or fully participative.

6 Conclusions and Future Work

Collaboration is one of the main strategies in e-learning and a long standing issue which has provided wide research over the last decades. As it has been reviewed in this chapter, there is a need for an intelligent support that facilitates collaboration management during the design, development and analysis of the collaborative learning experience. Thus, students and instructors alike will be benefited in their respective tasks. Research covered monitoring, analyzing and inferring methods that offered assessments on collaboration. Collaboration indicators are inferred from users' interactions. Further, collaboration assessment has been provided by machine learning and data mining technologies, which are appropriate in these contexts because they can be applied to support an automated process of analysis. In this way, instructors' workload can be reduced and meta-cognitive issues can be leveraged to improve students' engagement in collaborative learning.

In particular, to deal with the collaboration support in this chapter we have presented the design, deployment and a formative evaluation of a framework which is based on standards, run on a CSCL environment within an open LMS and considers foreseen and unforeseen situations. Three main aspects have been discussed: the component-based model to obtain these adaptive features, a particular instantiation of this framework using the open standard-based LMS called OpenACS/dotLRN together with the CLF, and a formative evaluation of a prototype during the Science Week 2009 in Madrid.

The results of this evaluation indicate that the CLF supports the collaboration among the members of a group interacting in a course. It is expected that the integration of the others components of the proposed framework will generate a complete adaptive system that manages collaboration activities with the ability to provide the personalization features required to adapt the learning experience to the student needs, which will enhance the collaboration objective and ultimately support a successful learning.

At the moment, the aDeNu research group is working in some of the components of the proposed framework to extend the support of the adaptive behavior. The steps to follow are to improve the CLF interface, use the User Model Server to store the learners' collaborative indicators, complete the integration with the

IMS-LD player, and improve the design and implementation of the TORMES Recommender System against the scenarios of the A2UN@ and EU4ALL projects.

Acknowledgments

Authors would like to thank the following colleagues from the aDeNu research group who in one or another way, had made possible the experience at the Week of Science 2009: Emmanuelle Raffenne, Héctor Romojaro, Jorge Granado, Emmanuelle Gutiérrez y Restrepo, Pilar Ulloa and Carmen Barrera. Moreover, authors would also like to thank the Spanish Ministry of Science and Innovation for funding the A2UN@ project (TIN2008-06862-C04-01/TSI).

References

1. Soller, A., Martinez, A., Jermann, P., Muehlenbrock, M.: From Mirroring to Guiding: A Review of State of the Art Technology for Supporting Collaborative Learning. International Journal of Artificial Intelligence in Education 15, 261–290 (2005)
2. Anaya, A.R., Boticario, J.G.: Application of machine learning techniques to analyse student interactions and improve the collaboration process. Expert Systems with Applications: Special Issue on Computer Sup-ported Cooperative Work in Design (2010)
3. Middleton, A.: Logical Framework Analysis: A Planning Tool for Government Agencies, International Development Organizations, and Undergraduate Students. Undercurrent 2(2), 41–47 (2005)
4. Johnson, R., Johnson, D.: Creativity and Collaborative Learning. In: Thousand, J., Villa, A., Nevin, A. (eds.), Brookes Press, Baltimore (1994)
5. Meier, A., Spada, H., Rummel, N.: A rating scheme for assessing the quality of computer-supported collaboration processes. In: Computer-Supported Collaborative Learning, vol. (2), pp. 63–86 (2006)
6. Perera, D., Kay, J., Yacef, K., Koprinska, I.: Mining learners' traces from an online collaboration tool. In: Workshop Educational Data Mining, Proceedings of the 13th International Conference of Artificial Intelligence in Education, Marina del Rey, CA. USA (July 2007)
7. Santos, O.C., Boticario, J.G.: Supporting a collaborative task in a web-based learning environment with Artificial Intelligence and User Modelling techniques. In: VI Simposio Internacional de Informática Educativa, SIIE 2004 (2004)
8. Magnisalis, I., Demetriadis, S.: Modelling adaptation patterns with IMS-LD specification: a case study as a proof of concept implementation. In: International Workshop on Adaptive Systems for Collaborative Learning. In conjunction with the INCoS 2009, International Conference on Intelligent Networking and Collaborative Systems (2009)
9. Hernández, L., Burgos, D., Tattersall, C., Koper, R.: Representing Computer-Supported Collaborative Learning macro-scripts using IMS Learning Design. In: 2nd European Conference on Technology Enhanced Learning, EC-TEL2007 (2007)
10. Santos, O.C., Boticario, J.G., Barrera, C.: Authoring a Collaborative Task Extending the IMS-LD to be Performed in a Standard-based Adaptive Learning Management System called aLFanet. In: ICWE Workshops, pp. 180–187 (2004)

11. de la Fuente, L., Pardo, A., Asensio, J.I., Dimitriadis, Y., Delgado, C.: Collaborative Learning Models on Distance Scenarios with Learning Design: A Case Study. In: International Conference on Advanced Learning Technologies, ICALT 2008, Santander, Spain (June 2008)
12. Koper, R., Tattersall, C. (eds.): Learning Design: A handbook on modelling and delivering networked education and training. Springer, Berlin (2005)
13. Santos, O.C., Boticario, J.G.: Modeling recommendations for the educational domain. In: Proceedings of the 1st Workshop 'Recommender Systems for Technology Enhanced Learning', RecSysTEL 2010 (2010) (in press)
14. Santos, O.C., Boticario, J.G.: Modeling recommendations for the educational domain. In: Santos, O.C., Boticario, J.G., Raffenne, E., Pastor, R. (eds.) Proceedings of the 1st Workshop 'Recommender Systems for Technology Enhanced Learning' (2007); Why using dotLRN? UNED use cases. In: Proceedings of the FLOSS (Free/Libre/Open Source Systems) International Conference 2007, pp. 195–212 (2007)
15. Russell, S., Norvig, P.: Artificial Intelligence: A Modern Approach, Englewood Cliffs, New Jersey. Prentice Hall Series in Artificial Intelligence (1995)
16. Meier, A., Spada, H., Rummel, N.: A rating scheme for assessing the quality of computer-supported collaboration processes. Computer-Supported Collaborative Learning 2, 63–86 (2007)
17. Kahrimanis, G., Meier, A., Chounta, I.-A., Voyiatzaki, E., Spada, H., Rummel, N., Avouris, N.: Assessing collaboration quality in synchronous CSCL problem-solving activities: Adaptation and empirical evaluation of a rating scheme. In: Cress, U., Dimitrova, V., Specht, M. (eds.) EC-TEL 2009. LNCS, vol. 5794, pp. 267–272. Springer, Heidelberg (2009)
18. Martínez, A., Dimitriadis, Y., Gómez, E., Jorrín, I., Rubia, B., Marcos, J.A.: Studying participation networks in collaboration using mixed methods. International Journal of Computer-Supported Collaborative Learning 1(3), 383–408 (2006)
19. Bratitis, T., Dimitracopoulou, A., Martínez-Monés, A., Marcos-García, J.A., Dimitriadis, Y.: Supporting members of a learning community using interaction analysis tools: the example of the Kaleidoscope NoE scientific network. In: Proc. of the IEEE International Conference on Advanced Learning Technologies, ICALT 2008, Santander, Spain, vol. 2008, pp. 809–813 (July 2008)
20. Gaudioso, E., Montero, M., Talavera, L., Hernandez-del-Olmo, F.: Supporting teachers in collaborative student modeling: A framework and an implementation. Expert Systems with Applications 36, 2260–2265 (2009)
21. Collazos, C.A., Guerrero, L.A., Pino, J.A., Renzi, S., Klobas, J., Ortega, M., Redondo, M.A., Bravo, C.: Evaluating Collaborative Learning Processes using System-based Measurement. Educational Technology & Society 10(3), 257–274 (2007)
22. Hong, W.: Spinning Your Course Into A Web Classroom - Advantages And Challenges. In: Hong, W. (ed.) International Conference on Engineering Education, Oslo, Norway, August 6 -10 (2001)
23. Daradoumis, T., Martínez-Mónes, A., Xhafa, F.: A Layered Framework for Evaluating OnLine Collaborative Learning Interactions. International Journal of Human-Computer Studies 64(7), 622–635 (2006)
24. Winter, M., McCalla, G.I.: An Analysis of Group Performance in Terms of the Functional Knowledge and Teamwork Skills of Group Members. Paper presented at the Workshop on User and Group Models for Web-based Adaptive Collaborative Environments, Int. Conf. on User Modeling (UM 2003), Johnstown, Pennsylvania (2003)

25. Axelrod, R.: The Evolution of Cooperation. Basic Books, New York (1984)
26. Marcos-García, J.A., Martínez-Monés, A., Dimitriadis, Y., Anguita-Martínez, R., Ruiz-Requies, I., Rubia-Avi, B.: Detecting and Solving Negative Situations in Real CSCL Experiences with a Role-Based Interaction Analysis Approach. In: Daradoumis, T., Caballé, S., Marquès, J.M., Xhafa, F. (eds.) Intelligent Collaborative e-Learning Systems and Applications. Studies in Computational Intelligence, vol. 246, pp. 129–146. Springer, Heidelberg (2009)
27. Romero, C., Ventura, S.: Educational data mining: A survey from 1995 to 2005. Expert Systems with Applications 33, 135–146 (2007)
28. Gómez-Sánchez, E., Bote-Lorenzo, M.L., Jorrín-Abellán, I.M., Vega-Gorgojo, G., Asensio Pérez, J.I., Dimitriadis, Y.: Conceptual framework for design, technological support and evaluation of collaborative learning. International Journal of Engineering Education 25(3), 557–568 (2009)
29. Muehlenbrock, M.: Formation of Learning Groups by using Learner Profiles and Context Information. In: Looi, C.-K., McCalla, G. (eds.) Proceedings of the 12th International Conference on Artificial Intelligence in Education AIED 2005, Amsterdam, The Netherlands (2005)
30. Duque, R., Bravo, C.: A Method to Classify Collaboration in CSCL Systems. In: Beliczynski, B., Dzielinski, A., Iwanowski, M., Ribeiro, B. (eds.) ICANNGA 2007. LNCS, vol. 4431, pp. 649–656. Springer, Heidelberg (2007)
31. Redondo, M.A., Bravo, C., Bravo, J., Ortega, M.: Applying Fuzzy Logic to Analyze Collaborative Learning Experiences in an e-Learning Environment. USDLA Journal (United States Distance Learning Association) 17.2, 19–28 (2003)
32. Talavera, L., Gaudioso, E.: Mining Student Data To Characterize Similar Behavior Groups In Unstructured Collaboration Spaces. In: Proceedings of the Workshop on Artificial Intelligence in CSCL. 16th European Conference on Artificial Intelligence (ECAI 2004), Valencia, Spain, pp. 17–23 (2004)
33. Anaya, A.R., Boticario, F.G.: Clustering learners according to their collaboration. In: Proceedings of the 13th International Conference on Computer Supported Cooperative Work in Design (CSCWD 2009), IEEE Computer Society Press, Los Alamitos (2009)
34. Brooks, C., Winter, M., Greer, J., McCalla, G.: The Massive User Modelling System (MUMS). In: Lester, J.C., Vicari, R.M., Paraguaçu, F. (eds.) ITS 2004. LNCS, vol. 3220, pp. 635–645. Springer, Heidelberg (2004)
35. Baldiris, S., Santos, O.C., Barrera, C., Boticario, J.G., Velez, J., Fabregat, R.: Integration of Educational Specifications and Standards to Support Adaptive Learning Scenarios in ADAPTAPlan. International Journal of Computer & Applications 5(1), 88–107 (2008)
36. Denaux, R., Aroyo, L., Dimitrova, V.: OWL-OLM: Interactive Ontology-based Elicitation of User Models. In: Workshop on Personalisation for the Semantic Web PerSWeb 2005 at 10th International Conference on User Modeling, Edinburgh, UK, July 2005,
37. Bull, S., Kay, J.: Metacognition and Open Learner Models. In: Roll, I., Aleven, V. (eds.) Proceedings of Workshop on Metacognition and Self-Regulated Learning in Educational Technologies, International Conference on Intelligent Tutoring Systems, pp. 7–20 (2008)
38. Bull, S., Gardner, P., Ahmad, N., Ting, J., Clarke, B.: Use and Trust of Simple Independent Open Learner Models to Support Learning Within and Across Courses. In: Houben, G.-J., McCalla, G., Pianesi, F., Zancanari, M. (eds.) User Modeling, Adaptation and Personalization, pp. 42–53. Springer, Heidelberg (2009)

39. Dimitracopoulou, A.: Computer based Interaction Analysis Supporting Self-regulation: Achievements and Prospects of an Emerging Research Direction. In: Kinshuk, M., Spector, D., Sampson, P. (eds.) Technology, Instruction, Cognition and Learning, TICL (2008)
40. Macarthur, V., Conlan, O.: Using Psychometric Approaches in the Modeling of Abstract Cognitive Skills for Personalization. In: Workshop Lifelong User Modelling, UMAP (2009)
41. Steffens, K.: Self-regulation and computer based learning. Anuarion de Psicología 32(2), 77–94 (2001)
42. Baghaei, N., Mitrovic, A.: From Modelling Domain Knowledge to Metacognitive Skills: Extending a Constraint-Based Tutoring System to Support Collaboration. In: Conati, C., McCoy, K., Paliouras, G. (eds.) UM 2007. LNCS (LNAI), vol. 4511, pp. 217–227. Springer, Heidelberg (2007)
43. Dillenbourg, P.: Introduction; What do you mean by Collaborative Learning? In: Dillenbourg, P. (ed.) Collaborative Learning. Cognitive and Computational Approaches, pp. 1–19. Elsevier Science Ltd., Oxford (1999)
44. Barkley, E., Cross, K.P., Major, C.H.: Collaborative Learning Techniques: A Practical Guide to Promoting Learning in Groups. Jossey Bass, San Francisco (2004)
45. Santos, O.C., Boticario, J.G.: Usability methods to elicit recommendations for Semantic Educational Recommender Systems. IEEE Learning Technology Newsletter 12(2) (April 2010)
46. Bayón, A., Santos, O.C., Boticario, J.G.: A component to carry out the Logical Framework Approach. In: dotLRN. 7th OpenACS /.LRN Conference (2008)
47. Witten, I.H., Frank, E.: Data Mining: Practical machine learning tools and techniques, 2nd edn. Morgan Kaufmann, San Francisco (2005)
48. Lafifi, Y., Halimi, K., Ghodbani, A., Salhi, N.: Learners Monitoring Based on Traces in CSCL System. INFOCOMP – Journal of Computer Science 8(2), 61–72 (2009) ISSN: 1807-545
49. Couchet, J., Santos, O.C., Raffenne, E., Boticario, J.G.: The Tracking and Auditing Module for the OpenACS Framework. In: 7th OpenACS /.LRN Conference (2008)
50. Cuartero, A., Santos, O.C., Granado, J., Raffenne, E., Boticario, J.G.: Management of standard-based User Model and Device Profile in OpenACS. In: OpenACS and.LRN Conference [International Conference and Workshops on Community based environments, Guatemala] (2008)
51. Felder, R.M., Silverman, L.K.: Learning and Teaching Styles In Engineering Education. Engr. Education 78(7), 674–681 (1988)
52. IMS Learner Information Package specifidation, `http://www.imsglobal.org/profiles/`
53. Santos, O.C., Granado, J., Raffenne, E., Boticario, J.G.: Offering recommendations in OpenACS/dotLRN. In: 7th OpenACS /.LRN Conference (2008)
54. Santos, O.C., Martin, L., del Campo, E., Saneiro, M., Mazzone, E., Boticario, J.G., Petrie, H.: User-Centered Design Methods for Validating a Recommendations Model to Enrich Learning Management Systems with Adaptive Navigation Support. In: Weibelzahl, S., Masthoff, J., Paramythis, A., van Velsen, L. (eds.) Proceedings of the Sixth Workshop on User-Centred Design and Evaluation of Adaptive Systems, held in conjunction with the International Conference on User Modeling, Adaptation, and Personalization (UMAP 2009), Trento, Italy, June 26, pp. 64–67 (2009)

Apt to Adapt: Micro- and Macro-Level Adaptation in Educational Games

Michael D. Kickmeier-Rust, Christina M. Steiner, and Dietrich Albert

Cognitive Science Section, Department of Psychology, University of Graz, Austria

Abstract. The popularity of computer games has lead to an increasing interest in educational games in research and development in the last decades. Educators as well as technicians are captivated of the idea of utilizing the motivational potential and the rich virtual worlds of today's computer games. To use the full educational potential of computer games a strong personalization and adaptation to the individual needs and preferences is needed. Conventional methods of educational adaptation, however, are oftentimes not suitable in the context of games, as they may force an interruption of the game experience and thus, destroy immersion and engagement of the player. In this paper we present approaches of educational adaptation in games that allow embedding instruction into the game experience and narrative, through non-invasive assessment of knowledge and motivation, the delivery of various types of adaptive interventions, and adaptive storytelling. The outlined approaches are focus of research, development, and evaluation in the context of the European research project 80Days.

Introduction

Educational (or serious) computer games are considered being a highly promising approach to make learning a more pleasant, engaging, satisfying, inspiring, and probably more effective task. Thus, it is not surprising that there is an increasing focus on the research on game-based learning and serious games per se. Many of the potential advantages of computer games (e.g., interactivity, feedback, situated learning) are considered being didactically important for successful and effective learning [1]. Moreover, games perfectly serve the demands of the so-called "digital natives" [2]. Whatever one might think about Marc Prensky's ideas, the key strengths of educational games are the intrinsic motivational potential and their potential to reach young people who actually are not necessarily willing or interested to learn (with conventional materials or within conventional settings). Since the 1990s research and development has increasingly addressed learning aspects of playing recreational games and also the realization of computer games for primarily educational purposes. Still, the idea of educational games is not fully mature yet and existing approaches oftentimes have certain disadvantages, for example, difficulties in providing an appropriate balance between gaming and learning activities or between challenge and ability, in aligning the game with national

T. Daradoumis et al. (Eds.): Technology-Enhanced Systems and Tools, SCI 350, pp. 221–238.
springerlink.com

curricula, or the extensive costs of developing high quality games [3]. From a global perspective we can say that most products can hardly compete with their commercial "non-serious" counterparts in terms of gaming experience, immersive and interactive environments, narrative, or motivational potential to play. Moreover, most educational games do not rely on sound instructional models, leading to a separation of learning from gaming; often such games provide gaming actions only as reward for learning. Provocatively speaking, many existing products do not differ significantly from other multimedia learning objects or applications.

To address specific existing pitfalls of educational games, some authors emphasize the importance of personalization and intelligent adaptation (e.g., [4]). The rational behind is simple; the big strength of educational games is their immersive and intrinsically motivational potential. Being motivated and engaged, however, strongly depends on personal preferences and abilities, both in a gaming-oriented sense as well as in an educational sense. Only if a game and the flow of the game suit the individual interests and preferences of the learners, they will be motivated and engaged and only if the educational requirements (the presented subject matter or problems/tasks that must be solved) neither overburden nor under-challenge the learners, they will be motivated and engaged.

Psycho-pedagogical and technical research, of course, addressed personalization and adaptation in educational systems (cf. [5] for an overview). Common techniques are adaptive presentation (i.e., adapting the look and feel of an educational environment to the preferences and needs of the learner) or adaptive curriculum sequencing (i.e., providing a specific sequence through the learning objects on an individual basis). The existing approaches and solutions, particularly those of adaptive curriculum sequencing, however, cannot be transferred easily to the genre of educational games. Usually, such methods require a continuous assessment of learning progress; most often realized by typical computer-supported queries and test items. To give an example, imagine a person is presented a specific learning object; to decide whether this person has understood the learning content and subsequently to decide which of the available learning objects is the most appropriate next one (which differs whether the previous learning object was understood and also with respect to the individual learning goals), the system needs to perform some kind of knowledge test. Conventional querying methods, for example a popping up multiple choice question, are not possible in the virtual environment of a game since it would immediately destroy the game's flow and, therefore, motivation, immersion, and engagement. Moreover, for the genre of educational computer games a pure sequencing of learning objects (or rather "learning situations") is problematic. On the one hand, games are usually (more or less) driven by a narrative. Sequencing and re-sequencing of learning objects, as it is done in traditional intelligent tutoring systems [5], would significantly compromise the storyline. On the other hand, in the rich virtual worlds of educational games, appropriate guidance and support is the more important and more effective type of adaptation [6]. To give an example, game-based instructional design preferably is realized in form of exploratory or experimental learning, strongly embedded in the game and the game's narrative. The aim of meaningful adaptation is to support the learners during exploration and experimenting and to provide them with hints, suggestions, warnings, or feedback [6].

Around an Inspiring Virtual Learning World

The idea of a non-invasive, continuous assessment of learning progress and motivational states of the learner and the provision of adequate responses in the context of competitive educational games is in the focus of the European research project 80Days (www.eightydays.eu), inspired by Jules Verne's novel "Around the world in eighty days". The project's (first) demonstrator game is teaching geography for a target audience of 12 to 14 year olds (see Fig. 1 for screenshots). The curriculum includes, for example, knowledge about the planet Earth, such as countries or cities but also aspects such as longitude or latitude. In the game the learner takes the role of a boy or a girl (depending on the learners' gender) at the age of 14. The story starts from an extraordinary event; a space ship is landing in the backyard and an alien named Feon appears. Feon turns out to be a friendly alien, being an alien scout who has to collect information about foreign planets, in this case planet Earth. The learner accompanies Feon and is having fun with flying a space ship and exploring interesting places on Earth. Feon creates a report about the Earth and its geographical features. This is accomplished with the help of the player by means of flying to different destinations on Earth, exploring them, and collecting and acquiring geographical knowledge. The goal is to send the Earth report as a sort of travelogue about Earth to Feon's mother ship. At a certain point in the game, however, the player makes a horrible discovery; the aliens are not really friendly but collect all the information about Earth to conquer the planet, lately. This discovery reveals the "real" goal of the game: The player has to save the planet and the only way to do it is to draw the right conclusion from the traitorous Earth report. The subject matter is embedded in the story and learning occurs at various events in the game. From a pedagogical point of view, learning occurs by receiving information (e.g., seeing/reading something in the game or hearing something from Feon or other game characters), problem solving (e.g., reducing the negative impacts of a flood by appropriate "terra-forming"), or imitation (e.g., watching other game characters and learning from their behaviors).

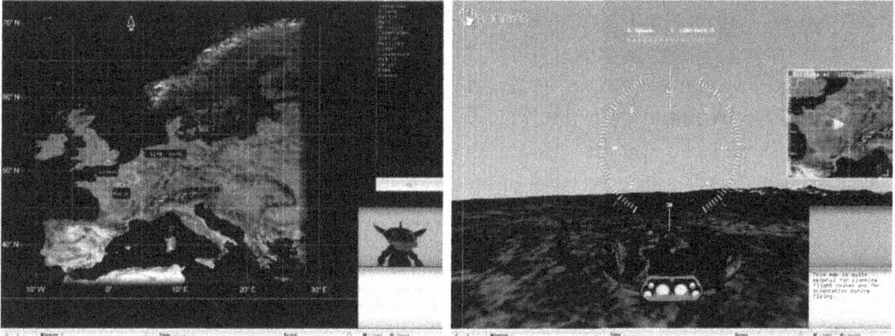

Fig. 1 Screenshots of 80Days' demonstrator game.

Micro Level Assessment

The very basis of a non-invasive, continuous assessment of learning progress and motivational states is to monitor and interpret the learner's behavior in the game. To achieve this, we utilize the formal framework of Competence-based Knowledge Space Theory (CbKST) [7, 8]. Originating from conventional adaptive and personalized tutoring, this set-theoretic framework allows assumptions about the structure of skills of a domain of knowledge and to link the latent skills with observable behavior. It provides an internal cognition-based logic that is quite similar to the logic of ontologies: well-defined entities (the skills) are in a well-defined relationship (a so-called prerequisite relation). The domain model, the set of meaningful skill states, and the resulting set of meaningful learning paths are combined with a model of tasks and problems within certain parts, so-called learning missions, of the game (equivalent to conventional "learning objects"), the so-called problem space (cf. [9]). A simple example for such mission might be the task to fly with the space ship to a certain city and to take a picture. The learning objective of this task might be (among others) to learn about the location of the city on the map. In this situation are various manipulable objects, for example the space ship. The learner can perform certain actions to achieve the goal, in this example primarily changing the directions of the flying space ship or controlling speed and altitude. The aim of micro level assessment is in the first instance to assign a problem solution state from the problem space to each action (e.g., pressing an arrow key). This mapping is done by classifying actions according to a set of rules. An example for such rule might be "the distance between space ship and target location is increasing". The second aim is to assign a set of available and a set of lacking skills to each problem solution state; for example, flying in the right direction indicates that the learner knows the wind direction towards the city. Of course, a single observation is not very convincing. Thus, CbKST provides a probabilistic approach to assessment. We have a probability distribution over all possible skill states and with each action we update the probabilities of those states that include the relevant skills and we decrease those states that include the lacking skills (for details on the probabilistic updating procedure refer to [10]). At the end of this procedure stands a more or less well-founded assumption about the skills the learners have, the skills they don't have, and their position in the problem solving process. Similarly, we can assign specific motivational assumptions to specific classes of actions, again based on a set of rules. An example is to interpret the density of actions, that is, the number of actions performed in a specific time interval.

The continuously gathered and updated assumptions on the skills and motivational state throughout the game serve the provision of adaptive interventions tailored to the learner's current needs.

Micro Level Interventions

As important as it is to avoid comprising the game's flow by assessing learning progress or motivational state, are interventions embedded in the game. Interventions are hints, suggestions, warnings, or feedback. For a systematic approach to providing interventions, we have elaborated a menu of adaptation for educational games that covers different types of psycho-pedagogically founded interventions.

Depending on the information stemming from the continuous, non-invasive assessment of a learner's currently available and lacking skills and current motivational state, from this menu of adaptation a system feedback in terms of an intervention in the game can be triggered (e.g. hints or suggestions through a non-player character, modification of display or interface) that is individually appropriate for the respective learner and situation. The interventions and their selection rules are defined based on sound cognitive, psycho-pedagogical, and motivational theories and considerations. All types of adaptive interventions have in common that they aim at supporting a beneficial game-based learning experience. In general, three broad categories of meta-cognitive, cognitive and motivational interventions can be distinguished. These can be related with the two perspectives of non-invasive assessment of competence and motivation: while meta-cognitive and cognitive interventions mainly address skill-related aspects and aim at supporting skill acquisition, motivational interventions naturally refer to motivational aspects and the final aim of enhancing or maintaining engagement.

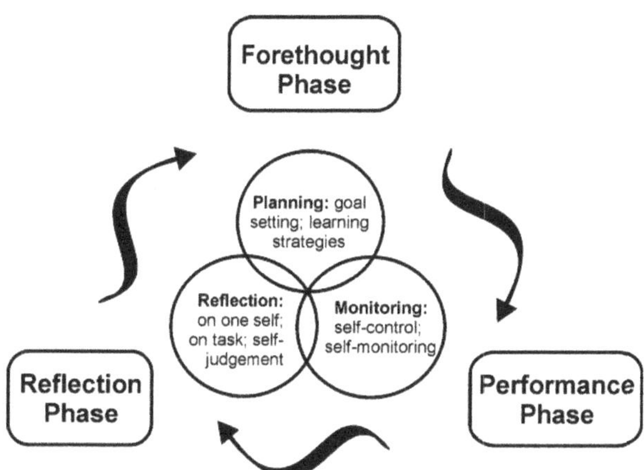

Fig. 2 The self-regulated learning cycle and involved meta-cognitive (sub)processes.

Meta-cognitive Interventions

Meta-cognition is basically knowledge about one's own knowledge/thinking – Flavell defines: "Meta-cognition refers to one's knowledge concerning one's own cognitive processes or anything related to them, e.g., the learning-relevant properties of information or data. For example, I am engaging in meta-cognition if I notice that I am having more trouble learning A than B; if it strikes me that I should double check C before accepting it as fact" ([11], p. 232).

Meta-cognitive interventions are supposed to provoke learners' reflection about their own abilities, thinking processes, solution behaviour, or confidence. This category of interventions is motivated by the psycho-pedagogical approach of self-regulated learning [12, 13, 14]. Meta-cognitive and reflective processes are critical for effective self-regulated learning; they are inherent part of all phases of the self-regulated learning process (see Fig. 2). The forethought phase is characterized by processes for planning and preparing for learning, like goal setting, strategic planning (learning strategies). The performance phase mainly involves monitoring one's own learning performance through self-control and self-observation, and the reflection phase involves reflective processes about oneself and the task as well as self-evaluation of performance [13, 15].

Meta-cognitive interventions firstly can be differentiated according to the phase of learning in which they are provided, aligned with the self-regulated learning cycle. As, however, the line between the single learning phases is blurred, there is also a smooth transition between the respective intervention types.

- **Meta-cognitive forethought interventions** mainly aim in promoting or supporting planning and preparatory thinking processes for actual learning. This may for example be a question like 'What are the strategies/tactics/principles appropriate for solving this problem?' posed by a non-player character.
- **Meta-cognitive performance interventions** are provided in the course of a learning or problem-solving process, aiming at fostering self-monitoring. This may for instance consist in asking the learner for self-recording of personal events in a diary or logbook.
- **Meta-cognitive reflection interventions** promote and support reflection on one self's performance and skills, e.g. by asking 'What did go wrong here?'

In addition to this differentiation there is another way of distinguishing different types of meta-cognitive interventions, depending on the nature of the intervention itself and the type of provoked meta-cognitive action. These types basically are orthogonal to the types mentioned above and constitute a second dimension of differentiating meta-cognitive intervention types.

- **Meta-cognitive questions** consist in interventions that, as the name implies, provide a question that may provoke or initiate meta-cognitive processes, such as 'Does this solution make sense?' These questions may
  - refer to the comprehension of a problem,
  - construct connections between previous and new knowledge elements,
  - ask for appropriate strategies for solving a problem,
  - or encourage reflection on the problem solving process and solution.
- **Meta-cognitive tasks** are interventions that provoke meta-cognitive thoughts through posing specific tasks or asking the learner to do something, like a prompt to write down one's thoughts in a diary.
- **Certainty questions** ask the learner for their certainty with a specific action or query the amount of certainty (which actually can also be related to the current motivational state w.r.t. confidence) – for example 'How sure are you about this?'

Cognitive Interventions

Interventions of this type strive to enhance cognitive abilities and to support the learner adaptively according to his/her task behaviour and underlying available or lacking skills. Consequently, these interventions target directly the learning objectives defined in terms of skills and foster their successful acquisition in terms of prompting reflection or assisting the learner. The cognitive interventions and their selection rules are defined based on the theoretical framework of Competence-based Knowledge Space Theory [7, 16] and the corresponding understanding of meaningful learning paths grounded on its evolving competence learning structures [17, 18] as well as referring to theoretically founded principles for the design of informative tutoring feedback [19].

The following cognitive intervention types can been distinguished, whereby partly no distinct line between the different types can be drawn:

- **Competence explication interventions** are realized in complex, simulation or experimentation situations. They are provided in case of correct learner actions leading to an increase of certain skill probabilities. This intervention type shall make explicit and reinforce the knowledge involved or underlying a certain action. In other words, if a learner has shown a 'correct' action, it shall be assured that he/she is aware of why this action was correct by explicating the underlying knowledge elements, principles, or rules.
- **Competence activation interventions** are applied if a learner gets stuck in a certain task, while foregoing assessment results led to the assumption that the learner possesses the necessary skills. In other words, this intervention type applies if the learner sticks in some area of the problem space and some skills are not used even though the system assumes that the learner possesses them. By the use of an appropriate intervention (e.g. 'We have come across this issue already before.') the temporarily 'inactive' skills are assumed to be stimulated and reactivated. Possible concrete interventions in this case may be
  - the reference to a specific skill of the learner
  - or a question or task covering the same knowledge but in a different manifestation.
- **Competence acquisition interventions** are selected when the system concludes that a learner lacks certain skills and thus, provide the required information – for example through a non-player character. This type of intervention is very similar (if not identical) to problem solving support (see below). Unlike the latter one, this intervention type is more common in a context that seems more detached from a concrete complex problem solving task.
- **Problem solving support** is provided in the context of an ongoing problem solving process and provides hints and indications for possible next problem solution steps in order to decrease the distance between the present solution state and the target state. Interventions of this type may be

- vague indications of possible next problem solution steps
- concrete hints about possible and promising next steps
- provide extra information, independent from the actual problem solving context
- suggest to visit or attend other/prior learning situations to acquire the necessary knowledge
- exemplify the solution for the current problem by a similar problem

- **Dissolving interventions** are a further form to present specific information to the learner. The purpose of this intervention type is to provide the solution of a problem/task if the learner was not able to show the required answer behaviour within a reasonable number of actions. Such interventions, ultimately, shall assure that the game can continue and thus the gaming and flow experience are kept going. Such intervention of course, for didactical reasons, might not be used or appropriate for all problems/tasks.
- **Progression feedback** is made up by interventions that provide the learner with information about the learning progress or the game – e.g. through a non-player character or different scoring mechanisms. This concrete feedback fosters cognitive abilities as well as monitoring and reflection on one's own performance, i.e. meta-cognition (because of it's concrete reference to skills or problems/tasks is categorized as cognitive intervention). Interventions of this type can be
  - feedback on responses/response actions: correctness, location of mistakes, elaborated explanation (e.g. why a certain answer is correct/incorrect)
  - goal-directed feedback: information about progress toward a desired goal, which has also a strong relevance for motivation as well
- **Cognitive assessment interventions** are a special form of intervention that is applied if the non-invasive assessment of skills led to unclear or ambiguous results after a certain number of actions. In order to gather additional information and improve the assessment this type of intervention is triggered. Typically this is realised by providing the learner with explicit questions or problems. As these interventions are strongly embedded into the game context and narrative, they differ significantly from conventional and possibly disrupting pop-up assessments known from 'traditional' learning systems. Assessment interventions may be realised by interactive dialogues with different answer options, which may not only refer to correct and incorrect responses but may also have a storytelling function and lead to different story strands depending on the learner's choice.

Motivational Interventions

Motivational interventions are supposed to enhance and retain the learner's motivation and engagement on a high level or to intervene when the system detects that the motivational state or certain aspects of it (potentially) decrease. On principle, these interventions may be triggered based on information stemming from the non-invasive assessment of a learner's motivational state (e.g. detection of a lack of attention) as well as of his/her skills (e.g. series of unpurposeful learner actions leading to continuous decrease of certain skill probabilities).

The differentiation of motivational intervention types is inspired and their selection rules are fed by theoretical elaborations of an advanced model of motivation for educational games [20], based on psycho-pedagogical theories on motivation and motivational design, such as the expanded model of motivation to learn [21], attribution theory [22] and the concept of self-efficacy [23], and Keller's ARCS model [24]. Motivational interventions may provide the learner with information about the learning progress or the game, provide or announce incentives or rewards, may address attention or confidence, but may also involve emotionally focused feedback. The following intervention types are distinguished:

- **Praising interventions** are used for congratulation in case of success. They are assumed to be incentives for learning and reinforcing motivation.
- **Encouraging interventions** are applied especially in case of failure in order to promote further trials and keep motivation.
- **Attributional interventions** go further than the previously mentioned interventions – they aim at fostering self-worth enhancing attributional styles for success and failure and are applied in case of lacking confidence or dysfunctional attributional styles. In general, the following factors affecting the attribution of success and failure of achievement are: ability, effort, task difficulty, and luck [22]. These factors can be classified according to stability (stable vs. variable) and locus of control (internal vs. external). This has been taken up in education for motivational training (e.g. [25]), by providing feedback in terms of attributions that are fostering self-worth and motivation. Such motivational training provides feedback that explains success first by effort (internal, variable factor) and later on by ability (internal, stable factor) and suggests attribution of failure to lacking effort (internal, variable factor) or bad luck (external, variable factor) [26]– for an overview see Fig. 3. Attributional interventions realise such a motivational training based on attribution theory – in form of feedback that directs attribution of success to internal factors (i.e. effort and ability) and attribution of failure to variable components (i.e. lack of effort and bad luck).
- **Incitation interventions** in general announce pleasing outcomes like rewards in order to foster motivation to carry on in the game or to proceed in a problem solving situation.
- **Affective interventions** address emotional-affective aspects of the game experience and social interaction with other game characters and are supposed to foster a positive affect.
- **Attention-catchers** are interventions that are applied if the system detects decreasing or lacking attention through the interpretation of the learner's actions. Such interventions constitute unexpected changes or incidents and in this way increase variability and further appeal of the game.

- **Motivational assessment interventions** are similar to their cognitive counterparts. They are utilised in case of inconclusive or contradicting inferences on the learner's motivational state based on the non-invasive assessment. For gathering further indications on the learner's current motivation assessment interventions realise an explicit questioning, usually in form of an interactive dialogue with a non-player character and with the answer options relating to certain aspects and states of motivation.

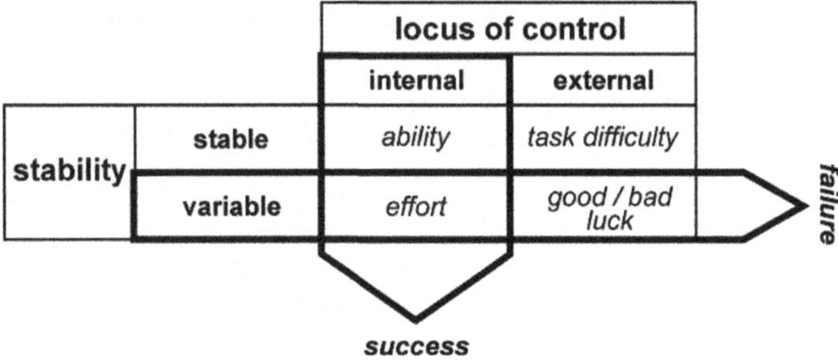

Fig. 3 Factors affecting attribution of success and failure and beneficial attributional styles.

All interventions of a game require a manifestation in form of game assets (e.g., a sound file with a specific sentence). Of course, not all possible interventions can be realized. In general, we pursue and propose an approach of using interventions conservatively or sparingly. We are perfectly aware that repeated inadequate interventions due to misinterpretations of a situation (e.g., assuming a lack of motivation on the basis of no actions for longer period of time while the learner just has gone to the toilette) are a significant harm to motivation, engagement, and the game flow. The conditions under which a certain adaptive intervention is given are to be developed on the basis of psycho-pedagogical rules.

The rules for triggering educational interventions and problem solving interventions, for instance, are defined based on cognitive psychological considerations in tight relation to the continuous assessment of skills in terms of CbKST. Through the definition of threshold values on skill probabilities an according intervention is prompted if the non-invasive monitoring procedures of the learner's actions provide substantial evidence for lacking skills or an increasing distance to the problem solution. Rules for motivational interventions, on the other hand, are developed grounding on motivational psychology, referring e.g. to attribution theory [22] and the ARCS model of motivational design [24]. Correspondingly, continuously unsuccessful behavior and unconfident reactions arguing for a decrease or lack of confidence and motivation will trigger an intervention fostering motivation and suggesting self-worth enhancing attribution styles (Fig. 4).

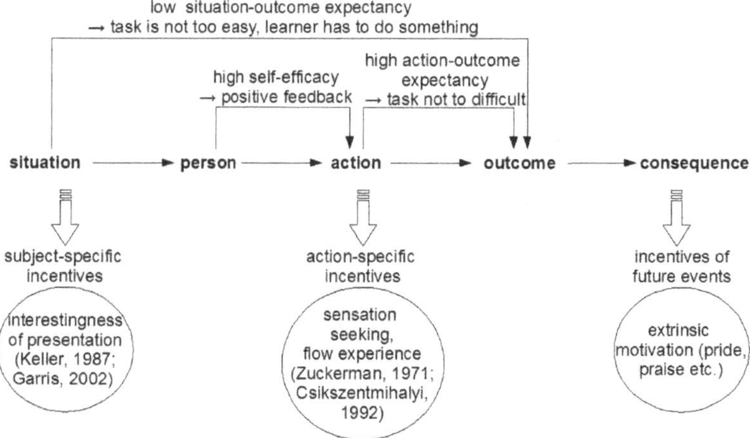

Fig. 4 Implications of motivational concepts.

Adaptation by Interactive Storytelling

The second core element of 80Days and in-game adaptation is – more globally – tailoring the game's storyline and pacing to the learners. Early story generation systems relied on non-dramatic models of narrative. In the spirit of Brenda Laurel's vision of interactive drama [27] implemented the first story generation systems. The most effective drama models developed for feature films by several authors (e.g., [28]) have since become global professional standards. Very few drama models developed for screenwriting have yet been integrated in story generation systems. In an educational game, adaptive and interactive digital storytelling basically serves two essential purposes:

- Support of personalized learning experiences by adapting the game's story to individual preferences and by providing the possibility of explorative learning processes. Additionally, it enables the learner to actively interfere with the game and its narrative. Such individual preferences in style and emotional quality are considered to be a crucial factor for facilitating learning and retaining motivation to play and learn. For that purpose, a specific authoring environment for DEGs has to be conceptualized, supporting the configuration of adapted story pacing, accounting for the integration of varied didactical drafts [29].
- Second, interactive and adaptive storytelling serves the re-usability of learning material by enabling the realization of different stories and entirely different games (even for different learning domains) based on more or less the same pool of atomic story units, patterns, and structures as well as learning and gaming concepts, elements, and objects.

The challenge of creating dynamic yet plausible adaptive narratives is not trivial and requires arduous manual editing of branching narratives. Experimental systems such as *Façade* [30] or *Virtual Human* [29] exemplify the high level of difficulty in creating adaptive storylines. There are different projects and initiatives targeting either at (interactive) storytelling issues as 'instrument' for virtual environments, training, and simulation or at educational games in general. However, the approaches did not converge yet and integrated solutions combining both interactive storytelling and gaming for learning and training purposes are lacking entirely. The question is how to transpose experimental approaches [29, 30] of interactive and adaptive storytelling to educational settings and educational games. From an application oriented point of view it is very exciting to combine the different disciplines, expectations, or typical workflows and analyze arising requirements and constraints for storytelling and game-based learning scenarios.

A crucial aspect of adaptive, interactive storytelling in adaptive educational games is to find an appropriate storyline on the basis of a pool of given scenes/game-based narrative learning objects. The key challenge in this context is to find a suitable and fair balance between the initially created story and 'exceptions' caused by user interactions (unforeseen or at least not intended by the author) or educationally inspired adaptations. Examples for such exceptions are wrong paths (not following the instructions of a virtual guide), skipped stations (passing artifacts without interacting), or too long/short interactions with artifacts (causing problems with external and internal time constraints). Moreover, the red thread through the story and therefore through the game must be in line with the learner's learning progress and goals. To accomplish this linkage, finding educationally meaningful, yet immersive and exciting storylines, formalisms and rules are required. This challenge is substantially more complex when focusing on collaborative learning in multiplayer environments. The following are the core constraints on which we have built rules for the adaptive story development in the 80Days' demonstrator game:

- *External constraints*: game design, learning progress, learning goal, and prior knowledge of the learner
- *Dramaturgic aspects and story models:* characteristics and heuristics of story model and narrative structures, e.g., timing of plot points, duration of story units, setting of story climax
- *Content and individual story elements:* classification of importance/weighting of individual story units for (i) the narrative, (ii) in the group context, (iii) from educational considerations

With respect to the last point and the importance of story elements from an educational and pedagogical viewpoint, [31] introduced a three level concept with a content level storing the learning units (learning objects = story objects), a learning level (where the learning objects and units are classified and assigned to didactic learning phases), and a story level (integrating the learning units into a story with plot points and an arc of suspense). On this basis it is easy to attribute individual narrative learning objects as well as a set of objects belonging to a didactic learning phase with an indicator of importance. On this basis, we introduced the notion of narrative learning objects, which are composed of

- a *narrative component*, distinguishing (i) prerequisites in the storyline (specifying which other objects must have been introduced before to understand an object), (ii) importance and weight, and (iii) dramaturgic quality (e.g., story climax), and (iv) contents of an object;
- an *educational component* on the basis of CbKST (as introduced above), distinguishing knowledge or skills that are (i) prerequisites to understand the object, (ii) that are (ii) taught be an object, and/or (iii) assessed;
- a *motivational component* specifying the motivational quality of an object;
- a *collaborative component* specifying the extent of cooperation that (i) is required or (ii) possible;

Similar to conventional adaptive tutoring systems we are now able to potentially adapt entire educational sequences to the individual learner, either in a single player or multiplayer environment. For example, learning elements/units are important for specific user groups and are skipped (to speed up a story) for other more advanced and experienced learners. In contrast, additional learning units

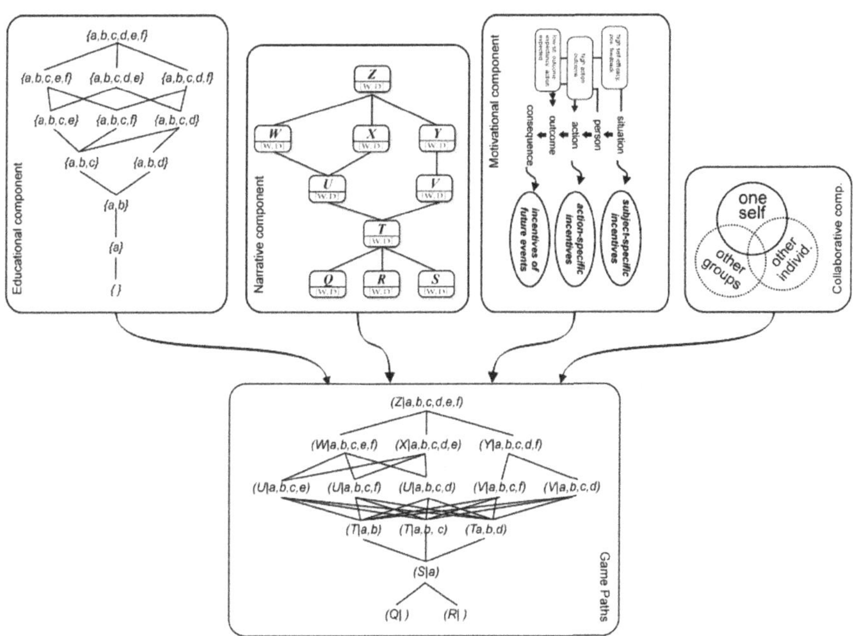

Fig. 5 Deriving game paths by merging competence spaces and story plots.

might be integrated in the course in order to 'keep fast learners busy' or provide further background information about a topic. Technically speaking, 80Days combines this story and learning by linking competence spaces with story plots (Fig 5), which generates game paths, possible and meaningful paths through the game accounting for story model, learning objectives, and pedagogical interventions. Similar to the competence-performance separation introduced in CbKST, we realize a competence-performance-story separation based on mathematical interpretation and representation functions. Therefore, from competence spaces and story plot we can derive a 'game space', the set of admissible and meaningful paths through story and game.

Collaborative Learning

As sketched above, the adaptation of educational sequences can on principle be realized for single as well as multiplayer games and thus, are applicable to both, individual and group learning scenarios. Entering the level of collaborative learning and trying to apply the principles of micro level assessment and interventions on the group level poses an additional research challenge. The principles for personalization and adaptation developed and implemented for individual learners cannot be adopted for groups of learners in an un-reflected manner. Rather, further research is needed in order to take care for the specific characteristics of collaborative learning – incorporating and synthesizing state of the art on computer supported collaborative learning, multiplayer games, group adaptation, and cognitive as well as social psychology of group learning and groups. Subsequently, first considerations on the integration of micro level assessment and adaptive interventions in educational multiplayer computer games for collaborative learning are presented, which we are going to be addressed and elaborated in our future work.

In a collaborative learning context, micro level assessment can serve different purposes. The continuous monitoring and interpretation of individual learners' behavior in the game can serve the adaptive formation of appropriate groups of learners (adaptive collaboration support, e.g. [32]). The aim hereby is to use the assessment results of different students to create a group of learners for different kinds of collaboration, like a collaborative problem solving process or finding a peer that could act as a tutor being competent on a certain topic and able to answer questions. The continuously gathered assumptions on individual learners' skills in terms of CbKST provide a detailed picture of their current knowledge and thus a sound basis for group formation at a proper moment of time. Furthermore, the non-invasive assessment can serve for establishing a group model (e.g., [33]) characterizing the competence of a group of learners. Micro level assessment is also usable for the purpose of monitoring and comparing performance of different learners (intelligent class monitoring, e.g. [32]). This can help learners to identify their position within the group and eventual knowledge gaps or leads. In addition to the consideration of individual assessment profiles for collaborative learning, the assessment procedure may be exploited to realize a continuous assessment on the group level itself.

Similar as in a scenario of individual learning, in a collaborative learning scenario the assessment results are utilized for the selection and presentation of appropriate interventions. The types of interventions presented before will in this case be chosen and adapted to the group model – e.g. provision of problem solving support if the group of learners turns out to be unable to progress in their problem solving state on a collaborative problem solving task. Moreover, there will be further, complementing types of interventions specifically suited for the context of collaborative learning, like group formation interventions suggesting to contact or consult specific peers. By a smooth integration of interventions into the game and story in this way a collaborative learning process can be enabled, embedded into a multiplayer game experience.

Conclusions and Outlook

This paper has elaborated on sound adaptation strategies and technologies for educational computer games, which allow tailoring the learning and gaming experience to the learner's needs without compromising game flow and motivation.

The aim of micro level assessment and adaptation is to enable an assessment of learning progress and motivational states in an educational game without compromising the game's flow. Moreover, the aim is to support the learner with psycho-pedagogical interventions in form of hints, suggestions, warnings, or feedback. The assessment process strongly relies on the combination of formal skill structures and problem spaces. Thus, the assessment is strongly associated with ideas of problem-based, exploratory learning.

The probabilistic assessment is not perfect, thus it is reasonable to strengthen the conclusions by "harder" test items, for example the accomplishment of a certain task in order to reach a new level of the game. On the basis of this assessment, non-invasive adaptive interventions can be triggered in order to support the learning process.

The mechanisms described here, of course require the game system to have significant understanding of the domain (the skill and skill structures), of the game (the problem spaces), the learner (the learning progress and motivational states), and the rules that glue all the information together. From a technical perspective, we realized this data storage in form of (OWL) ontologies [34]. The game system has reasoning services that query the ontologies in real time to interpret the learner's behavior. The ontologies at the same time provide a storage medium to link rules and available interventions.

In the context of 80Days and its predecessor project ELEKTRA (www.elektra-project.org) we started to conduct empirical investigations and evaluation studies to investigate the educational effectiveness of adaptive features in assessment and interventions. Early analyses revealed that adaptive features result in better learning performance and also superior gaming experience than non-adaptive control groups [6]. However, micro adaptivity is still in an early stage of research and development. The underlying framework uses some simplifying assumptions like the identity of properties and position categories and actions. Based on our experiences, the framework must be generalized within and beyond the domain of game-based learning. Future work may also address a stronger integration of meta-cognitive aspects such as confidence ratings into the assessment procedure. A strong focus of 80Days lies on complementing the micro level adaptation with a macro level, which essentially means integrating adaptive storytelling and ideas of adaptive curriculum sequencing.

So far, the focus of our work was on finding a suitable method to provide the learner with psycho-pedagogical guidance and support, strongly embedded in the game and with the major objective of not compromising the gaming experience. The basis for this attempt is an approach to non-invasively monitor and interpret the learner's behavior in the game. In a future step we will focus on multiplayer learning games, which are even more challenging in terms of assessment and interventions on group level.

Acknowledgments. The research and development introduced in this work is funded by the European Commission under the sixth framework programme in the IST research priority, contract number 027986 (ELEKTRA, www.elektra-project.org) as well as under the seventh framework programme in the ICT research priority, contract number 215918 (80Days, www.eightydays.eu).

References

[1] Merrill, M.D.: First principles of instruction. Educational Technology, Research and Development 50, 43–59 (2002)

[2] Prensky, M.: Digital game-based learning. McGraw-Hill, New York (2001)

[3] Van Eck, R.: Digital game-based learning: It's not just the digital natives who are restless. Educause Review 41, 16–30 (2006)

[4] Peirce, N., Conlan, O., Wade, V.: Adaptive Educational Games: Providing Non-invasive Personalised Learning Experiences. In: 2008 Second IEEE International Conference on Digital Games and Intelligent Toys Based Education, pp. 28–35 (2008)

[5] De Bra, P.M.E.: Adaptive hypermedia. In: Adelsberger, H.H., Kinshuk, Pawlowski, J.M., Sampson, D. (eds.) Handbook on Information Technologies for Education and Training, pp. 29–46. Springer, Berlin (2008)

[6] Kickmeier-Rust, M.D., Marte, B., Linek, S.B., Lalonde, T., Albert, D.: The effects of individualized feedback in digital educational games. In: Conolly, T., Stansfield, M. (eds.) Proceedings of the 2nd European Conference on Games Based Learning, October 16-17, pp. 227–236. Academic Publishing Limited, Barcelona (2008)

[7] Albert, D., Lukas, J.: Knowledge spaces: Theories, empirical research, and applications. Lawrence Erlbaum Associates, Mahwah (1999)

[8] Doignon, J.-P., Falmagne, J.-C.: Spaces for the assessment of knowledge. International Journal of Man-Machine Studies 23, 175–196 (1985)

[9] Newell, A.: Unified Theories of Cognition. Harvard University Press, Cambridge (1990)

[10] Augustin, T., Hockemeyer, C., Kickmeier-Rust, M.D., Albert, D.: Individualized skill assessment in educational games: Basic definitions and mathematical formalism. Submitted for Publication to IEEE Transactions on Learning Technologies (2009)

[11] Flavell, J.H.: Metacognitive aspects of problem solving. In: Resnick, L.B. (ed.) The Nature of Intelligence, pp. 231–236. Erlbaum, Hillsdale (1976)
[12] Zimmerman, B.J.: Self-regulated learning and academic achievement: An overview. Educational Psychologist 25, 3–17 (1990)
[13] Zimmerman, B.J.: Becoming a self-regulated learner: An overview. Theory into Practice 41, 64–70 (2002)
[14] Puustinen, M., Pulkkinen, L.: Models of self-regulated learning: A review. Scandinavian Journal of Edcuational Research 45, 269–286 (2001)
[15] Aviram, A., Ronen, Y., Somekh, S., Winer, A., Sarid, A.: Self-Regulated Peronalized Learning (SRPL): Developing iClass's pedagogical model. eLearning Papers (9) (2008), `http://www.elearningeuropa.info/files/media/media15974.pdf` (retrieved September 21, 2009)
[16] Doignon, J.-P., Falmagne, J.-C.: Knowledge spaces. Springer, Heidelberg (1999)
[17] Heller, J., Steiner, C., Hockemeyer, C., Albert, D.: Competence-based knowledge structures for personalised learning. International Journal on E-Learning 5, 75–88 (2006)
[18] Korossy, K.: Organizing and controlling learning processes within competence–performance structures. In: Albert, D., Lukas, J. (eds.) Knowledge Spaces: Theories, Empirical Research, and Applications, pp. 157–178. Lawrence Erlbaum Associates, Mahwah (1999)
[19] Narciss, S., Huth, K.: How to design informative tutoring feedback for multi-media learning. In: Niegemann, H.M., Leutner, D., Brunken, R. (eds.) Instructional Design for Multimedia Learning, pp. 181–195. Waxman, Munster (2004)
[20] Mattheiss, E., Kickmeier-Rust, M., Steiner, C., Albert, D.: Motivation in Game-Based Learning: It's More than, Flow'. In: Schwill, A., Apostolopoulus, N. (eds.) Lernen im Digitalen Zeitalter, Potsdam, pp. 77–84 (2009)
[21] Heckhausen, J., Heckhausen, H.: Motivation und Handeln: Einführung und Überblick [Motivation and action: Introduction and overlook]. In: Heckhausen, J., Heckhausen, H. (eds.) Motivation und Handeln [Motivation and action], pp. 1–8. Springer, Heidelberg (2006)
[22] Weiner, B.: Achievement motivation and attribution theory. General Learning Press, Morristown (1974)
[23] Bandura, A.: Self-efficacy: Toward a unifying theory of behavioral change. Psychological Review 84, 191–215 (1977)
[24] Keller, J.M.: Development and use of the ARCS model of motivational design. Journal of Instructional Development 10(3), 2–10 (1987)
[25] Rheinberg, F., Krug, S.: Motivationsförderung im Schulalltag [Motivation encouragement in school practice]. Hogrefe, Göttingen (1999)
[26] Dresel, M., Ziegler, A.: Langfristige Förderung von Fähigkeitsselbstkonzept und impliziter Fähigkeitstheorie durch computerbasiertes attributionales Feedback [Long-term encouragement of self-concept of ability and implicit ability theory through computer-based attributional feedback]. Zeitschrift für Pädagogische Psychologie 20, 49–63 (2006)
[27] Mateas, M.: Interactive drama, art and artificial intelligence. Technical Report CMU-CS-02-206. School of Computer Science, Carnegie Mellon University (2002)
[28] McKee, R.: Story: Substance, structure, style, and the principles of screenwriting. Harper Collins, New York (1997)

[29] Göbel, S., Iurgel, I., Rössler, M., Hülsken, F., Eckes, C.: Design and narrative structure for the Virtual Human scenarios. International Journal of Virtual Reality 5(3), 1–10 (2006)

[30] Crawford, C.: Chris Crawford on Game Design. New Riders Press, Indianapolis (2003)

[31] Crawford, C.: Chris Crawford on Interactive Storytelling. New Riders Press, Indianapolis (2004)

[32] Brusilovsky, P.: Adaptive and intelligent technologies for Web-based education. Künstliche Intelligenz 4, 19–25 (1999)

[33] Jameson, A., Smyth, B.: Recommendation to groups. In: Brusilovsky, P., Kobsa, A., Nejdl, W. (eds.) Adaptive Web 2007. LNCS, vol. 4321, pp. 596–627. Springer, Heidelberg (2007)

[34] Kickmeier-Rust, M.D., Albert, D.: The ELEKTRA ontology model: A learner-centered approach to resource description. In: Leung, H., Li, F., Lau, R., Li, Q. (eds.) ICWL 2007. LNCS, vol. 4823, pp. 78–89. Springer, Heidelberg (2008)

A Collaborative and Adaptive Design Pattern of the Jigsaw Method within Learning Design-Based E-Learning Systems

Maria Kordaki and Haris Siempos

Department of Computer Engineering and Informatics, University of Patras,
26500, Rion Patras, Greece
kordaki@cti.gr, siempos@ceid.upatras.gr

Abstract. This chapter presents an innovative description of the Jigsaw collaboration method, in the form of an online, adaptive collaborative design-pattern that has been constructed taking into account adaptation techniques, within the context of open-source learning design-based environments such as LAMS. This method is described with special reference to the learning of essential issues in Computer Science and especially in the area of programming languages. These issues include an understanding of: (a) basic elements of structured programming languages, (b) the rapid evolution of the area of programming languages, (c) the learning of programming languages' levels and techniques. The innovative description of the Jigsaw collaborative method within LAMS is based on the fact that: (a) the tasks assigned to the expert groups consist of investigation of real world scenarios and not merely the study of learning material as is usually proposed, (b) adaptive techniques are integrated with the method and (c) for the design of the collaborative learning activity, an intuitive learning design tool like LAMS is used.

1 Introduction

Nowadays, e-learning is widely accepted as a promising approach in education, one which encourages new forms of learning, performs various innovative types of interactions, and enjoys virtual communication and collaboration all over the world while providing flexible opportunities for learners to overcome time and space constraints in terms of their learning (Harasim, Hiltz, Teles & Turoff, 1995; Van Eijl & Pilot, 2003; Pallof & Pratt, 2004; Roberts, 2005; Diggelen and Overdijk, 2009). Most significantly, e-learning demands careful planning of all learning activities considered necessary during a lesson or a course. Indeed, within e-learning contexts, teaching is performed as a conscious and carefully-planned procedure, not as a spontaneous activity.

With learning being a subjective activity (von Glasersfend, 1990), in terms of individual learner differences in knowledge, skills, goals and preferences, each individual learner needs specific support within e-learning environments. Learners

T. Daradoumis et al. (Eds.): Technology-Enhanced Systems and Tools, SCI 350, pp. 239–255.
springerlink.com

could therefore be better assisted in understanding the learning concepts in question -when e-learning is coupled with adaptation techniques- than when they participate in e-learning systems providing learning opportunities that do not take into account the learner's individual characteristics (Brusilofsky, et al., 1996). In fact, the use of adaptation techniques within e-learning systems can support each individual learner, taking into account some of their individual characteristics, .i.e.: learning styles, background knowledge including her/his misconceptions, experience with the knowledge in question, goals and preferences. In general, the architecture of adaptive e-learning systems consists of the 'learners' model', the 'subject matter model' - or expert model - and the 'learning model'. The latter contains the previously mentioned individual characteristics for each learner, the subject-matter model contains the knowledge viewed appropriate for learning by students and the learning model consisting of the pedagogical methods –including adaptation strategies- proposed as appropriate for the learning of the subject matter, e.g. the use of specific collaboration strategies.

Research in e-learning shows that involving learners in online collaborative learning activities could provide them with essential opportunities to,: motivate active engagement in their learning (Scardamalia & Bereiter, 1996), extend and deepen their learning experiences, try new ideas and improve their learning outcomes (Picciano, 2002; Pallof & Pratt, 2004), trigger their cognitive processes (Dillenbourg, 1999), enhance their diversity in terms of the learning concepts in question (Johnson and Johnson, 1994) and interact socially,developing a sense of community and belonging online (Haythornthwaite, Kazmer, Robins, & Shoemaker, 2000). Although. on the whole, computer-supported collaborative learning has been recognized as an emerging paradigm of educational technology (Kosschmann, 1996),many teachers remain unsure of why, when, and how to integrate collaboration into their teaching practices in general, let alone their online classes (Panitz, 1997; Brufee, 1999).

At this point, it is worth differentiating collaborative from cooperative learning situations. In cooperative settings, the task is split into subtasks and each participant is responsible for solving a portion of the problem at hand, whereas in collaborative situations, participants are mutually involved in shared activities and must coordinate their efforts if they are to solve problems together. In cooperative settings, learners usually produce separate solutions, while in collaborative learning, it is essential to construct a shared solution (Liponen, 2002). So as to encourage teams to achieve effective collaboration, some structuring may be necessary (Lehtinen, 2003; Lipponen, 2002), for example through the use of computer-supported collaborative design patterns. A pattern is seen as something that will not be reused directly but can aid the informed teacher in building up their own range of tasks, tools or materials drawing on a collected body of experience (McAndrew, Goodyear & Dalziel, 2006).

The concept of specific collaborative patterns could be well integrated into 'learning design'-based e-learning environments, with a 'learning design' defined as the description of the teaching-learning process that takes place in a unit of learning, e.g. a course, a lesson or any other designed learning event, such as a specific collaboration structure (Koper & Tattersall, 2005). An important aspect of this is that pedagogy is conceptually abstracted from context and content, so that excellent pedagogical models can be shared and reused across instructional contexts and subject domains. Specifically, best pedagogical practices can be reflected in the formation of context-free 'design patterns' which could be shared and reused across instructional contexts and essentially assist online learning. The key principle in 'learning design' is that it represents the learning activities that need to be performed by learners and teachers within the context of a unit of learning. Within the context of "learning design', the role of collaborative design patterns is to indicate clearly the flow of collaboration activities using specific collaboration structures.

The IMS Learning Design (LD) specification aims to represent the design of units of learning in a semantic, formal and machine-interpretable way (LD, 2003). Although various examples of e-learning environments close to the LD specification have appeared in the literature, authoring using LD is not a simple task for teachers, their being unfamiliar with both the use of the tools provided and the underlying concepts of the LD modeling language to be taken into account when planning educational activities. Nevertheless, involving teachers in not only the implementation but also the design of their teaching sessions is thought to be essential (Griffiths and Blat, 2005). To this end, the essential role of suitably-designed tools in supporting teachers in their mindful and appropriate 'learning design' has been acknowledged by many researchers (Lloyd & Wilson, 2001; Babiuk, 2005; Kordaki & Daradoumis, 2009). It would appear clear that teachers need high level tools in order to understand learning design and it is likely that specialized tools for a particular pedagogic context would be easier to use (Griffiths & Blat, 2005). To this end, it is important to note that the type of editor usually required by classroom teachers should be similar to the authoring environment provided by LAMS (Dalziel, 2003), a well-known integrated e-learning system that effectively supports the idea of 'learning design' and whose tools have recently been used to construct a number of collaboration design patterns (Kordaki & Siempos, 2009; 2010; Kordaki, Siempos & Daradoumis, 2009).

The teaching of programming is one of the main challenges in Computer Science, despite it being not only an essential topic proposed for a K-12 curriculum, and a fundamental subject in studying Computing at Tertiary level, but also a 'mental tool' of general interest (Dagdilelis, Satratzemi & Evaggelidis, 2002) where problem-solving skills can be encouraged in learners. In truth, programming is more a mental skill than a body of knowledge (Hadjerrouit, 2008). It is a complex task, including understanding of the task at hand, method finding, coding, testing and debugging of the resulting program (Brooks, 1999). It is essential to note here that students encounter serious difficulties in performing any or all of the above (Allwood, 1986; Du Boulay, 1986; Soloway and Spohrer, 1989; Winslow, 1996; Lemone & Ching, 1996; Christiaen, 1998; Robbins, Rountree and Rountree, 2003; Hadjerrouit, 2008; Pacheco, Gomes, Henriques, de Almeida, Mendes, 2008). It is also worth noting that, ever since the early 1970s, there has been strong interest in the adoption of effective methods to improve the ability of students to comprehend and solve computational problems requiring programming solutions (Dijkstra, 1969; Gries, 1974; Soloway and Spohrer, 1988; Wegner et al.,

1996; McCracken et al., 2001; Robins et al., 2003). In light of this, a framework of collaborative activities is proposed that can be applied to the vast majority of programming languages. To ensure greater effectiveness, it is proposed that a programming language supported by a web compiler could be used.

Taking into account all the above, we have attempted to form the 'Jigsaw' collaborative method (Aronson, 1971; Aronson, Blaney, Sikes, Stephan & Snapp, 1978) as an adaptive collaborative design pattern within the context of LAMS to construct a sequence of learning activities for the learning of essential issues in Programming such as: (a) basic elements of structured programming languages, (b) the rapid evolution of the area of programming languages, (c) the learning of programming languages' levels and techniques. Such a sequence of online, adaptive and collaborative learning activities for the learning of Programming- using the Jigsaw method within LAMS - has not yet been reported.

The essential features of LAMS are briefly presented in the following section of this paper, followed by a description of the Jigsaw collaboration method. Next, a sequence of online, adaptive, collaborative learning activities using Jigsaw-within-LAMS with special reference to the aforementioned issues of learning programming is demonstrated. In closing, the design of this sequence is discussed and conclusions and future research plans are drawn.

2 The Rationale

2.1 LAMS

LAMS (Learning Activity Management System; http://www.lamsfoundation. org/) is an open-source tool for designing, managing and delivering online collaborative learning activities. In fact, LAMS offers a set of predefined learning activities, shown in a manner comprehensible to teachers that can be graphically dragged and dropped in order to establish a flow chart of sequence of activities. When using LAMS, teachers gain access to a highly intuitive, visual authoring environment for the creation of sequential learning activities. These activities may be individual tasks, small group work or whole class activities. LAMS is based on the belief that learning does not arise simply from interacting with content but from interacting with teachers and peers. The creation of content-based, self–paced learning objectives for single learners is now well-understood in the field of e-learning. However, the creation of sequential learning activities which involve groups of learners interacting within a structured set of collaborative environments - referred to as 'learning design' - is less common; LAMS allows teachers both to create and to deliver such sequences. In essence, LAMS provides a practical way to describe multi-learner activity sequences and the tools required to support these. Furthermore, LAMS provides tools that support various activities such as communication, presentation of information, writing and sharing resources, as well as posing and answering questions.

LAMS also offers to the designers of educational activities specific tools that support grouping and conditional branching. In fact, the grouping can be random or it will be based on learner's choice or author's choice. Additionally, the students can be directed in different sequences of activities depending on the group they belong (grouped branching) or based on what the learner has contributed in a specific activity (tool output branching). LAMS can make branching decisions based on criteria like the number of correct answers in a questionnaire, the certain words that a learner has or has not typed into an activity like chat, forum or survey. In any case, the author of the learning activity can assign students manually in any branch he likes.Nevertheless, Dalziel (2003) has commented on the absence of tools supporting broader ranges of collaborative tasks. In fact, despite the availability of all the tools mentioned above, sequences of learning activities for the performance of the Jigsaw collaboration method within LAMS –using adaptation techniques- for the learning of specific CS concepts have not yet reported.

The said sequence of collaborative activities was implemented using specific tools provided by LAMS: http://wiki.lamsfoundation.org/display/lamsdocs/Home. These tools are demonstrated in its interface and are briefly presented below:

The *Assessment tool* allows authors to create a series of questions with a high degree of flexibility in total weighting

The *Chat Activity* runs a live (synchronous) discussion for learners

The *Chat and Scribe Activity* combines a *Chat* Activity with a *Scribe Activity* for collating the chat group's views on questions posed by the teacher

The *Forum Activity* provides an asynchronous discussion environment for learners, with discussion threads initially created by the teacher

The *Forum and Scribe Activity* combines a *Forum* Activity with a *Scribe Activity* for collating Forum Postings into a written report

The *Mindmap activity* allows teachers and learners to create, edit and view mindmaps in the LAMS environment. Mindmaps allow for the organizing of concepts and ideas, and exploring how these interact

The *Multiple Choice* activity allows teachers to create simple automated assessment questions, including multiple choice and true/false questions

The *Notebook Activity* is a tool for learners to record their thoughts during a sequence of activities

The *Noticeboard Activity* provides a simple way to supply learners with information and content. The activity can display text, images, links and other HTML content.

The *Question and Answer Activity* allows teachers to pose a question or questions to learners individually and, after they have entered their response, to see the responses of all their peers presented on a single answer screen.

The *Share Resources tool* allows teachers to add content to a sequence, such as URL hyperlinks, zipped websites, individual files and even complete learning objects.

The *Submit Files Activity* allows learners to submit one or more files to the LAMS server for review by a teacher.

The *Survey Tool* presents learners with a number of questions and collects their responses. However, unlike Multiple Choice, there are no right or wrong answers.

The *Wiki Tool* allows authors to create content pages that can link to each other and, optionally, allow learners to make collaborative edits to the content provided.

2.2 The Jigsaw Collaborative Method

The *Jigsaw* method was originally proposed by E. Aronson (1971) at the University of Texas and the University of California. Hundreds of schools have employed Jigsaw-based activities in their classrooms with much success (see http://www.jigsaw.org). Jigsaw has been seen as a method that can support both cooperative learning (Johnson & Johnson, 1992) and collaborative situations (Silverman, 1995). Gallardo (2003) also thought that this method could sit well within the constructivist framework of learning. In addition, many researchers have proposed the implementation of this method within the online context (Gallardo et al. 2003; Hernandez-leo et all; Kordaki, Siempos and Daradoumis, 2009; Kordaki and Siempos, 2010), despite the fact that Jigsaw was originally proposed for face-to-face education (Aronson & Patnoe, 1997). The *Jigsaw* method is a co-operative/collaborative learning strategy which enhances the process of listening, commitment to the team, interdependence and team work.

Each member of the team has to excel in a well-defined subpart of the educational material, undertaking the role of expert. The experts form a different group to discuss the nuances of the subject and later return to their teams to teach their colleagues. The ideal size of teams is 4 to 6 members. Specifically, the implementation of the Jigsaw method could be realized through the following *process*: 1) Divide the problem into sub-problems, 2) Create heterogeneous groups, 3) Assign roles and material to each student, 4) Form a group of experts, 5) Let experts study the material and plan how to teach their colleagues, 6) Let experts teach in their groups, 7) Assess students.

Through Jigsaw, the following goals could be achieved: 1) Building of interpersonal and interactive skills, 2) Ensuring that learning revolves around interaction with peers, 3) Holding students accountable among their peers, 4) Encouraging active student participation in the learning process.

In the next section of this paper, the set of collaborative learning activities for the learning of the aforementioned essential issues in CS using the Jigsaw-within-LAMS design pattern is reported.

2.3 Adaptation

For the design of an adaptive system, a four-stage process has been proposed (Brusilovsky, 2003): (i) design of the 'knowledge-base', including a hierarchy of learning goals and specific learning topics, (ii) design of the 'learner's model', including her/his individual learning characteristics and preferences. In fact, the learner's profile in terms of her/his knowledge-background and experience, goals, preferences and learning style must be investigated. To this end, the learner's knowledge has to be diagnosed before they can be characterized as novices, intermediate or experts. The learners' background and experience with regard to the knowledge in question is useful to explore, because learners with different backgrounds need different treatment by the system. Learner's goals should also be examined. For example, the goal of some learners may be to acquire some information about the learning topics in focus, while for other learners it is probably to gain knowledge that will facilitate problem solving. Exploration of learners' preferences is also significant, as the basic learning style of each individual learner plays a basic role in finding ways to support them in their learning. In terms of learners' learning styles, various classifications have been proposed. Some important classifications view learners as holistic-analytic and verbalizers-imagers (Riding and Cheema, 1991), some others separate learners as field-dependent (F/D) and field-independent (F/I; Witkin, et al., 1977), while other classifications sort learners into activists, pragmatists, reflectors and theorists (Honey and Mumford, 1992). The exploration of these individual characteristics is usually performed through a questionnaire that could be completed just after their entrance into the e-learning system. (iii) design of the 'media space', including various materials and topics which are interconnected with the topics included in the previously mentioned knowledge-base, and (iv) design of the 'adaptation model', including the rules for the selection of appropriate topics - from both the knowledge-base and the media space - taking into account each learner's individual characteristics as these emerged from the relevant 'learner model'.

To this end, various ways of adaptation could be used by the system to support learners in their learning, namely: (a) *adaptive sequencing curriculum*, where sequences of educational materials are formed and proposed to the learner by the system according to her/his individual characteristics (Brusilofsky & Pesin, 1994), (b) *adaptive presentation*, and *adaptive navigation* techniques (Kay and Kummerfeld, 1997; De Bra & Calvi, 1998). These techniques are usually proposed for the design of adaptive hypermedia educational materials, where sequences of web pages are created and the adaptation could be implied at both content level and link level, (c) *problem solving support*. Here, too, three modes of support have been reported: (i) intelligent solution analysis, where the ideal solution of a problem is compared with the solution provided by the learner and appropriate feedback is given by the system - after the problem-solving process has been completed - regarding her/his mistakes, (ii) interactive problem solving support. Here, the system monitors the learner's problem-solving path and provides appropriate feedback during the problem solving process, and (iii) example-based problem solving (Brusilofsky, 1996), where a repository of examples regarding the problem solving in focus is provided by the system, to support each learner's problem-solving actions, and (d) *collaboration support*, where the system can use the learners' personal characteristics to support the creation of appropriate groups for collaboration and communication in order to face suitable learning activities (Brusilofsky, 1999).

Adaptive techniques are also useful for the design of tests used for the assessment of learners' knowledge throughout a learning experiment. Such tests are generated by the system according to each individual learner's knowledge. For example, when a learner successfully answers a set of questions - appropriate for the assessment of a piece of knowledge - then the system provides questions aiming to assess another, probably more complicated, piece of knowledge. Contrarily, when a learner does not succeed in answering a set of questions, the system provides her/him with easier questions and various kinds of help.

In the next section of this chapter, the design of the Jigsaw adaptive online activity is presented.

3 Design of the Jigsaw Adaptive Online Learning Activity

The design of the adaptive online learning activity consists of the following phases: 1) Introduction to the learning activity, 2) Formation of the original groups, 3) Formation of the expert-groups, 4) Return to the original groups, 5) Group report submission, 6) Group report presentation, 7) Assessment. The implementation of these phases within the context of LAMS is diagrammatically represented - as an 'adaptive design pattern' - in Figure 1. The proposed activity can be used in environments of synchronous and asynchronous collaborative learning. The only necessary modification is the replacement of the "chat" tool with the "forum" tool and vice versa.

Phase 1. Introduction to the learning activity

The main aim of this educational activity is encouragement of learners, through their interaction within an adaptive collaborative learning environment, to explore fundamental issues concerning Computer Science, and especially programming languages. Intermediate goals are: 1) the comprehension of basic elements of procedural programming languages such as C and Pascal, 2) the learner's familiarization with the issues of programming levels and techniques, e.g. visual programming, and 3) the ascertainment of the rapid evolution of Computer Science through the study of programming languages development. Additionally, through the learners' efforts to fulfill the educational objectives, some secondary skills are developed, e.g. a) the practice of word processing and presentation software, and b) the practice of web searching techniques. The learning activity aims to highlight the value of collaborative learning as a modern method of teaching.

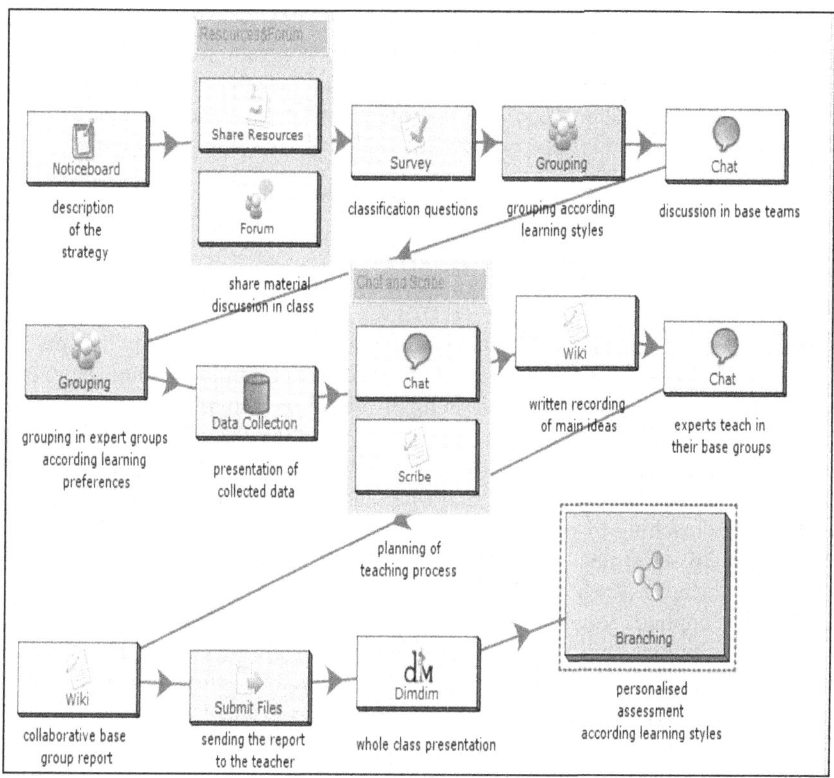

Fig. 1 A diagrammatic representation of the Jigsaw, online adaptive collaborative design pattern within LAMS

The proposed design utilizes the tool of 'Survey' for the investigation of the students' main learning styles. Next, the teacher has to group the students in base groups according to their main learning styles as it emerged from their answers in this survey. In addition, the teacher has to group students in expert groups according to their learning preferences as these can also be diagnosed through the related answers given to the aforementioned survey. Here, it is worth noting that, the work of grouping was assigned to the teacher, since LAMS does not support effectively tool output grouping capabilities. In fact, there is an available option of tool output branching but the routing of students in different branches would have made difficult to group them since the created groups couldn't be related with the initial sequence. This lack of connection between the base and experts group would have made impossible to apply the Jigsaw collaborative learning scenario. Another reason for the decision to involve the teacher in student's grouping relies on the fact that, in the case of the possible presence of more than four students with the same learning style, it is necessary to encharge the teacher the task to decide in which group the redundant students have to be assigned.

In this learning context, learners must study issues concerning the fundamental structures of programming languages, the levels of programming languages, the recent approaches of programming techniques and the main advantages and disadvantages of each approach. According to this classification, the learners can be separated into four expert groups, which will study the aforementioned issues:

a) Levels of programming languages Jigsaw group
b) Programming techniques Jigsaw Group 1
c) Programming techniques Jigsaw Group 2
d) The fundamental algorithmic structures Jigsaw Group

The usual process of data collection in school environments is the study of a given literature or web searching. To perform their duties successfully, students can collect data from various and significant areas of life where programming languages are used, such as: (a) a financial organization, e.g a bank (b) a web technologies software company, (c) a mobile communications company, (d) a university lab specializing in artificial intelligence, and e) a university department observing the structure of the teaching of programming languages. In fact, Pragmatists can learn through practical activities – e.g. in a mobile communications company and in a financial organization - where they can observe how programming theories and techniques are applied; activists can gain knowledge by being involved in interactive learning activities (eg. in a software company), reflectors can learn through various programming examples – e.g. reflecting in their experience within a university programming lab - and theorists can absorb knowledge through exploring theoretical materials available in a university Computer Science department. Other appropriate types of content could also be created for online study.

In this phase of the Jigsaw activity, students are informed - using a Notice board - about the whole context of the activity, including its aims and the specific issues of programming languages that have to be studied during this activity as well as the various places where they could collect appropriate data. Students should exchange ideas on the whole procedure of the activity using a whole-class Forum or a whole class Chat-room.

Phase 2. Jigsaw: *Original group creation*

The students are assigned a questionnaire, specially designed by the teacher, to explore their personal learning characteristics in terms of: learning styles, background knowledge and experience regarding programming as well as their goals and preferences. Based on students' preferences and learning styles, the system can propose they form 4 groups of 4 students – by appropriately using the question tool in combination with the branching tool and the Grouping tool- so that each group will consist of students who prefer to study all the issues mentioned in the previous section and share the same learning style. Ideally, the members of each original group should have the same main learning style. In fact, all members of an original group should have different preferences but the same main learning styles. Initially, each original group– using a group chat-room or a group forum - discusses the issues presented in the introduction, striving to form a commonly acceptable framework of ideas.

Phase 3. Jigsaw: *Creation of expert groups*
Next, every member of each group gains expertise in a specific issue of the proposed learning activity through their participation in specific expert-groups. The formation of the expert groups could be supported by the system –using the grouping tool in combination with the branching tool- which already knows the students' preferences in terms of the aforementioned learning issues for programming. In case of dispute, the students who prefer to be in another group could ask the educator to assign them accordingly. It is worth mentioning that each expert group consists of students of different learning styles, as it contains one member from each original group. Each expert group must visit the specific areas of life mentioned in the 'Introduction' to the activity, where programming languages are used, to collect specific data. The system can advise the students of each expert group on the selection of these areas by using the data referring to their main learning styles. Each expert group has to fulfill a well-defined task, as described in the next section.

The programming language levels Jigsaw Group has to note how the programming languages are grouped into categories according to the level of distraction in relation to details of the computer architecture. It is proposed that the historical evolution of the programming languages and the circumstances that led to the need for multiple levels to exist be studied. Workspaces such as companies in the computer architecture or telecommunications systems fields are suitable for information search.

The programming techniques Jigsaw Group 1 will cope with the presentation of the two main categories of programming techniques: procedural programming and object-oriented programming. Issues like variables, scope, functions, objects and classes can be studied, in combination with comprehension of the advantages and disadvantages of each approach. A visit to an IT company can help students to clarify the differences between these programming techniques.

The programming techniques Jigsaw Group 2 should study less known programming techniques like functional, logic or visual programming. Since these techniques are not used widely in the software industry, the students should find more information in an academic environment where research for subjects like artificial intelligence or theoretical computer science is taking place.

The algorithmic structures Jigsaw Group should deal with the key ideas of algorithms and data structures. An interview with a software engineer could help students to clarify many questions regarding the inner details of software design.

The data collected by each expert group should be categorized using specific criteria and questions they themselves have formed and those suggested by their teacher. Here, the use of the 'Data collection" tool will be useful. To this end, appropriate learning materials can be used for further understanding of the experimental activity of each expert group.

Besides data collection and processing, the expert groups have to organize an interesting and efficient teaching process to share their knowledge and experiences in base groups. The exchange of ideas and proposal for the teaching process to be followed can be supported by the use of chat and forums tools. There follows a template of possible actions that can be followed by the expert group students:

1) They should try as much as possible to comprehend the deeper meaning of the data they have collected and the materials they have studied. If necessary, they could ask their teacher for help.

2) It is important to emphasize the value of commenting on the key ideas of each specific issue at hand.

3) They should research alternative and interesting learning scenarios in order to provide a pleasant teaching experience for their colleagues. To this end, the teaching process can comprise a variety of learning representations: e.g., photographs, videos, simulations, charts, Power-Point presentations etc. This variation in learning materials is very important because the members of each group have to teach all the original groups (each member of an expert group teaches her/his original group), which consists of students of different learning styles. Different types of learning activities should also be designed to teach the participants in the original groups, taking into account their different learning styles. The experts should also not forget the importance of stimulating their colleagues' interest and motivating them to participate in a constructive thinking process, the result of cultivating discussion with the other students.

4) Using a wiki, they should provide their colleagues - in their original groups - with appropriate presentations and activities that could help them to absorb and better comprehend the knowledge offered.

5) Using a wiki, they should concentrate on the knowledge acquired during their experimentation to design a representative questionnaire reflecting the critical - and not the memorizing - skills of learners.

Phase 4. Jigsaw: ***Back to the original group***

Each expert, on returning to their original group, should propose alternative ways to present the knowledge they acquired during their participation in the experimentation performed within a specific expert group. Here, the members of the original groups could be provided with essential activities, so that every student can participate actively in their learning experience. Each expert should also encourage their colleagues to better comprehend the knowledge provided. Chatrooms or forums could be used by each expert to teach their original groups.

Phase 5. Jigsaw: ***Group Report formation***

Each group has to prepare a presentation about the total knowledge acquired during their learning process. To form this report, the use of a wiki would be useful. The use of the 'Submit Files' activity could be used to send the reports to the teachers.

Phase 6. Jigsaw: ***Group Report presentation***

Here, it would be useful to provide students with some recommendations as to how to prepare and deliver a good presentation. Some useful guidelines for the former are: (a) The presentation must begin with the main idea of the subject, (b) only the key points of the subject have to be presented, (c) On every slide, only 4-5 key points should be presented, (d) A uniform style of presentation must be followed (unnecessary effects must be avoided since these distract the learner from the key concepts), (e) The duration of each presentation should be around 10 minutes (for synchronous presentation using a chat-room) since there is always the danger the students may get bored. There will be additional time to further discuss the learning material.

Some essential guidelines that can be given to students about their actual online presentation are: (a) Students have to be careful not to overstep the time limit given, (b) The presentation slides are a reference for further development of the subject and not a paper for reading, (c) It is advisable to prepare the presentation before their group, in order to evaluate the time needed and obtain experience in speaking in public, (d) It is very important to keep a steady pace in presentation; the audience is not as well informed as they are. (e) It is better to give less information well-presented than large amounts that are incomprehensible.

Online presentations could also be performed by each group, using a whole-class chat or forum. During the online presentation, the teacher can initiate a 'question and answer' session to encourage experts to present their area of study in greater detail.

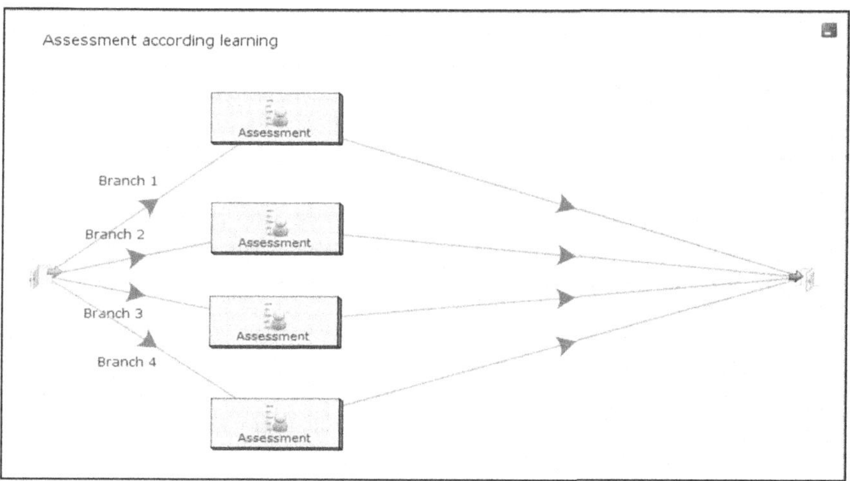

Fig. 2. A diagrammatic representation of the assessment phase in Jigsaw adaptive collaborative pattern within LAMS

Phase 7. Jigsaw: *Assessment*

In this final phase, each student should be set an adaptive quiz - once the learning activity is concluded - for purposes of assessment. The students cannot help each other during the testing process. The educator could use the tool "assessment" in combination with branching techniques - based on tool output branching capabilities of LAMS tools - to design suitable questions for students of different levels of knowledge and of different learning goals. A diagrammatic representation of the assessment phase within the Jigsaw adaptive collaborative pattern within LAMS is presented in Figure 2. The question types can be multiple choice, true or false and open types. The assessment may include the tracking of errors in source code or the execution of a given program in a web compiler.

4 Summary and Future Research Plans

This chapter proposed an innovative approach to the Jigsaw collaborative method taking into account adaptation techniques, in the context of online learning-design based learning. In fact, an adaptive online collaborative design pattern of the Jigsaw method has been formed within LAMS, a well-known open source learning-design based system. The design of this pattern was also presented through a specific collaborative Jigsaw-activity for the learning of essential issues in the area of programming languages. Basic educational goals of this activity include: (a) an understanding of basic elements of structured programming languages, (b) the rapid evolution of the area of programming languages, (c) the learning of programming languages' levels and techniques. The innovative description of the adaptive online Jigsaw collaborative method within LAMS is based on the fact that: (a) the activity takes place in the online context (b) the tasks assigned to expert groups consist of the investigation of real world scenarios and not merely the study of learning material as is usually proposed, (c) adaptive techniques are integrated with the method and (d) for the design of the collaborative learning activity, an intuitive learning design tool like LAMS is used. For the evaluation of the proposed adaptive, online, investigative, collaborative Jigsaw design pattern, further research with real online learners is necessary, and this forms the basis for our future research plans. Finally, it is also proposed to develop tool output grouping capabilities in LAMS learning management system. This development could transform LAMS in a platform with rich adaptive capabilities.

References

Allwood, C.: Novices on the computer: a review of the literature. International Journal of Man-Machine Studies 25, 633–658 (1986)

Aronson, E.: History of the Jigsaw Classroom (1971), `http://www.jigsaw.org/history.htm` (retrived from The Jigsaw Classroom)

Aronson, E., Patnoe, S.: The jigsaw classroom: Building cooperation in the classroom. Longman, NY (1997)

Babiuk, G.: A Full Bag of "Tech Tools" enhances the reflective process in Teacher Education. In: Crawford, C., et al. (eds.) Proceedings of Society for Information Technology and Teacher Education International Conference 2005, pp. 1873–1877. AACE, Chesapeake (2005)

Brooks, R.: Towards a Theory of the Cognitive processes in Computer Programming. Int. J. Human-Computer Studies 51, 197–211 (1999)

Brusilovsky, P.: Methods and techniques of adaptive hypermedia. User Modeling and User-Adapted Interaction 6(2-3), 87–129 (1996)

Brusilovsky, P., Pesin, L.: An intelligent learning environment for CDS/ISIS users. In: Levonen, J.J., Tukianinen, M.T. (eds.) Proc. of the Interdisciplinary Workshop on Complex Learning in Computer Environments (CLCE 1994), Joensuu, Finland, May 16-19, pp. 29–33 (1994)

Brusilovsky, P.: Adaptive and Intelligent Technologies for Web-based Education. Künstliche Intelligenz (4), 19–25 (1999), http://www2.sis.pitt.edu/~peterb/papers/KIreview.html

Brusilovsky, P.: Developing adaptive educational hypermedia systems: from design models to authoring tools. In: Murray, T., Blessing, S., Ainsworth, S. (eds.) Authoring Tools for Advanced Technology Learning Environment, pp. 377–409. Kluwer Academic Publishers, Dordrecht (2003)

Brufee, K.A.: Collaborative Learning: Higher Education Interdependence, the authority of knowledge. The John Hopkins University, Baltimore (1999)

Christiaen, H.: Novice Programming Errors: Misconceptions or Mispresentations? SIGCSE Bulletin 20(3), 5–7 (1998)

Dagdilelis, V., Satratzemi, M., Evangelidis, G.: Introducing secondary education to algorithms and programming. Education and Information Technologies 9(2), 159–173 (2004)

Dalziel, J.: Implementing Learning Design: The Learning Activity Management System (LAMS). In: Proceedings ASCILITE 2003, Adelaide, Interact, Integrate, Impact, December 7-10, pp. 593–596 (2003), http://www.ascilite.org.au/conferences/adelaide03/docs/pdf/593.pdf

De Bra, P., Calvi, L.: AHA! An open Adaptive Hypermedia Architecture. The New Review of Hypermedia and Multimedia, 115–139 (1998)

Diggelen, W.V., Overdijk, M.: Grounded design: Design patterns as the link between theory and practice. Computers in Human Behavior (2009), doi:10.1016/ j.chb, 01.005

Dijkstra: On the cruelty of really teaching computing science. Communication of the ACM 32(12), 1398–1404 (1989)

Dillenbourg, P.: Introduction: What do you mean by collaborative learning? In: Dillenbourg, P. (ed.) Collaborative Learning: Cognitive and Computational Approaches, pp. 1–19. Pergamon, Oxford (1999)

Du Boulay, B.: Some difficulties in Learning to Program. Journal of Educational Computing Research 2(1), 57–73 (1986)

Gallardo, T., Guerrero, L.A., Collazos, C., Pino, J.A., Ochoa, S.: Supporting JIGSAW-type Collaborative Learning. In: System Sciences, Hawaii, p. 8 (2003)

Gries, D.: What should we teach in an introductory programming course? In: Proceedings of the Fourth SIGCSE Technical Symposium on Computer Science Education, Detroit, Michigan, USA, pp. 81–89 (1974)

Griffiths, D., Blat, J.: The role of teachers in editing and authoring Units of Learning using IMS Learning Design. Advanced Technology for Learning 2(4) (2005), http://www.actapress.com/Content_Of_Journal.aspx?JournalID=63 (retrieved June 10, 2009)

Hadjerrouit, S.: Towards a Blended Learning Model for Teaching and Learning Computer Programming: A Case Study. Informatics in Education 7(2), 181–210 (2008)

Harasim, L., Hiltz, S.R., Teles, L., Turoff, M.: Learning Networks: a field guide to Teaching and Learning Online. MIT Press, Cambridge (1995)

Haythornthwaite, C., Kazmer, M.M., Robins, J., Shoemaker, S.: Community development among distance learners: temporal and technological dimensions. Journal of Computer-Mediated Communication 6(1) (2000), http://www.ascusc.org/jcmc/vol6/issue1/haythornthwaite.html

Hernández-Leo, D., Villasclaras-Fernández, E.D., Asensio-Pérez, J.I., Dimitriadis, Y., Jorrín-Abellán, I.M., Ruiz-Requies, I., Rubia-Avi, B.: COLLAGE: A collaborative Learning Design editor based on patterns. Educational Technology & Society 9(1), 58–71 (2006)

Honey, P., Mumford, A.: The manual of learning styles. Peter Honey, Maidenhead (1992)

Johnson, D.W., Johnson, R.T.: Learning together and alone: Cooperative, competitive, and individualistic learning, 3rd edn. Allyn and Bacon, Boston (1999)

Kay, J., Kummerfeld, B.: User models for customized hypertext. In: Nicholas, C., Mayfield, J. (eds.) Intelligent Hypertext. LNCS, vol. 1326, pp. 47–69. Springer, Heidelberg (1997)

Koper, R., Tattersall, C.: Learning Design: A handbook on modeling and delivering networked education and training. Springer, Berlin (2005)

Kordaki, M., Daradoumis, T.: Critical Thinking as a Framework for Structuring Synchronous and Asynchronous Communication within Learning Design-Based E-Learning Systems. In: Daradoumis, T., Caballé, S., Marquès, J.M., Xhafa, F. (eds.) Intelligent Collaborative e-Learning Systems and Applications. SCI, vol. 246, pp. 83–98. Springer, Heidelberg (2009)

Kordaki, M., Siempos, H.: Encouraging collaboration within learning design-based open source e-learning systems. In: Dron, J., Bastiaens, T., Xin, C. (eds.) Proceedings of World Conference on E-Learning in Corporate, Government, Healthcare & Higher Education (E-Learn 2007), Vancouver, Canada, USA, October 26-30, pp. 1716–1723. AACE, Chesapeake (2009)

Kordaki, M., Siempos, H.: The JiGSAW Collaborative Method within the online computer science classroom. In: 2nd International Conference on Computer Supported Education, Valencia, Spain, April 7-10 (2010) (accepted)

Kordaki, M., Siempos, H., Daradoumis, T.: Collaborative learning design within open source e-learning systems: lessons learned from an empirical study. In: Magoulas, G. (ed.) E-Infrastructures and Technologies for Lifelong Learning: Next Generation Environments. IDEA-Group Publishing, USA (2009) (to appear)

Koschmann, T.: CSCL: Theory and practice of an emerging paradigm. LEA, Mahwah (1996)

LD. IMS Learning Design. Information Model, Best Practice and Implementation Guide, Version 1.0 Final Specification IMS Global Learning Consortium Inc. (2003), http://www.imsglobal.org/learningdesign/ (retrieved February 29, 2010)

Lehtinen, E.: Computer-supported collaborative learning: an approach to powerful learning environments. In: de Corte, E., Verschaffel, L., Entwistle, N., van Merrieboer, J. (eds.) Powerful Learning Environments: Unravelling Basic Components and Dimensions, pp. 35–54. Pergamon, Amsterdam (2003)

Lemone, K.A., Ching, W.: Easing into C: Experiences with RoBOTL. SIGCSE Bulletin 28(4), 45–49 (1996)

Lipponen, L.: Exploring foundations for computer-supported collaborative learning. In: Stahl, G. (ed.) Proceedings of the Computer-supported Collaborative Learning 2002 Conference, Computer Support for Collaborative Learning: Foundations for a CSCL Community, pp. 72–81. Erlbaum, Hillsdale (2002)

Lloyd, G., Wilson, M.: Offering Prospective Teachers Tools to Connect Theory and Practice: Hypermedia in Mathematics Teacher Education. Journal of Technology and Teacher Education 9(4), 497–518 (2001)

McAndrew, P., Goodyear, P., Dalziel, J.: Patterns designs and activities: unifying descriptions of learning structures. International Journal of Learning Technology 2(2-3), 216–242 (2006)

McCracken, M., Almstrum, V., Diaz, D., Guzdial, M., Hagan, D., Kolikant, Y.B.-D., Laxer, C., Thomas, L., Utting, I., Wilusz, T.: A multinational, multi-institutional study of assessment of programming skills of firstyear cs students. In: ITiCSE-WGR 2001, Canterbury, UK, pp. 125–180 (2001)

Pacheco, A., Gomes, A., Henriques, J., de Almeida, A.-M., Mendes, A.-J.: Mathematics and Programming: Some studies. In: International Conference on Computer Systems and Technologies - CompSysTech 2008, pp. V15-1–V15-6 (2008)

Palloff, M.R., Pratt, K.: Learning together in Community: Collaboration Online. In: 20th Annual Conference on Distance Teaching and Learning, Madison, Wisconsin, August 4-6 (2004), http://www.uwex.edu/disted/conference/Resource_library/proceedings/04_1127.pdf

Panitz, T.: Faculty and Student Resistance to Cooperative Learning. Cooperative Learning and College Teaching 7(2) (Winter 1997)

Picciano, A.G.: Beyond student perception: Issues of interaction, presence and performance in an online course. Journal of Asynchronous Learning Networks 6(1), 21–40 (2002)

Riding, R., Cheema, I.: Cognitive styles - an overview and integration. Educational Psychology 11(3-4), 193–215 (1991)

Roberts, T.S.: Computer-supported collaborative learning in higher education: An introduction. In: Roberts, T.S. (ed.) Computer-Supported Collaborative Learning in Higher Education, pp. 1–18. Idea Group Publishing, Hershey (2005)

Robins, A., Rountree, J., Rountree, N.: Learning and teaching programming: A review and discussion. Journal of Computer Science Education 13(2), 449–450 (2003)

Van Eijl, P., Pilot, A.: Using a virtual learning environment in collaborative learning: Criteria for success. Educational Technology 43(2), 54–56 (2003)

Scardamalia, M., Bereiter, C.: Computer support for knowledge-building communities. In: Koschmann, T. (ed.) CSCL: Theory and Practice of an Emerging Paradigm, pp. 249–268. Erlbaum, Mahwah (1996)

Silverman, B.G.: Computer Supported Collaborative Learning (CSCL). Computers Education 25(3), 81–91 (1995)

Soloway, E., Spohrer, J.C.: Studying the novice programmer. Erlbaum, Hillside (1989)

von Glasersfeld, E.: An Exposition of Constructivism: Why Some Like It Radical. In: Davis, R.B., Maher, C.A., Noddings, N. (eds.) Constructivist Views on the Teaching and Learning of Mathematics, pp. 1–3. N.C.T.M, Reston (1990)

Wegner, P., Roberts, E., Rada, R., Tucker, A.B.: Strategic directions in computer science education. ACM Computing Survay 28(4), 836–845 (1996)

Winslow, L.E.: Programming Pedagogy. SIGCSE Bulletin 28(3), 17–22 (1996)

Witkin, H.A., Moore, C.A., Goodenough, D.R., Cox, P.W.: Field-dependent and field-independent cognitive styles and their educational implications. Review of Educational Research 47(1), 1–64 (1977)

Recommendation of Learning Material through Students´ Collaboration and User Modeling in an Adaptive E-Learning Environment

Daniel Lichtnow[1,3], Isabela Gasparini[2,3,4], Amel Bouzeghoub[4], José Palazzo M. de Oliveira[3], and Marcelo S. Pimenta[3]

[1] Centro Politécnico
Universidade Católica de Pelotas - UCPEL, Pelotas, Brazil
[2] Departamento de Ciência da Computação
Universidade do Estado de Santa Catarina - UDESC, Joinville, Brazil
[3] Instituto de Informática
Universidade Federal do Rio Grande do Sul – UFRGS, Porto Alegre, Brazil
[4] Département Informatique
TELECOM & Management SudParis, Paris, France
{dlichtnow,igasparini,mpimenta,palazzo}@inf.ufrgs.br,
Amel.Bouzeghoub@it-sudparis.eu

Abstract. In this chapter, we present an approach for recommendation of learning materials to students in an e-learning environment. Our aim is to increase the current system's personalization capabilities for students in different scenarios making use of recommendation techniques. The recommendation is produced considering learning materials' properties, student's profile and the context of use. In addition, the process of recommendation is improved through students´ collaboration. In the context of this work, a learning material is a link to a Web page or a paper available on the Web and previously stored in a private repository. The process of collaboration occurs during student's evaluations of the recommendations. These student´s evaluations are used by the system to produce new recommendations for other students. The main features of the recommendations aspects are described and some examples are also used to discuss and illustrate how to provide this personalization.

1 Introduction

A Web-based e-learning environment (ELE) is used by a wide variety of students with different skills, background, preferences, and learning styles. Thus, one of the most desired characteristics of ELEs is being adaptive and personalized [7]. Adaptive educational systems adjust the content presentation and navigation to a student's model. Personalization (or adaptation) is the process of adapting a computer application to the needs of specific users and takes advantage of the acquired knowledge about them. In fact, the use of personalization techniques improves

T. Daradoumis et al. (Eds.): Technology-Enhanced Systems and Tools, SCI 350, pp. 257–278.
springerlink.com

ELE usability since a personalized system customizes the user interface considering the user profile (usually called student model) and each user has a perception that the system was designed specifically for him/her. One aim of our research is to investigate approaches putting the users' profile and contextual knowledge into practice in the development process of real ELEs – in particular of the adaptive environment application for Web-based learning AdaptWeb® (Adaptive Web-based learning Environment) [25], [36] whose goal is to adapt the content, the presentation and the navigation according to the student model. AdaptWeb® is an open source environment and in actual use in different universities.

In the present work we are particularly interested in a strategy for adaptive recommendation of a specific kind of learning objects –Web links to additional information (Web pages, papers, etc). Thus, we use the term *learning material* to make reference to this type of learning objects.

Typically in most of ELEs, a professor (author) suggests learning materials, usually in an unsystematic manual procedure, although some automatic mechanisms have been created. The problem of adequacy evaluation of a suggested Web link (herein called `quality evaluation problem´) appears because most of these learning materials are not adapted to student´s profile, and in many cases could be inadequate to the student in some situations. This may happen for example when the student has not enough knowledge to understand the concepts developed in a learning material suggested as reference, or when the content is written in a language in which the student doesn't have sufficient proficiency.

In these situations, a professor must review the material in advance for each student. In huge e-learning groups, this task is harder and very time-consuming. Thus, it is necessary to create some mechanisms allowing evaluation of learning material and addition information of quality evaluation of learning material.

The most common contents of students' models for e-learning are: students' interests, knowledge, background and skills, experiences, goals, behavior, interaction, preferences, individual traits and learning styles. However, ELEs may be dynamically adjusted not only according to the student's model but also depending on a richer notion of context.

A contextualized ELE provides the learner with exactly the material he needs, and appropriate to his/her knowledge level and which makes sense in a special learning situation, called a scenario in our work [9].

The goal of this chapter is to define an approach to make recommendation of learning materials (Web links or papers in our work) that are most suitable for students´ profile and current tasks (tasks currently being done) in the context of a specific ELE. The present work results of authors' previous experiences with AdaptWeb® [9] [25] [36] and with Recommender Systems [18] [19].

In order to provide a basic summary of recommendation techniques, section 2 presents an overview of aspects related to Recommender Systems. Some recommendations techniques that could be used in the context of an ELE are then identified. In section 3 the architecture and some limitations of actual version of AdaptWeb® are presented. The section 4 describes details about our proposal, that involves the use of an ontology and a collaborative evaluation process to recommend learning materials to students (users of AdaptWeb®). Finally, the section 5 discusses some final remarks and perspectives of future work.

2 Related Works

There is an explosive growth of the volume of information: users should be able to make choices without knowing all alternatives. In this case, user´ expectation is a personalized assistance service - a recommendation. A recommendation looks a sentence like this: "*Customers who bought this CD also bought: Rush – Moving Pictures*". Recommender Systems have been introduced for sifting through very large sets of items selecting those items that are relevant for a determined user.

Recommendations change the way people interact with the Web. Nowadays recommenders help people to choose between diverse products and complex information by providing a more personalized access experience. Recommenders Systems are usually adopted in a great variety of Web sites from e-commerce Web sites like *Amazon.com* to news and information sites like *Digg* and *Slashdot*.

One problem is the complex definition of these recommendation mechanisms due to the lack of semantic representation of the web content. More than this, the quality must be evaluated considering multi-criteria and contextual aspects of the students and can vary from one situation to another for the same student. For example, let´s suppose a student John is studying a new course about Artificial Intelligence (AI). As an activity of AI course, he has to do a survey paper about "search methods". Considering this situation, the system has to select the best learning material in that moment: the material adequate for his particular context (task: do a survey). If we suppose that another student David is trying to understand some basic concepts of AI, the system has to provide totally different materials for him.

A Recommender System is basically a system that try do discover user's interests. The *Tapestry*, considered to be the first Recommender System [28], was created to reduce problems related to the information overload generated by increasing number of electronic messages in a corporate environment. Basically, *Tapestry* filtered the electronic messages based on messages content and in previous evaluation that the first readers have done. Since the system considered the ratings of several users, the process was called *Collaborative Filtering* [13].

Although there are many different approaches to produce the recommendation, *Content-Based* and *Collaborative Recommender Systems* are considered the basic ones [2]. In the *Content-based* approach the user's profile is compared with items to be recommended. In this approach, in general, the user's profile is represented by a set of keywords and the items to be recommended consist of a set of textual documents [4]. The recommendation is produced comparing the user's profiles with documents (the approach employed by many techniques of *Information Retrieval*).

In the *Collaborative* approach the user´s profile is represented by a set of ratings related to specific items that user has previously evaluated. The recommendation is produced comparing the user's ratings to find out which users have a similar profile (Pearson correlation is generally used here). The aim is recommending items to users that others users with similar preferences have evaluated well [2].

It is possible to combine these two approaches to reduce some of problems of each approach [4]. For example, in the case of *Content-based*, the quality of an item is not considered, since it is only based on similarity measures [31]. In the case of *Collaborative* approach, new items will be recommended just after the first evaluation. New users will not be receiving good recommendations since there is not enough information about their profile and preferences (The *cold start* problem) [31].

Many Recommender Systems use users' profiles representation and items semantically poor (e.g. a set of keywords, a set of ratings). In order to make these representations richer and solve some problems present in many Recommender Systems, some researchers have proposed to use ontologies to represent user profiles and items [24] [38] [30] [17]. For example, Vincent Schickel-Zuber and Boi Faltings [30] define a similarity measure used with ontology to reduce problems of absence of ratings related to new items or new users. The idea is that items with similar properties tend to have similar evaluation compared to a previously evaluated item. Thus, a rating given to an item is propagated to new items that belong to the same concept or similar concepts.

Another point is related to the fact that knowledge about users' tasks can allow to produce better recommendations [14] [22] [23]. These users' tasks represent the context of use of an item. Considering the goal of our work there are some interesting proposals regarding to use of Recommender Systems in e-Learning environments. There are different scenarios of traditional Recommender Systems for example, in *e-Commerce*. In their work Drachsler, Hummel, and Koper [8] and Santos [29] discuss aspects related to use of Recommender Systems in *e-Learning* environments. Firstly in an e-Learning environment the recommendation must consider pedagogical aspects rather than just users' preferences (a movie recommender will try to recommend a movie according user's preference). Besides, Drachsler emphasizes that "*the cognitive state of learner and the learning content may change over time and context*". Another point is that beginner's learners could be helped by information given by advanced learners (*What is the best learning material for this task? Which materials my classmates have used before?*).

Considering these aspects, with respect to Recommender Systems and according to [29] "*there is not a single recommendation strategy to apply (due to the diversity of needs and situations)*", we developed an hybrid approach.

In our approach, the process of recommendation combines four aspects: (i) the users' model (e.g. cognitive style, knowledge about language, etc - these data are previously added by system's administrator); (ii) the content and properties of the learning material (Web pages), (iii) the context (tasks related to students), (iv) and finally the quality of recommendation is improved using users' ratings.

Our approach is predestined to undergraduate students. It is not our intent to retrieve automatically material from Web. We believe this is a nice feature suitable for e-learning environments where students have more knowledge and could evaluate the accuracy of the material (some postgraduate courses, for example). In the case where material is retrieved from Web [32], it is important to note that they only use a repository that contains papers that have been previously reviewed by a professor. We believe this kind of materials is not adequate for novice students who are starting to study a subject (especially in the first classes). Thus, in the context of our work, all materials must be reviewed by authors/professors before being available to students.

A mentioned problem of Recommender Systems is related to *cold start* – that means it is not possible to produce recommendation because there is no information about a new user. To solve this problem, some Recommender Systems ask new users to access and evaluate some items. Schickel-Zuber et. al. [30] try to solve or reduce this problem using an ontology. In a similar way, our recommendation process considers user's model, task taxonomy and an ontology to organize learning materials. Our learning materials ontology defines the aspects of quality (quality = "*fitness to use*" [35]) of each learning material. Using this information it is possible to produce recommendation for new users.

Finally, we consider that students can help each other in a collaborative way of recommending material. Students give feedback about their perception of learning materials´ quality. Each rating is related to a specific context- task and it is considered in the student´profile.

3 AdaptWeb® Environment

The aim of our approach is to improve the personalization process in the context of AdaptWeb® using an ontology and recommendation techniques. This section presents the architecture and the use of the AdaptWeb®.

The AdaptWeb® environment is an adaptive hypermedia system providing the same content adapted to different students groups. AdaptWeb® it is an open source environment in operation on different universities. The Fig. 1 shows the Architecture of AdaptWeb®.

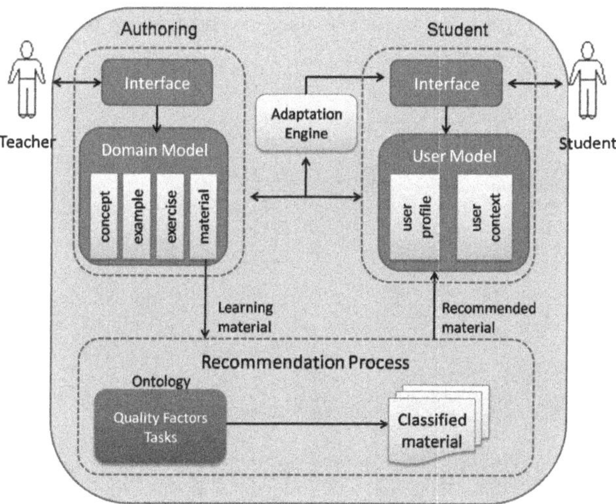

Fig. 1 Architecture of AdaptWeb® with Recommendation Module

Following the architecture, the system is composed of an authoring environment where the professor organizes and creates the structure of content of their courses adapted to degree programs (for example: Engineering, Computer Science or Mathematics), and by an environment for students that produces the personalization.

Thus, using the authoring environment, the professor starts defining the concepts related to each subject. These concepts are organized hierarchically (Fig. 2). This structure is stored in XML (Extensible Markup Language) format in a private repository present in the AdaptWeb®.

Database Systems

1 Introduction to Database Systems

1.1 The Evolution of Database Systems

1.2 Overview of a DBMS

...

2 The Relational Model

...

Fig. 2 Concepts of Database Course

The authoring environment also allows authors to include learning objects. In this sense, we adopt a learning object broad definition - according to IEEE - as "*any digital entity which can be used, reused or referenced during technology supported learning*"[1].

For each concept there may be a list of examples, exercises and learning materials (Fig. 1). We use the term learning materials to refer to a specific kind of learning objects – the content present on the Web accessible via a Web link (e.g. Web page, papers etc). Fig 3 shows the interface used by professors to include learning materials.

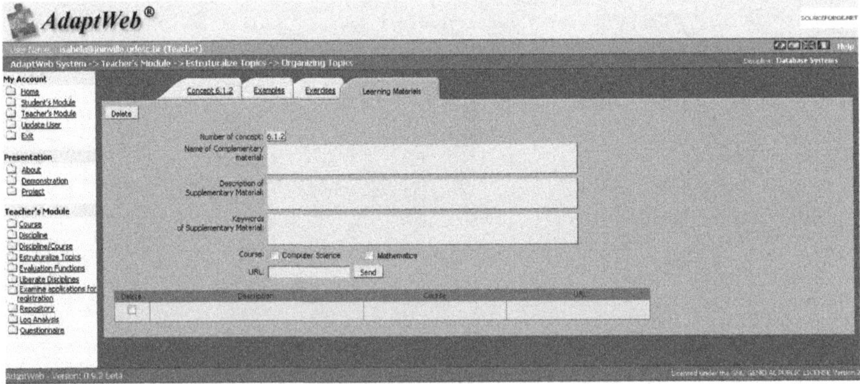

Fig. 3 Interface to include learning materials

[1] http://ltsc.ieee.org/wg12/files/LOM_1484_12_1_v1_Final_Draft.pdf

Examples and exercises are defined by professors using the authoring environment and are stored in a private repository as XML files. Learning materials are not produced by professors – They only propose Web links that are stored in AdaptWeb® repository.

Regarding to implementation aspects, two DTD (Document Type Definition) were defined for the XML files creation. A DTD depicts the hierarchical structure of concepts and the other DTD describes the specific content of each concept. An algorithm to store the content structured was defined based on DTD. This algorithm creates an XML file for each subject with its respective concepts structure and the features of each concept and an XML file for each concept, which includes tags for concepts, examples, exercises and learning material in its body.

Storing files in XML format makes it possible to structure data in a hierarchical way, because there is always a single XML file with the structure of concepts of the subject and as many XML files as concepts defined. The XML files generation is validated through a parser that scans the documents.

The Adaptation Engine combines the student information with the structure of concepts defined by the professor using authoring environment. In the user model is defined the user profile (e.g. navigation mode preference, the learning style and the language skill) and user context (e.g. the history of navigation, task that user is doing). Student's information are also stored in XML files in the AdaptWeb's repository. The Adaptation Engine generates an instance of an XML file adapted to student and the presentation is dynamically generated. In the interface, the adaptation occurs in the links that are available for the student. For example, if the concept involved in the prerequisites was studied by the student, then the concept that depends on it can be enabled. Thus a concept can assume three categories: 1) "studied", it means that the student has already accessed the concept; 2) "under study", it refers to the concept that is being accessed, also named current; 3) "not studied", a concept can be in this category for two reasons, (a) the concept was not enabled because its prerequisites were not studied yet; or (b) the concept was not studied yet, but it is enabled after prerequisites were studied [36].

Regarding to the navigation mode, the adaptation can work in two ways: tutorial mode or free mode. In the tutorial mode (guide tour), prerequisites criteria among concepts determine the student's navigation, and navigation adaptation is based on the register of concepts studied ("studied", it means that the student has already accessed the concept). In the free mode, the student can study any concept available in the navigation menu. These aspects are presented with more details in some previous publications [25], [36] and [9].

One of the limitations that we observed using AdaptWeb® is the lack of adaptation when proposing learning materials to students. These limitations are due to the fact that learning materials are shown to any student anytime (the only aspect that is considered is the suitable concept). Aspects related to students' profiles and tasks are not considered.

Thus, our proposal focused on the definition of recommendation mechanisms to propose learning materials to students in a more adequately way. As consequence, professors must fill values of learning material's properties (see section 4). In addition, professors must assign each learning material to a specific task, previously identified. A quality measure is also used to classify the learning materials according to quality factors. These aspects, related to recommendation process and techniques, are described in the next section.

4 Integrating Recommendation Mechanism to AdaptWeb®

This section describes our approach to recommend learning materials to students into AdaptWeb®. The recommendation process uses an ontology to classify the learning materials according students' needs and feedback to indicate learning material with more quality to them ("quality" means more appropriate for students). In the present section, initially the ontology that contains learning material's properties and classes is described. This ontology is used to classify a learning material. After, we describe the process of collaboration that takes into account students' evaluations to improve the quality of recommendation. Finally, we discuss about some scenarios of use and aspects related to implementation.

4.1 Multi-Criteria Ontology for Learning Materials

The Recommender System aims to recommend items that are likely to be adequately to users. Quality, in many works, is related to *"fitness for use"* [35]. Our approach defines an ontology that represents quality properties of learning materials. This quality dimensions are described in works that defines some quality content factors. Although there is not an agreement on the quality factors definition, many works agree with Wang and Strong [35] in terms of categories of quality dimensions:

- *Intrinsic*: independent of the user's context, emphasizes that data have quality in their own right.
- *Contextual*: emphasizes that quality must be considered within the context of the task at hand.
- *Representational*: emphasizes aspects related to data format.
- *Accessibility*: emphasizes aspects related to availability and security.

For each of these categories, some quality dimensions must be considered. These quality dimensions are selected considering the requirements of our work and chosen based on their importance considering Bizer [5] and Knight and Burn [15]. Quality dimensions examples related to intrinsic category are Wang and Strong [35] and Pipino, Lee, and Wang [27]:

- *Accuracy*. The extent to which data are correct, reliable, and certified free of error (reflects real world).
- *Objectivity*. The extent to which data are unbiased (unprejudiced) and impartial.
- *Believability*. The extent to which data are accepted or regarded as true, real and credible.

In the context of our work the quality factors related to *Intrinsic* category are assured by authors (the professor). Our assumption is that the author has enough knowledge to select web learning content that have these qualities. Therefore the problem is how to associate a learning material with a context (student and task). This problem is related to *Contextual, Representational* and *Accessibility* dimensions. Contextual quality dimensions considered are [35]:

- *Timeliness*. The extent to freshness of data is appropriate for the task at hand.
- *Relevancy*. The extent to which data are applicable and helpful for the task at hand.
- *Amount of Data*. The extent to which the quantity or volume of available data is appropriate.

Regarding to *Representational* and *Accessibility* categories two quality dimensions are considered:

- *Understandability*. The extent to which data is easily comprehended [27].
- *Accessibility*. The extent to which data is available or easily and quickly retrievable [35].

The ontology represents some classes where the learning materials is classified according to some characteristic. The Fig. 4 shows the classes hierarchy of our ontology. Table 1 presents the properties used in our ontology and the quality dimensions related to each property.

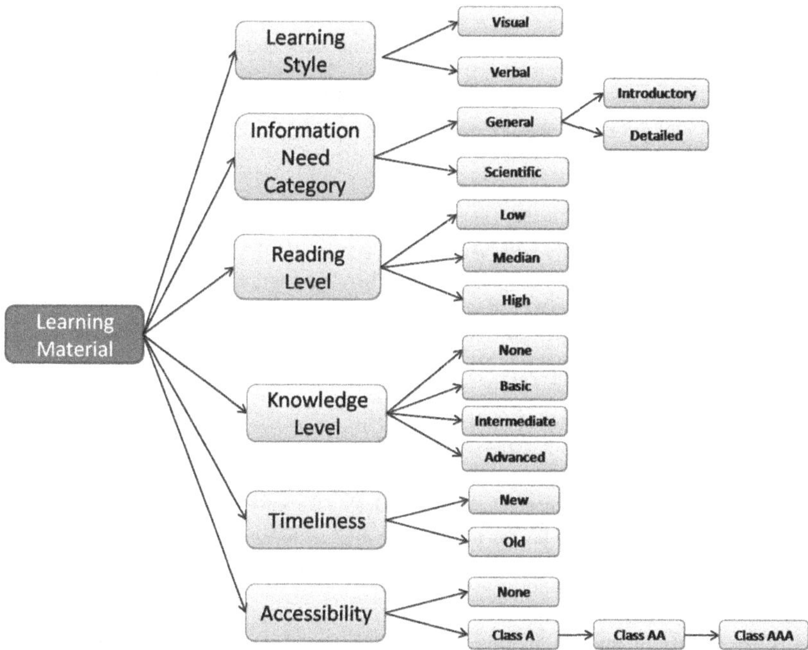

Fig. 4 Ontology´s classes hierarchy

Each Learning Material's subclasses shown in Fig. 4 (*Learning Style, Information Need Category, Reading Level, Knowledge Level, Timeliness and Accessibility*) are related to specific quality aspects. In the ontology *Learning Material*'s subclasses are not disjoints. Consequently, a specific learning material is assigned to each one of subclass. However, the subclasses of these classes (*Learning Style, Information Need Category, Reading Level, Knowledge Level, Timeliness and Accessibility*) are disjoint. Each learning material will be classified according to properties values (see Table 1).

To classify a learning material, in these classes, authors must fill metadata entries before storing a learning material in the repository. Details about how the

Table 1 Learning Material Properties

Property	Description and Quality Dimensions
knowledgeLevelRequired	Refers to a knowledge level (knowledge that a user must to have to understand the content). Can be none, basic, intermediate or advanced. It is related to subject property. Quality Dimension: Understandability.
created	Date of creation of learning material on the Web. It is not related to time when the learning material has been added in AdaptWeb. Quality Dimension: Timeliness
modified	Date on which the learning material was changed on the Web. It is not related to time when the learning material has been updated in AdaptWeb. Quality Dimension: Timeliness
numberOfWords	Can be few or many. Quality Dimension: Amount of Data.
publishedInJournal	Yes or no. Quality Dimension: Understandability.
publishedInConference	Yes or no. Quality Dimension: Understandability.
contentPage	Refers to the type of the content (writing, pictures, graphs, etc.). Quality Dimension: Understandability.
language	Refers to language of the learning material (e.g., English, Spanish, Portuguese). Quality Dimension: Understandability.
readability	Refers to reading ease. It is depending on the language. There are some readability formulas (e.g. SMOG). These formulas estimates the years of education needed to completely understand a text. In the context of AdaptWeb® these property is related to learning material written in a language that is not student's mother tongue. Quality Dimension: Understandability
subject	Refers to some concept of hierarchical structure cited (see Fig. 2). Quality Dimension: Relevancy
textEquivalent	Refers to the fact that the Web page has a text equivalent for every non-text element (e.g., via "alt"). Quality Dimension: Accessibility.
sufficientContrast	Refers to the fact that in the Web pages foreground and background color combinations provide sufficient contrast. Quality Dimension: Accessibility.
acronymExpansion	Refers to the fact that there is a expansion of each acronym in the Web page. Quality Dimension: Accessibility.

professor informs this metadata are presented in section 4.3.1. We present in the following sub-section the ontology's classes, properties and conditions that associate a learning material to specific classes.

4.1.1 Ontology Classes

Knowledge Level. A learning material will be classified in one of subclasses of this class according to students' knowledge level. Thus, the professor must indicate the knowledge level required to use a learning material. For the professor accomplish this, he assigns one specific value to *knowledgeLevelRequired* property. This value can be *none*, *basic*, *intermediate* or *advanced.* According to this value a learning material is associated to a specific class. There are 4 classes (the names of some classes are the same names of values to property):

- *None.* Learning material for students who just start a discipline or a discipline's subject (*∀knowledgeLevelRequired None*[2]).
- *Basic.* Learning material for students with minor knowledge, just attend initial classes (*∀knowledgeLevelRequired Basic*).
- *Intermediate.* Learning material for students that have some knowledge, attend a good percentage of classes (e.g. more than 50%) related to a subject (*∀knowledgeLevelRequired Intermediate*).
- *Advanced.* Learning material for students at the end of course or students that finished a specific subject (*∀knowledgeLevelRequired Advanced*).

Information Need Category. This class is related to idea that for each category of information there are different quality factors and indicators that have to be considered and it is useful to specific demands (based in [12]). The subclasses are:

- *General.* Learning material has not been published in a conference or journal as research paper *(∀publishedInJournal No*) and *(∀publishedInConference No*). This class has two subclasses *Detailed* and *Introductory. Introductory* refers to the information whose amount of information is smaller than *Detailed* information Thus a learning material is classified as *Detailed* (*∀numberOfWords Many*) or *Introductory* (*∀numberOfWords Few*).
- *Scientific.* Refers to publications produced specially in academic environments. This kind of publication was published in a Journal *(∀publishedInJournal Yes*) or in a Conference *(∀publishedInConference Yes*).

Learning Style. The learning style is a characteristic defined by the way people prefer to learn. This feature is widely used in adaptive educational systems. A learning-style model classifies students according to how they fit in a number of scales representing how they receive and process information. Several models and frameworks for learning styles have been proposed. The Felder-Silverman's model classifies students according to the way that each one receives and processes the information considering the styles as skills that can be developed [10]. Felder-Silverman's model categorizes students as sensitive/intuitive, visual/verbal,

[2] The Protégé´s ontology representation is used in this paper to describe the ontology.

active/reflective, and sequential/global, depending on how they learn. The visual-verbal dimension of Felder´s learning style model, differentiates learners who remember best what they have seen, e.g. pictures, diagrams and flow-charts, and learners who get more out of textual representations, regardless of the fact whether they are written or spoken [10]. Considering this fact, a learning material can be classified into two classes:

- *Visual.* Learning material that contains pictures or graphs *(∃contentPage some Picture or Graph).*
- *Verbal.* Learning material that hasn't pictures or graphs *(∀contentPage Text).*

Timeliness. A learning material is classified as *New* or *Old* according to freshness. Here timeliness is related to data of creation of a learning material, it is not related to date of inclusion of a learning material in environment. This can be useful for some task where it is important to consider the learning material age (e.g. a student is searching about recent advances related to a research area):

- *New.* Refers to learning material that was created or modified recently *(∀created Actual or ∀ modified Actual).*
- *Old.* Refers to learning material that was created or modified some time ago *(∀created Outdate and ∀ modified Actual).*

It is not easy to know the meaning of *Actual* and *Outdate*. We consider a learning material is actual if it is created or modified in the last year from the date of use. Again is related to date of a learning material, this it is not related to time when the learning material has been added in our environment.

Reading Level. The subclasses are used to indicate the difficulty's degree related to reading. It is important to identify difficulty's degree of learning material written in a language that is not student's mother tongue. For doing this, the author assigns one specific value to *readability* property. This value can be *low*, *medium* or *high*. According to this value a learning material is associated to a specific class like in the case of subclasses of *Reading Level* class (section 4.3.1 discusses details about use of SMOG or Flesh-Kincaid).

Accessibility. The subclasses are used to verify how much a webpage is accessible for users, in W3C terms [33], where we classify a page by none, when a page or a learning material is not accessible, A, when a page has certain proprieties agreeing recommendation, and so on. It is a hierarchy because a Website that fills the requirements of accessibility of *ClassAAA* also fills requirements of *ClassAA*, and *ClassAA* fills requirements of *ClassA*. This accessibility hierarchy can be studied in W3C 1999 [34]. This class is associated to user, not so much with task. Aim to illustrate, the Table 1 contains three properties related to accessibility (*textEquivalent*, *sufficientContrast* e *acronymExpansion*) but there are others. The *textEquivalent* is related to conformance level *A*, *sufficientContrast* is related to conformance level *AA* and *acronymExpansion* is related to conformance level *AAA*.

4.1.2 Taxonomy of Tasks

One important point is to identify which kind of learning material is more suitable to some specifics tasks. Thus, it is necessary to identify tasks and assign each task to ontology's classes. This tasks' list is not exhaustive. Examples of tasks are:

- *Studying Basic Concepts.* At any time (especially in the beginning) a student may need to study or review (in case of doubts) some basic aspects related to a topic. In this case, the student needs learning material that belongs to *Information Need Category -> General -> Introductory* class.
- *Doing a Final Work.* A final work may be writing a program or solving more complex exercises. In this case, the student needs learning material that belongs to *Information Need Category -> General -> Detailed* class.
- *Fulfill a Survey.* This activity requires reading papers about a specific topic. The student needs learning material that belongs to *Information Need Category -> Scientific* class.
- *Search for Recent Advances.* The student needs recent papers about a topic. Therefore, the student needs learning material that belongs to *Information Need Category -> Scientific* class and *Timeliness->New.*
- *Studying for final exam.* Happens generally at the end of a semester or a scholar year. The student needs to review the key subjects and concepts. *Information Need Category -> General -> Detailed* class.

4.2 The Collaboration Process

Although the learning material has been adapted to student's profile and to specific task, the process of recommendation could be improved considering student´s opinion, especially when there is a great number of learning materials.

The evaluation process take in care the context of use: a student gives his opinion about a learning material considering one specific task that he is doing or he has done.

We define a rating function *R* on the space *Student×Learning Material×Task* specifying how much students liked learning material *lm* on working in a task *t*. Thus, a student evaluate a paper about "XML databases", that he used to *Fulfill a Survey* (see section 4.1.2) as "3". Our approach is similar to Adomavicius and Tuzhilin [1] related to classical *OLAP - Online Analytical Processing* model in databases. Our predicted score for a specific learning material *lm* considering a task *t* is given by (1):

$$\overline{rlmt} = \frac{1}{n}\sum_{i=1}^{n} rlmt_i \tag{1}$$

Where $rlmt_i$ is a rating (range of 0 to 4) given by a student to a learning material (*lm*) considering a specific task *t*. The same learning material can be evaluated with distinct ratings by the same student - one rating for each task. This score is used to generate ranking of learning materials to students in the recommendation process (see section 4.3.3).

In Recommender Systems context, an important point is the collaborative filtering where users who gave similar ratings for the same items are identified. This similarity is measured using Pearson coefficient. After, considering the most similar users (*neighborhoods*) items that do not have rating by a specific user are presumed [2]. This process could be used in our environment, but it seems to be a technique with high computational cost, e.g., if we have only 4 student´s review, we get 6 combinations, but if we have 5 student´s review, the number increases to 10 combinations and so on. The fact of learning materials, tasks and users profiles are richer represented become possible to use another alternative, simpler and with a lower computational cost. Besides, problems related to *cold start* could be decreased.

The techniques described in this section are based in a previous work [11]. Our proposal is also supported by a recent work that shows improvement of search results by users' feedback [3].

4.3 Scenario of Use

The next sections describe the use of the system and some aspects related to its implementation. The start point is the inclusion of new learning materials. After, to explain adaptations provided, we start by describing some learning situations and then we detail how those situations trigger the corresponding adaptation process in AdaptWeb®. Finally the collaborative evaluation of learning material is illustrated.

4.3.1 Including a New Learning Material

The process of content creation is described in the section 3. However, considering the use of ontology in the recommendation process, a new interface for include a learning material must be used. The new proposal interface allows to professor inform data related to ontology (section 4.1).

Regarding to properties presented in Table 1, it could be possible in some cases to extract part of this information automatically [37]. However, sometimes it is too difficult to extract some of these properties' values. This work does not emphasize these aspects, but we have identified some possibilities.

Considering that a learning material could be a Web page, the first problem is that sometimes a Web page may contain other information (e.g. navigation menus, user comments, advertisings, snippet previews of related documents, legal disclaimers etc.). The first step is to identify the main content of a Web page; some recent works address this problem [16].

After identifying the main content of a Web page, it is easy to obtain the number of words and calculate the level of readability. In the case of readability, it is important to consider that Flesh-Kincaid formulas, for example must be adapted to a specific language. In the case of Portuguese language, for example words contain in average a higher number of syllables than English [21].

If there is metadata present following some pattern (e.g. Dublin Core[3]) it is easier to extract properties' values such as "*created*" and "*modified*". However in most of content there isn't any associated metadata, in this case this extract become more difficult, even impossible. In this situation the author must manually enter this information.

It is more difficult to extract other properties' values, such as the subject. In previous works [18] and [19] a technique to identify the text' subject was presented. Following this technique, each subject known in AdaptWeb® is associated to a list of terms and their respective weights.

The presence of terms in a text indicates (with some probability) a specific subject. Weights are used to state the relative importance (or the probability) of the term for identifying the subject in a text (e.g. the term *neural* is associated to *Artificial Intelligence* subject). The relation between concepts and terms is many-to-many, that is, a term may be present in more than one concept and a concept may be described by many terms.

Thus, the text mining method (a kind of classification task) evaluates the relationship between a text and a subject using a similarity function that calculates the distance between the two vectors. One vector represents texts of main content and the other, representing a specific subject, is composed of a list of terms with a weight associated to each term.

The identification of the presence of pictures (*contentPage* properties) could be done by analyzing the *HTML* code of a Web page (*e.g. <img src="url" />*). Regarding to language it is possible to consider metadata (if there is) or methods like proposed in Martins and Silva [20]. The most difficult property's value to extract is related to *knowledgeLevelRequired* that refers to the knowledge level that a user must have to understand the content of a subject.

Considering the context of our work, in general, any learning material aims to help students in a specific subject. When they start their study, they need more simplified learning material than when they are deeply studying for a test, for example. Thus, the default value of this property will be *none* for learning material that does not have been published in a conference or in a journal. When the learning material is related to a Journal or a conference the default value of this property will be *high*. For all the other cases, this information must be indicated by the author (professor) in authoring phase.

Regarding the accessibility´s degree over a webpage, using some automatic tool which supports WC3 patters, as *A-Checker*, *A-Prompt*, *Bobby*, *Hera*, *TAW*, *WAVE*, and others [33]. Obviously, this property is related specially to Web pages. Finally, it is possible to extract some other information about Web page using Google API, for example. One example is the snippet that contains a small description about the content of a Web page.

Only the professor can include a learning material. Another point is that the professor can import and export a set of learning materials to use in other moment with other group of students. There is an important restriction here: the concepts structure (e.g. Fig. 2 – section 3.1) must be equal. The possibility of import/export

[3] http://dublincore.org/

learning materials can be usefully to reduce the *cold start* problem, because students can use evaluations of others students who finished their course before.

4.3.2 Using the Ontology to Produce Recommendation

We show some examples in a Database Systems course context where the professor provided a set of links learning materials with diverse content about database system, for example: History and motivation for database systems, Components, DBMS functions, Database architecture and data independence, etc. For a simplification purpose, we have a few variables over student´s model: student's course and background, student's knowledge, learning style, subject, task, language, language level and country.

In our example, Mike is a computer science student that lives in England and his mother language is English. He is at the end of Database´ course and his task is to fulfill a final work for his grade (*Doing a Final Work*). He has no knowledge in the specific research theme (*XML databases*), but he has good superficial understanding of the overall of Database. One of the course´ tasks is to investigate and use some XML database. Clearly for this, Mike needs a deep understanding of the XML Databases. He has visual learning style, and he has basic language skills in Portuguese.

In this moment, Mike is using AdaptWeb® to access to his task description in order to start. There are some available links learning materials. Only the recommended links are presented to him, according to his profile. Notwithstanding the recommendation module provides user with the best possibilities at this moment, the environment don´t discard the others learning materials links, and if the user wants to see the whole content, he can do it (in "*more materials*", Fig. 5).

A more rigorous representation of this situation is given as follows (according to notation defined in Eyharabide et. al. [9]):

Situation 1 =
{(Student.Mike, **isStundentof**, *Grade.ComputerScience),*
(Student.Mike, **isCoursing**, *Course.DatabaseSystems),*
(Course.DatabaseSystems, **hasSubject**, *Subject.XMLDatabase),*
(Student.Mike, **isLearning**, *Subject.XMLDataBases),*
(Student.Mike, **hasUserKnowledge**, *UserKnowledge.bad),*
(Course.DataBasesystems, **hasLearningMaterial**, *Language.english),*
(Course.DataBasesystems, **hasLearningMaterial**, *Language.spanish),*
*(Course.DataBasesystems,***hasLearningMaterial**, *Language.portuguese),*
(Student.Mike, **hasUserTask**, *UserTask.finalWork),*
(Student.Mike, **hasStyle**, *LearningStyle.visual),*
(Student.Mike, **hasMotherTongue**, *Language.english),*
(Student.Mike, **hasLanguageSkill**, *Language.english),*
(Student.Mike, **hasLanguageSkill**, *Language.portuguese),*
(Student.Mike, **hasEnglishLanguageLevel**, *LanguageLevel.high),*
(Student.Mike, **hasPortugueseLanguageLevel**, *LanguageLevel.low),*
(Student.Mike, **isCitizenOf**, *Country.England)}*

Thus, the recommender mechanism suggests the material according of these features, and classifies them. First, the recommender removes links of learning material with Database introductory pages. Second, it matches the study subject with the pages contents, eliminating pages with no corresponding subject. It verify the content and the student' learning style.

As Mike has visual learning style (based on Felder and Silverman dimensions), the recommender mechanism is going to prioritize learning material related to this learning style. After, it classifies links more relevant to less relevant. The ordering results are presented in Fig. 5 (The data and the interface are just illustrative). Basically, more relevant items have properties according the context (task and user' profile). Considering Mike's task and profile the more important classes of ontology are *Learning Style -> Visual*; *Information Need Category -> General -> Detailed* and *Knowledge Level->Advanced.* Note that *Timeliness* is not too important here. Regards to *Reading Level*, learning material written in Portuguese will be recommended only with readability is low (*Reading Level->Low*).

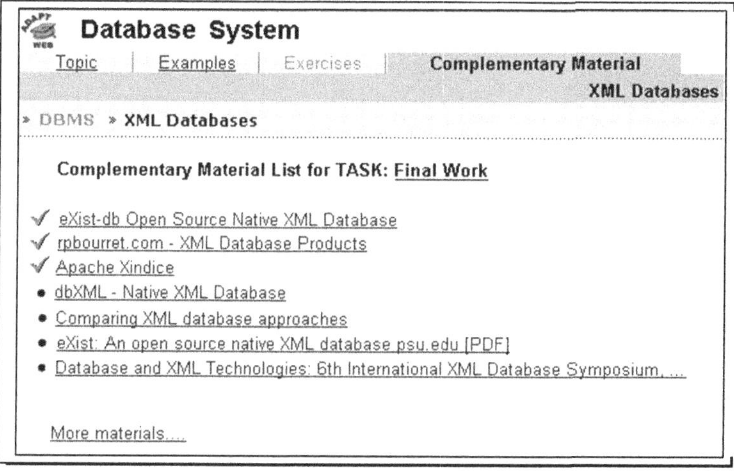

Fig. 5 Learning Material adapted to Mike in situation 1

4.3.3 Collaborative Evaluating of Learning Material

The last aspect of our approach is related to collaborative evaluation of learning materials recommended to students. This process of evaluation starts when a student receives a recommendation that is produced using the ontology (see section 4.3.2).

Thus, regarding to collaborative aspects of our approach, considering the situation 1, Mike will be invited to give a grade for each learning material that he had accessed to perform his task (final work). The data and figures showed here are only to illustrate our approach. The interfaces are being implemented. The process of evaluation will occurs following these steps:

1. Student receives recommendations of learning material (Fig. 6);
2. Student accesses some learning materials (Fig. 6);
3. The AdaptWeb® registers student's access;

4. The AdaptWeb® requires a rating for each accessed learning material. Each rating is represented by a numerical value: 0-Irrelevant, 1-Partial Irrelevant, 2-Neutral, 3-Partial Relevant, 4-Relevant;
5. The AdaptWeb® registers student's ratings.

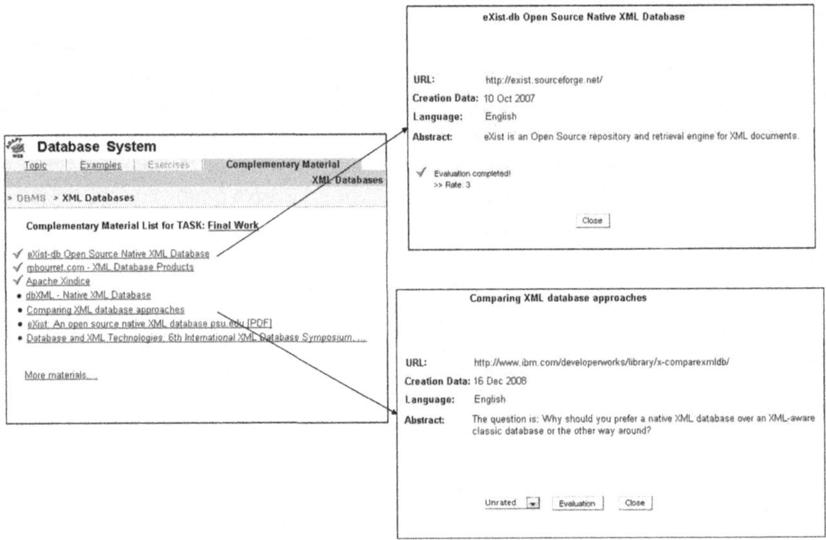

Fig. 6 In situation 1, Mike evaluates the quality of recommendation.

The recommendation process occurs in similar situations to those described in the section 4.3.2 and we combine students' ratings to improve the quality of recommendation. The new recommendation process consists of following steps:

1. The learning materials are recommended as described in the section 4.4.2.
2. The ratings are retrieved. The system considers only the student's ratings related to specific task (see section 4.2).
3. The average rating is calculated (more details in section 4.2).
4. The ranking is generated.

A problem found here is related to *cold start* (see section 2): the learning materials without ratings or with a few ratings will be not recommended to student.

To reduce this problem our approach is to present all learning material to student in a distinguished way into the interface. Thus the system will show learning materials without ratings, with few rating (only 10% of students had evaluated) and the rated ones (the ranking generated using this strategy) in a different manner.

In the case of learning materials without ratings or with few ratings the system will show these learning materials in a random order.

5 Conclusions

As e-learning systems become more sophisticated, new opportunities and new challenges are emerging. One meaningful example is the need to deal with recommendation of diverse learning materials, and in different scenarios.

We proposed the use of ontology to evaluate the quality factor of these learning materials according to different student´s profiles and tasks. This chapter presents how this ontology is integrated in the ELE AdaptWeb®, whose objective is to adapt the content, the navigation and the interface for each student.

In addition, we propose the use of students' evaluation to improve the quality of recommendation. In these sense we are considering some techniques used in the Collaborative Recommender Systems context. Regarding to use of collaborative recommendation techniques, firstly, collaborative filtering techniques, in general, do not consider contextual aspects [2]. Furthermore, the use of items' properties, users' profiles' and task's taxonomy helps recommendation even when there are no ratings - as others works we use an ontology to reduce the *cold start* problem. About *cold start*, we decided to present the learning materials evaluated and learning materials without ratings as well.

Our aim is to improve even more the student´s learning, by giving them the best available learning materials, where the notion of "best" is totally oriented by multi criteria recommendations.

The present work results of experiences of authors with AdaptWeb® and with Recommender Systems. This chapter describes a work that is being developed. We intend to implement and incorporate all these features in the actual version of AdaptWeb®. In addition, our future works include:

- Building more complete task´s taxonomy similar to one present by Broder [6] in the Web Search Goals field. In our case we focus on learning tasks;
- Improving the ontology using others quality metrics, e.g. considering some metrics presents in [26].
- Reducing professors' work. A problem found is related to add new learning material. The professor has to inform all properties values in the authorship phase. Some possible solutions are described in section 4.3.1, but they are not implemented yet.
- Carrying out experiments tests, with a variety of actual students in order to validate our proposal.

Acknowledgments. This work is partially supported by CNPq, Conselho Nacional de Desenvolvimento Científico e Tecnológico, Brazil and CAPES, Coordenação de Aperfeiçoamento de Pessoal de Nível Superior, Brazil.

References

1. Adomavicius, G., Tuzhilin, A.: Multidimensional recommender systems: A data warehousing approach. In: Fiege, L., Mühl, G., Wilhelm, U.G. (eds.) WELCOM 2001. LNCS, vol. 2232, pp. 180–192. Springer, Heidelberg (2001)

2. Adomavicius, G., Tuzhilin, A.: Toward the next generation of recommender systems: A survey of the state-of-the-art and possible extensions. IEEE Trans. on Knowl. and Data Eng. 17(6), 734–749 (2005), doi: `http://dx.doi.org/10.1109/TKDE.2005.99`
3. Agrahri, A.K., Manickam, D.A.T., Riedl, J.: Can people collaborate to improve the relevance of search results? In: RecSys 2008: Proceedings of the 2008 ACM Conference on Recommender Systems, pp. 283–286. ACM, New York (2008), `http://doi.acm.org/10.1145/1454008.1454052`
4. Balabanović, M., Shoham, Y.: Fab: content-based, collaborative recommendation. Commun. ACM 40(3), 66–72 (1997), `http://doi.acm.org/10.1145/245108.245124`
5. Bizer, C.: Quality-Driven Information Filtering. In: The Context of Web-Based Information Systems. VDM Verlag, Saarbrücken (2007)
6. Broder, A.: A taxonomy of web search. SIGIR Forum 36(2), 3–10 (2002), doi: `http://doi.acm.org/10.1145/792550.792552`
7. Brusilovsky, P., Peylo, C.: Adaptive and intelligent Web-based educational systems. International Journal of Artificial Intelligence in Education 13(2), 159–172 (2003)
8. Drachsler, H., Hummel, H., Koper, R.: Recommendations for learners are different: Applying memory-based recommender system techniques to lifelong learning. In: Proceedings of the EC-TEL Conference, Crete, Greece (2007)
9. Eyharabide, V., Gasparini, I., Schiaffino, S.N., Pimenta, M.S., Amandi, A.: Personalized e-learning environments: Considering students' contexts. Education and Technology for a Better World 302, 48–57 (2009)
10. Felder, R.: Learning and Teaching Styles in Engineering Education. Journal of Engineering Education 78(7), 674–681 (1988)
11. Gheller, L.F.M., Pimenta, M.S.: An aditional user interface layer for search mechanisms (uma camada de interfaces adicional para mecanismos de busca). In: S.B. de Computao SBC (ed.) Proceedings of V Symposium on Human Factors in Computer Systems (IHC 2002), vol. 1, pp. 366–370 (2002) (in Portuguese)
12. Glover, E.J., Lawrence, S., Gordon, M.D., Birmingham, W.P., Giles, C.L.: Web search—your way. Commun. ACM 44(12), 97–102 (2001), doi: `http://doi.acm.org/10.1145/501317.501319`
13. Goldberg, D., Nichols, D., Oki, B.M., Terry, D.: Using collaborative filtering to weave an information tapestry. Commun. ACM 35(12), 61–70 (1992), doi: `http://doi.acm.org/10.1145/138859.138867`
14. Herlocker, J.L., Konstan, J.A.: Content-independent task-focused recommendation. IEEE Internet Computing 5(6), 40–47 (2001), doi: `http://dx.doi.org/10.1109/4236.968830`
15. Knight, S., Burn, J.: Developing a Framework for Assessing Information Quality on theWorld Wide Web. Informing Science: International Journal of an Emerging Transdiscipline 8, 159–172 (2005)
16. Kohlschütter, C., Nejdl, W.: A densitometric approach to web page segmentation. In: CIKM 2008: Proceeding of the 17th ACM Conference on Information and Knowledge Management, pp. 1173–1182. ACM, New York (2008), doi: `http://doi.acm.org/10.1145/1458082.1458237`
17. Lee, T., Chun, J., Shim, J., Lee, S.G.: An ontology-based product recommender system for b2b marketplaces. Int. J. Electron. Commerce 11(2), 125–155 (2006), doi: `http://dx.doi.org/10.2753/JEC1086-4415110206`

18. Loh, L., Lichtnow, D., Kampff, A.C., de Oiveira, J.P.M.: Recommendation of complementary material during chat discussions. Knowledge Management & E-Learning 2(4) (2010) (to be appear)
19. Loh, S., Garin, R.S., Lichtnow, D., Borges, T., Rodrigues, R., Simões, G., Amaral, L., Primo, T.: Analyzing web chat messages for recommending items from a digital library. In: ICEIS, vol. (4), pp. 41–48 (2004)
20. Martins, B., Silva, M.J.: Language identification in web pages. In: SAC 2005: Proceedings of the 2005 ACM Symposium on Applied Computing, pp. 764–768. ACM, New York (2005), doi: http://doi.acm.org/10.1145/1066677.1066852
21. Martins, T., Ghiraldelo, C., Nunes, M., Oliveira Jr, O.: Readability formulas applied to textbooks in brazilian portuguese. Notas do ICMSC-USP, Série Computação 28, 11 (1996)
22. Mcnee, S.M., Konstan, J.A. (Adviser): Meeting user information needs in recommender systems. Ph.D. thesis, University of Minnesota, Minneapolis, MN, USA (2006)
23. McNee, S.M., Kapoor, N., Konstan, J.A.: Don't look stupid: avoiding pitfalls when recommending research papers. In: CSCW 2006: Proceedings of the 2006 20th Anniversary Conference on Computer Supported Cooperative Work, pp. 171–180. ACM, New York (2006), doi: http://doi.acm.org/10.1145/1180875.1180903
24. Middleton, S.E., Shadbolt, N.R., De Roure, D.C.: Ontological user profiling in recommender systems. ACM Trans. Inf. Syst. 22(1), 54–88 (2004), doi: http://doi.acm.org/10.1145/963770.963773
25. de Oliveira, J.P.M., de Lima, J.V., Gasparini, I., Pimenta, M.S., Brunetto, M.A.C., Proença Jr., M., Faggion, R.: Adaptive multimedia content delivery in adaptweb. In: XIII Taller Internacional de Software Educativo TISE, vol. 4, pp. 23–39 (2008)
26. Olsina, L., Rossi, G.: Measuring web application quality with webqem. IEEE MultiMedia 9(4), 20–29 (2002), doi: http://dx.doi.org/10.1109/MMUL.2002.1041945
27. Pipino, L.L., Lee, Y.W., Wang, R.Y.: Data quality assessment. Commun. ACM 45(4), 211–218 (2002), doi: http://doi.acm.org/10.1145/505248.506010
28. Resnick, P., Varian, H.R.: Recommender systems. Commun. ACM 40(3), 56–58 (1997), doi: http://doi.acm.org/10.1145/245108.245121
29. Santos, O.C.: A recommender system to provide adaptive and inclusive standard-based support along the e-learning life cycle. In: RecSys 2008: Proceedings of the 2008 ACM Conference on Recommender Systems, pp. 319–322. ACM, New York (2008), doi: http://doi.acm.org/10.1145/1454008.1454062
30. Schickel-Zuber, V., Faltings, B.: Inferring user's preferences using ontologies. In: AAAI 2006: Proceedings of the 21st National Conference on Artificial Intelligence, pp. 1413–1418. AAAI Press, Menlo Park (2006)
31. Shardanand, U., Maes, P.: Social information filtering: algorithms for automating "word of mouth". In: CHI 1995: Proceedings of the SIGCHI Conference on Human Factors in Computing Systems, pp. 210–217. ACM Press/Addison-Wesley Publishing Co., New York, NY, USA (1995), http://doi.acm.org/10.1145/223904.223931
32. Tang, T., McCalla, G.: Evaluating a smart recommender for an evolving e-learning system: A simulation-based study. In: Tawfik, A.Y., Goodwin, S.D. (eds.) Canadian AI 2004. LNCS (LNAI), vol. 3060, pp. 439–443. Springer, Heidelberg (2004)

33. W3C: Complete list of web accessibility evaluation tools WAI (Web Accessibility Initiative) (2010), http://www.w3.org/WAI/RC/tools/complete
34. W3C: Web Content Accessibility Guidelines 1.0 (1999), http://www.w3.org/TR/WAI-WEBCONTENT
35. Wang, R.Y., Strong, D.M.: Beyond accuracy: what data quality means to data consumers. J. Manage. Inf. Syst. 12(4), 5–33 (1996)
36. Bry, F., Kraus, M.: Adaptive hypermedia made simple with HTML/XML style sheet selectors. In: De Bra, P.M.E., Nejdl, W. (eds.) AH 2004. LNCS, vol. 3137, p. 472. Springer, Heidelberg (2004)
37. Warpechowski, M., Souto, M., de Oliveira, J.P.M.: Techniques for metadata retrieval of learning objects. In: SW-EL International Workshop on Applications of Semantic Web Technologies for E-Learning, with AH 2006. Citeseer, Dublin (2006)
38. Ziegler, C.N., McNee, S.M., Konstan, J.A., Lausen, G.: Improving recommendation lists through topic diversification. In: WWW 2005: Proceedings of the 14th International Conference on World Wide Web, pp. 22–32. ACM, New York (2005), doi: http://doi.acm.org/10.1145/1060745.1060754

Modeling Adaptation Patterns in the Context of Collaborative Learning: Case Studies of IMS-LD Based Implementation

Ioannis Magnisalis and Stavros Demetriadis

Aristotle University of Thessaloniki, Greece
P.O. Box 114, 54124, Thessaloniki, Greece
{imagnisa,sdemetri}@csd.auth.gr

Abstract. Research on collaborative learning has emphasized the need for providing flexible yet supportive tools to teachers in order to design collaborative learning tasks. In our work we present a next step in our pattern-based approach demonstrating how educators' ideas can provide the basis for adaptation patterns which, in turn, can be expressed in IMS-LD modeling language. In this paper we present representative and selective design case studies exemplifying the implementation of the core specification of an Adaptation Pattern (Input, Rules, Model and Output) on the basis of using tools compliant to IMS-LD. We analyze what is necessary for implementing an adaptation pattern and discuss the benefits of the pattern-based approach. Finally, we highlight what issues would be important toward integrating the adaptation pattern capabilities in LD compliant tools for collaborative learning design.

1 Introduction

The design and development of adaptive systems for collaborative learning (ASCL) emerges currently as a significant issue at the crossroad of adaptive educational hypermedia and CSCL research traditions (see, for example, [1], [2]).

From our point of view, we have emphasized the need for a generalized conceptual framework of adaptive scripting, relevant to all types of collaboration scripts, as a basis for formalizing the design of flexible adaptive interventions to support group learning [3]. Research has consistently emphasized that collaborating students might fail to engage in productive learning interactions when left without teachers' support (e.g. [4]). Consequently, collaboration scripts have been proposed as a means to structure the collaborative activity by didactic scenarios and engage all students in fruitful learning interactions (e.g. [5], [6]).

Nevertheless, adjusting the script level of granularity and flexibility emerges as an important issue that affects the outcome of scripted collaboration ([7]). We have argued elsewhere ([3], [8], [9]) that a solution to the script flexibility issue could be the integration of adaptive characteristics to systems for scripted collaboration by means of integrating "Adaptation Patterns" (APs) to the design. An AP

T. Daradoumis et al. (Eds.): Technology-Enhanced Systems and Tools, SCI 350, pp. 279–310.
springerlink.com

captures some core idea of pedagogical value on how to adapt the collaborative learning activity when specific conditions occur. Therefore, an adaptation pattern is essentially an abstraction based on teachers' key ideas regarding adaptivity and flexibility during collaborative learning. We envision a situation where teachers would be able to define and enact the type of adaptivity supported by adaptation patterns in a CSCL system, both during the design (foreseen situations) and run-time (unforeseen situations) of the learning activity.

In this work we present specific design case studies (as a proof of concept), exemplifying how the key issues of the adaptation pattern approach can be expressed using the IMS-LD modeling language. As background we discuss mainly the Learning Design (LD) modeling language, relevant tools and specifications and literature identifying hitherto their advantages and limitations. We next present the methodology of extracting adaptation patterns together with how we model an adaptation pattern in the context of a collaboration script, following what we call 'IRMO' specification (section 2). In section 3, we demonstrate how three specific APs are expressed in terms of IMS-LD and are embedded in the core design of a collaboration script. In section 4 we summarize our experience from implementing APs with IMS-LD. Finally, in section 5 we discuss how a software component can be engineered to facilitate the application/design of such APs in IMS-LD format.

2 Background

2.1 *Adaptation Patterns: Adaptivity in Systems for Collaborative Learning*

In the context of technology-enhanced learning, system designers have tried to systematically exploit the modeling potential of computers and develop systems that support learners through adaptive operation. An adaptive educational system (AES) is mainly a system that aims to adapt some of its key functional characteristics (for example, content presentation and/or navigation support) to the learner needs and preferences [10]). Thus an adaptive system operates differently for different learners, taking into account information accumulated in the individual or group learner models.

Introducing adaptive characteristics gave birth to the strand of Adaptive Hypermedia Systems (AHS), a significant subset of which is Adaptive Educational Systems (AES) with systems like AHA, InterBook and WebCOBALT [10]. Respectively the strand of Intelligent Tutoring Systems (ITSs) appeared with systems like ELM-ART, KBS-Hyperbook and SQL-Tutor [10]). According to Brusilovsky and Peylo [10], ITS traditionally focused on Curriculum Sequencing, Intelligent Solution Analysis & Problem Solving Support, while AES focused strongly on Adaptive Presentation & Navigation Support. The above approaches aim principally on helping the individual learner. Recently research efforts have focused on introducing adaptivity and intelligence in the context of computer-supported collaborative learning (CSCL) bringing together AESs and ITSs on one hand and

CSCL systems on the other. There is strong evidence that adaptation advances the learning effects of CSCL (e.g. [11], [12]).

The pedagogical roots of collaborative learning are to be found in Vygotsky's work [13] who extended Piaget's constructivist perspective toward the social field, that is, the dialogue between learners who interact developing a shared understanding of a problem and its solution process. So, collaborative learning brings together social and construction elements of the learning process and CSCL aims to efficiently introduce technologies capable of supporting both these components ([14], [15]). However CSCL does not simply imply using technology for communication purposes. Successful CSCL applications aim to capture and model information and knowledge of group activity and use it to achieve a more effective group monitoring and support [16], thus leading to the development of adaptive collaborative learning support (ACLS) systems. Developing this type of systems instantiates two key notions of the CSCL conceptual framework, namely, distributed cognition and the zone of proximal development [17]. In relation to the former, ACLS systems aim to capture the complexity of interactions in the collaborative learning setting and transform it to understandable and useful representations, thus offloading the teacher and the learners from respective cognitive overhead. In relation to the second, the ambition of system designers is that their tools exhibit a supportive behavior similar to that of a helpful experienced partner who intervenes unobtrusively and "just in time" to support group learners in achieving a productive level of interaction and accomplishing their task.

Brusilovsky and Peylo in [10] identify at least three distinct technologies that implement some type of adaptation regarding the collaborative learning activity: adaptive group formation and peer help, adaptive collaboration support, and virtual students. Efforts to implement adaptive techniques in CSCL in order to improve the learning experience have been systematically reported in the literature providing encouraging evidence on the impact of adaptive methods to enhance student learning (Ronen and Kohen-Vacs [18], Furugori et al. [19], Walker et al.[20], Miao and Hoppe [21]). Adaptively supporting the group of learners employing techniques such as prompting, for example, has attracted recently the attention of several researchers. For example, Gweon et al. in [1] provide research-based evidence in favor of adaptive collaboration support through scripting, when learning in an on-line collaborative environment. The authors use the term "scripting" to refer to the provision of support to collaborative students in the form of prompts (not necessarily within the framework of a collaboration script). In their study, they show how students increase their contribution over time when they adaptively receive prompts indicating ways of improving their within group interactions. From a similar perspective Tsovaltzi et al. in [11] use the term "adaptive scripting" to describe the situation in which a system wizard, who observes the students as they collaborate, provides adaptive support via prompts sent to the students, to promote explanations, reflection, and help giving/receiving. This type of "wizard-of-Oz" research paradigm explores the impact of adaptive design; that is, the adaptive system is not actually built but some human (a teacher) acts "behind the scenes" simulating the behavior of the system. This means that building an adaptive system for collaborative learning (ASCL) is far from being trivial.

Walker, Rummel and Koedinger [20] propose a two dimensional design space which explores alternative methods of adaptive assistance that are implicit, indirect, or both. Moreover, Walker at al. [22] emphasize that few ACLS systems have been implemented (and even less evaluated) and one major reason for this is the difficulty to effectively model the partners dialog and provide feedback. Also, Baghaei et al., [23] developed a collaboration system with an adaptive support mechanism which is based on the components of an individual tutoring system. This study showed that the learners both acquired declarative knowledge about effective collaboration and also collaborated more effectively than the learners who used the relevant individual tutoring system. Despite promising evidence, however, it is a fact that the systems that adaptively support collaboration are at an early stage.

From our point of view, we have emphasized so far the need for a generalized conceptual framework of adaptive scripting, relevant to all types of collaboration scripts, as a basis for formalizing the design of flexible adaptive interventions to support group learning (Demetriadis & Karakostas in [8]). This framework should not only consider the learner's (or group) characteristics but also the specific characteristics of the implemented script. We suggest that an adaptive system for collaborative learning should satisfy at least three criteria: (a) it is a CSCL system, i.e. it somehow supports collaborating groups of students; (b) it includes a user model (learner's cognitive characteristics and preferences), a group model (data relevant to the synthesis and the dynamics of the group) and a script model (computer-based script representation comprising information on specific script characteristics); (c) it comprises also an adaptation model; i.e. a set of rules to initiate some adaptation pattern based on available input.

An adaptation pattern is a process which takes into account the user, a group and/or script model (or other modeled entity) and adjusts certain aspects of the collaborative activity in order to maximize student engagement, satisfaction and, consequently, the learning outcomes (Karakostas & Demetriadis, in [3]). For an adaptation pattern at least three issues should be defined: (a) conditions of initiation, (b) aspects of script to be adapted, and (c) processes to be executed. An adaptation pattern essentially is the reification of key ideas regarding adaptivity and flexibility, strongly connected to anticipated situations where an appropriate strategy would be the enactment of adaptive system behavior. For example a system in a group of learners can adapt the difficulty level of a task for the advanced learner -for example, providing more demanding material and/or assigning a more demanding role to the advanced learner-, thus, making the activity more interesting for him/her. The adaptation pattern may adjust also the guidelines offered to and the role assigned to the novice partner(s) making the activity more beneficial for all learners of the group.

Adaptation patterns' rationale is presented extensively by Karakostas & Demetriadis in [24]. They are supporting together social and construction elements of the learning process [13] and are in general independent and abstracted from the details of a specific learning process. Thus, adaptation patterns contribute in developing ACLS systems as they instantiate two key notions of the CSCL conceptual framework, namely, distributed cognition and the zone of proximal development [17]. Thus, it is out of the scope of the current work to deal with specific learning processes, theories and models as adaptation patterns intentionally stay at an abstract level simply not to limit their applicability. On the other hand applying an adaptation pattern implies using it along a specific learning process or ideally a specific "CLFP" (collaborative learning flow pattern). CLFPs are best practice learning designs, i.e. learning designs that when applied under certain circumstances may lead to a successful CSCL process ([25], [26]) and examples of them are JigSaw, Pyramid, TAPPS etc [27], [28].

In general, we envision a situation where teachers would be able to select and implement the type of adaptivity they deem necessary in any demanding situation during collaboration. Of course, this generalization leads to the question of how to define what a demanding situation is and how to develop accordingly the needed adaptation patterns. We have proposed and exemplified elsewhere (Karakostas & Demetriadis, in [3]) a design methodology (DeACS) for identifying adaptation patterns to be embedded in adaptive scripting systems.

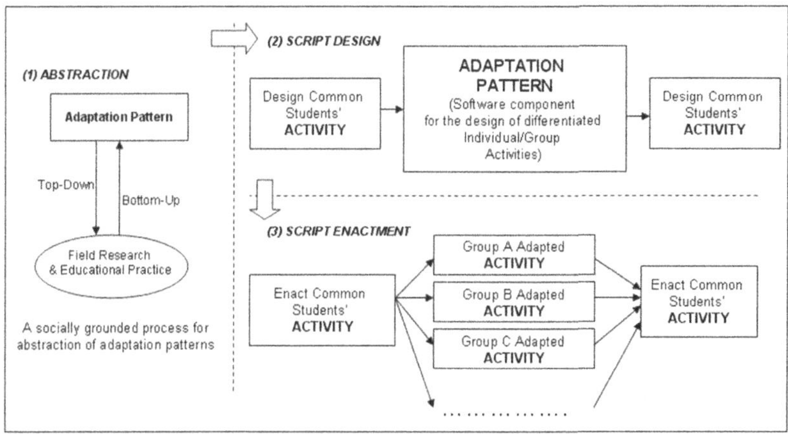

Fig. 1 Left: Abstraction of adaptation patterns through applying the DeACS methodology. Right: Integration of adaptation patterns in script design process and enactment.

The DeACS methodology proposes three major processes: (a) a top-down process: integration of identified adaptation patterns in the ideal script (the form of the script that the teacher initially wishes to put into practice) based on particular activation conditions, (b) a bottom-up process: identification of adaptation patterns that emerge from students' needs for help, support, adjustments, etc. during script runtime, (c) an evaluation process aiming to assess the added value of the adaptation patterns in the previous two categories. If the evaluation of patterns reveals beneficial impact on student learning then these patterns can become part of the computerized script representation embedded to the collaboration support system (see figure 1).

Naturally, the important technical challenge is how to link the core non-adaptive pedagogical design of the script (currently supported by various non-standardized script editors) with adaptive design functionalities. Our position on this is that adaptation patterns can be built either as software add-ons or web services that are invoked by a script editor when available (i.e. the software extends its functionalities depending on the available add-ons library or list of web services). The teacher then could integrate the selected adaptation pattern at the appropriate point of the computerized script representation and parameterize the properties and methods of the pattern as desired (figure 1). In this way the "adaptive logic" can reside at a separate software component (outside IMS-LD manifest) and pedagogically effective UoLs are decoupled from the flexibility it is desired to have under certain circumstances. This way also we take advantage of modifying the adaptive strategy without touching the original pedagogy pattern expressed with LD.

We also maintain that not any script feature can be candidate for adaptation. The script "intrinsic" constraints (Dillenbourg & Tchounikine in [7]), that is, the core features that give to the script its specific pedagogical character and value, should not be adapted in any way. For example, if in a script a teacher decides that students must follow the Jigsaw script then both groups of experts and (subsequently) Jigsaw groups should be formed, since this is an essential feature of the Jigsaw activity ([28]).

On the contrary, extrinsic features should be susceptible to adjustment. Extrinsic constraints can be considered as belonging to either of two categories: (a) "*Non-pedagogical*", that is constraints without any pedagogical relevance. These can be altered by the teacher and/or the students simply to make the script to better accommodate the conditions of the specific implementation (for example, extending the duration of a phase because of a learner's temporal inability to meet a deadline). (b) "*Pedagogical*" constraints that can (should) be adapted in order to provide a well suited learning experience (for example, increasing the level of support when diagnosing learners' misconceptions). For example, the *extrinsic non-pedagogical* features of a script (which can be considered as not affecting the quality of learning interactions), such as: a) a teacher wishes to be able to increase the number of group participants if asked for it, b) a group asks for deadline extension and the system adapts accordingly, c) a late student asks to be included in an already active group.

Computerized script representations should clearly define intrinsic and extrinsic (also pedagogical and non-pedagogical) script features, and adaptation patterns should affect only those features characterized as extrinsic. For example, a teacher may decide that working in dyads is an essential script condition. Consequently, he/she should be able to identify this attribute as an intrinsic aspect of the script not to be affected by any adaptation pattern.

Having said the above, it is clear that a number of issues should be considered when different types of adaptation need to be supported with some formalization method, such as LD. In the following sections (4 & 5) we discuss what the Learning Design (LD) standardization can offer to formally express the adaptive design of collaborative learning activities.

2.2 Modeling Adaptation Patterns

An adaptation pattern is a core idea of how to adapt the collaborative learning activity when specific conditions occur. By contrast to a design pattern (which prescribes a course of action as a solution to a commonly occurring problem) an AP suggests a valuable alternative (to the whole or part of the solution) depending on conditions. We argue that introducing adaptation patterns can help reusable knowledge on common and pedagogically valuable adaptations to become part of the design process. Moreover, APs could be integrated to authoring tools, much like some script editors (e.g. Collage [29]) encourage editing of a whole design pattern (or CLFP) (for example, collections of adaptation techniques could become available in the form of an 'adaptation toolbox'). Additionally, teachers and designers may become familiar and reflect on the use of valuable adaptations during collaboration and transfer research-based conclusions on adaptation to everyday educational practice.

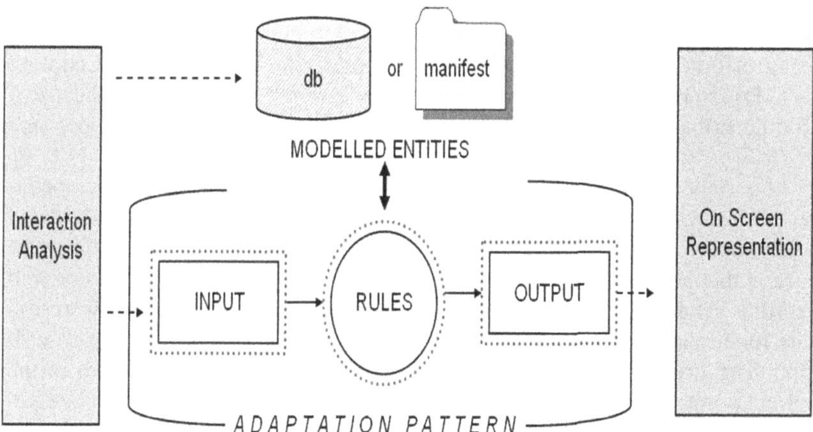

Fig. 2 The IRMO specification for defining the structure of an adaptation pattern.

However, although an AP is eventually experienced as a specific adaptation of the collaborative activity, it is essentially more than that. An AP needs to be somehow modelled in order to become a reusable software component. Thus, what differentiates the adaptation pattern approach from simply introducing hard-coded possibilities for adaptation to a CSCL system, is the need for modelling the patterns at a more abstract level.

In our work so far, we have argued that the structure of an AP can be modelled through four major components, namely: Input, Rule(s), Model(s) and Output [9] ('IRMO' specification, figure 2). Input refers to one or more parameter(s) which are monitored by the AP during runtime and trigger the enactment of the adaptation (these could be, for example, a student assessment outcome, a group deliverable, the synthesis of a group, etc.). Rule(s) implies input processing: one (or more) rules (of the form: "IF Input satisfies condition THEN the Output is

ADAPTED") are applied to input. The Model part defines which (one or more) entities of the collaborative activity are necessary to be modelled in order for the AP to function properly (these entities could be learner or group characteristics, collaboration script aspects, activity phases, material, etc.). Finally, Output refers to the result produced when applying Rule to Model according to some Input. The Output could be, for instance, a change in the synthesis of the group, the material provided to individual learners, the sequence of the activity phases, the roles of the learners, etc. In general, the Output results to an updated representation (internal and/or external) of the activity.

Put briefly, the IRMO specification suggests that for constructing an AP one has to: a) define monitored parameters (e.g. from interaction analysis tools) to be used as Input, b) construct Rules, c) decide which Model characteristics, in various databases and even the manifest of an IMS-LD script, are to be affected by Rules, and d) define Output (form, content, etc.).

However, if an AP is to be reusable it has to be expressed using a common modelling language 'understandable' by CSCL systems. Our next step, therefore, is to explore how the IRMO modelled structure of an AP can be expressed using the IMS Learning Design specification, which is reckoned as one of the most promising efforts to aid CSCL activity design and play by a machine. Learning Design (LD) is primarily a modeling tool which uses the metaphor of a theatrical play for describing a teaching-learning process (Halm et al. [30], Koper and Olivier [31]).

The important technical challenge is how to link the core non-adaptive pedagogical design of the script (currently supported by various non-standardized script LD editors, such as LAMS [32]) with adaptive design functionalities. Our approach is that adaptation patterns can be built either as software add-ons or web services that are invoked by a script editor when available (i.e. the software extends its functionalities depending on the available add-ons library or list of web services). The teacher then could integrate the selected adaptation pattern at the appropriate point of the computerized collaborative activity representation and parameterize the properties and methods of the pattern as desired. In this way the "adaptive logic" can reside at a separate software component (outside the core IMS-LD manifest) and pedagogically effective units of learning (UoLs) are decoupled from the flexibility it is desired to have under certain circumstances. This way also we take advantage of modifying the adaptive strategy without touching the original pedagogy pattern expressed with LD.

2.3 The Basics of LD: Adaptivity Capabilities and Limitations

Learning Design (LD) is primarily a modeling tool which uses the metaphor of a theatrical play for describing a teaching-learning process [30]. Its main components are: metadata, roles, acts, environment, role-part (i.e. activities of actor, who does what, when and how), sequence of activities within a role-part, conditions and notifications (interactivity and control over a live learning design as a form of event driven messaging system within an LD player). Through LD tool we formally express a unit of learning (UoL), that is, a complete, self-contained unit of education or training, such as a course, a module, a lesson etc.

To be usable by computers, Learning Design has to be given a concrete syntax and semantics. Thus, we come to Learning Design specification ([31], [33]). LD specification consists of three levels of implementation and compliance and each level is mapped to separate XML Schemas:

(a) Learning Design Level A: contains all the core vocabulary needed to support pedagogical diversity. (b) Learning Design Level B: adds Properties and Conditions to level A, which enable personalization and more elaborate sequencing and interactions based on learner portfolios. (c) Learning Design Level C: adds Notification to level B, much like an event-driven messaging system, which provides more interactivity and control during CSCL script runtime.

The approach taken in LD specification is therefore not to define a single large schema with a core of mandatory elements and numerous optional elements, but rather to define a complete core that is yet as simple as possible, and then to define two levels of extension that capture more sophisticated features and behaviors. Analyzing the LD structure Burgos et al. [34] identify three levels of support that the specification can offer to various types of adaptation: (a) well supported (for learning flow, content, evaluation and interactive problem solving support), (b) partially supported (for user grouping, interface adaptation, adaptive evaluation and full modification of a course on-the-fly), and, finally, (c) no support (for dynamic modification of learning structure and method in run-time, and adaptive information filtering and retrieval).

In the following, we criticize LD on the basis of what can be modeled by the specification. We also state some conclusions as to what could be done in order to overcome these limitations.

At first, a framework should be established on how pure decentralized P2P interaction models can be incorporated inside LD. This should be done in contrast with centralized P2P models ([31]). Put more clearly, LD should permit to declare whether the LD player itself is web-based and called from a server or it can be on a client machine, runs independently and when in need for communication then it uses networking. What could run on an individual's machine? For instance, could a machine host dynamically a synchronous VOIP meeting? Could all necessary interaction information of an individual be stored locally and then, when network permits, be transmitted on a server centrally? These and similar questions are interrelated with the idea of ubiquitous learning. That is everywhere –even in everyday workplace– a learning activity occurs; do learners have the chance to learn everywhere in a CSCL sense? Can ubiquitous learning have collaborative characteristics whether synchronous or asynchronous? LD specification at its current version does not incorporate elements that could model such situations. This obviously suggests a possible LD specification enhancement.

LD should facilitate representation of "loose" scripts where persons and groups have the opportunity to self-plan, just like a good actor/actress does in a movie. This requirement emerges for instance from the need to model/incorporate games within LD. An educational game is a medium for content rather than the content itself. As a quick answer one could say that here exists a possible role for a game meta-language variant of LD (possibly RDF or OWL/OWL-S), perhaps with extensions of the semantic web variety that will allow a wider range of both representations and operations on those representations (Richards in [35]).

In literature we find criticism for the capabilities of LD to express adaptive behavior. Paramythis ([36]) concludes that LD offers: (1) No support for modeling groups, (2) no support for modeling artifacts (e.g. a vote, an argument, an answer etc), (3) poor support for dynamic features modeling, (4) poor support for modeling complicated control flow, (5) poor support for modeling social interaction, (5) no exchange of information across UoLs, (6) poor modeling of services and their characteristics (additional services maybe "name-spaced" into the LD specification), (7) acts within plays cannot be re-sequenced or structurally modified. Due to above limitations, LD cannot support alternative policies for role playing, for example assigning more than one roles to the same learner to be played during a specific script phase. Another issue is that we cannot use IMS-LD to maintain collaboration activity data and support the identification of group activity patterns in semantically meaningful way.

More limitations are identified by Towle and Halm in [37] including: (1) difficulty of supporting multiple rule interactions (e.g. student profile with multiple characteristics); (2) lack of user/group driven activity ordering; LD is agnostic to the eventual user/group experience (e.g. users' capability to perform selected activities based on their preferences is not supported); (3) manifest-centered vs. server-centered (LD is a manifest-based representation, so once delivered cannot be changed on the fly); (4) knowledge is embedded in manifest and can not be accessible through metadata for use in new arbitrary strategies.

However, some newer research has proposed SLD 2.0 [38] an extension to IMS-LD (and also to tools widely used in the community like Collage) that enables to specify several characteristics of the use of tools that mediate collaboration. This specification would offer much more possibilities in term of learning activities. SLD 2.0 promises to rethink the learning design in the LMS's context while keeping the most essential features of LD like its capacity to express collaborative learning activities. Though this research attempt is very fresh and has not yet gained maturity and acceptance in the CSCL community, it deserves a researcher's attention.

3 IMS-LD and APs Case Studies: APs Expressed in IMS-LD

In our research we have conducted research using IRMO and LD tools and players in order to gradually move from personalized learning and Design-Time Adaptation, towards "groupalised" learning and finally reach Run-Time (or on the fly) Adaptation (see figure 3).

Three APs, among others (see Karakostas & Demetriadis, [24]), are chosen in the following for implementation with IMS-LD. These are:

P1. "Advance the Advanced" AP. This pattern aims at offering a more challenging version of the task for the advanced learner and, also, an adapted version of the task for the novice learner exploiting the partnership with the advanced learner

P2. "Lack of confidence" AP. This AP expresses the simple idea that a novice learner in a specific known group needs support taking into consideration the context of the group (i.e. other learners' domain knowledge) learner belongs in a specific CSCL setting. Therefore, this AP moves up the axis of "groupalized" learning.

P3. "Group heterogeneity based on Prior Domain Knowledge" AP and more specifically "Group Heterogeneity based on Prior Domain Knowledge". This is a complex AP focusing on group synthesis. It expresses the idea that forming groups of mild heterogeneity creates favourable conditions for peer interaction

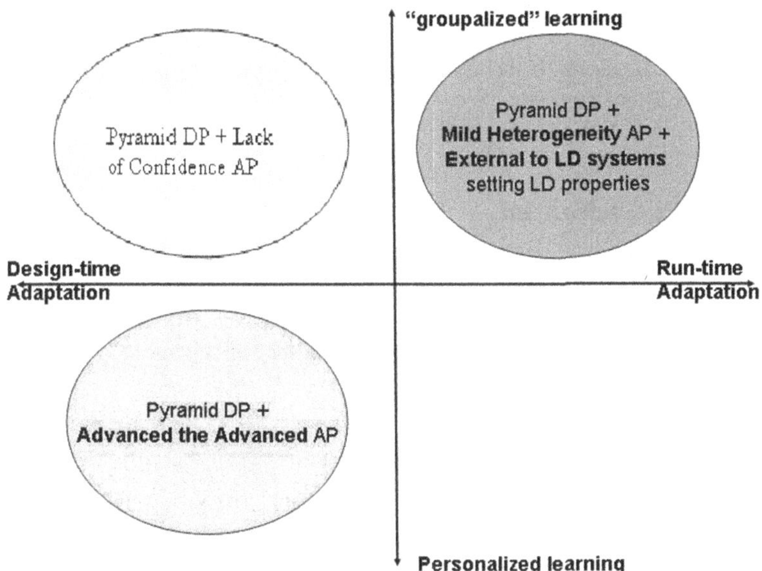

Fig. 3 2 Dimensions characterizing adaptation patterns.

P1 "Advance the Advanced" above is considered as an AP more focused on personalized learning (see axis of personalized vs. "groupalized" learning) and providing adaptation defined at design-time of a CSCL script. Similarly P2 "Lack of confidence" AP moves up towards "groupalized" learning as for this pattern we take into consideration the whole group's context and knowledge level as a metric, but we still stay at adaptation decided at design-time. The scenario described by P3 "Group heterogeneity based on Prior Domain Knowledge" AP and more specifically "Group Heterogeneity based on Prior Domain Knowledge" is demanding more dynamic forms of adaptation from a script, adaptations that can not be prescribed at design-time. Thus, the last AP is considered as a representative one belonging at the upper right quarter shaped by the two axes in figure 3, i.e. focusing on "groupalised" learning and providing adaptive capabilities not introduced or foreseen at design-time.

Thus, in this section we present three design cases where the above adaptation patterns are integrated in an activity structured as a 'pyramid type' collaboration script. Notice that APs are independent of the script they are used in. However, when an AP is implemented it has to be applied along with a specific script. In our cases we use the 'pyramid type' collaboration script, but any other script could be equally applied. First we use the Collage editor [29] to produce the initial structure of a 'pyramid-type' collaboration script. What we call design pattern (DP) is called elsewhere CLFP (Hernández et al. in studies [25] and [26]). For instance, in Collage tool we can implement such a DP as an option from a library of available DPs. We used Collage in order to produce an initial structure of a Pyramid Script. A miniature of the whole 'adaptation pattern' process is showcased. We define an AP, formalize it, develop a computer representation of it and through a user interface we enact and evaluate the process.

The above introduces us to the following showcase (s), where we implement APs with IMS-LD instruments. We start with the "Advance the Advanced" pattern.

3.1 Advance the Advanced

AP Name: Advance the Advanced

Key-idea: Modify the level of task difficulty and guidelines for the partners in order to offer a challenging learning experience also for the advanced peer in the group.

Activation Conditions: When an advanced student participates in a group (regardless of the group size).

What to model: (a) learners' prior domain knowledge, (b) group synthesis, (c) learning material and/or aspects of the task (various characteristics, for example level of difficulty).

What to adapt: the difficulty level of the task for the advanced learner. This includes, for example, providing more demanding material and/or assigning a more demanding role to the advanced learner. It is noteworthy that the adaptation pattern may adjust also the guidelines offered to and the role assigned to the novice partner(s).

The expected result of implementing this adaptation pattern is a more challenging version of the task for the advanced learner and, also, an adapted version of the task for the novice learner exploiting the partnership with the advanced learner.

The pattern 'advance the advanced' expresses the simple idea that when an advanced learner participates in a group (especially when this learner is the sole advance learner in a small group) then the task needs to be modified so that (a) the advanced learner(s) is engaged in a more challenging task and (b) the novice(s) get some benefits from working with the advanced peer(s) as depicted by Karakostas and Demetriadis in [39].

According to the IRMO specification this AP is described as follows:

1. *Input*: the outcome of an appropriate instrument (for example a prior knowledge questionnaire) for measuring learners' expertise and classify them as "Advanced" or "Novice".
2. *Rule*: "IF Learner is Advanced THEN Adapt the Task according also to the Group Synthesis" (the working hypothesis is that the default design of the collaborative activity corresponds to Novice-Novice group synthesis).
3. *Model*: Learner's prior knowledge (for example Advanced/Novice), Group Synthesis and Version of the Task (for example, model the learning material and guidelines for partners in a dyad, as appropriate for Advanced-Novice or Advanced-Advanced group synthesis).
4. *Output*: provide the appropriate version of the task (for example the appropriate learning material and guidelines) to peers.

Next, we design, "run" the adaptive form of the collaboration script and draw conclusions from this effort.

3.1.1 Implementing the Adaptation Pattern

As already mentioned, first we used the Collage editor [29] to produce the initial structure of a 'pyramid-type' collaboration script. After the pyramid script was codified, we integrated in it the code of the "Advance the Advanced" AP. The key idea in this pattern is to provide a more suited form of the task for the learners in a group according to their prior domain knowledge. This is done by modeling the learners' prior domain knowledge individually and then in comparison with other group members. Then a rule is defined to enact the adapted behavior of the system (e.g. provide challenging material for the advanced learner and specific instructions for the novice one).

In the following 3 tables: a) we ask the learner to answer a prior domain knowledge questionnaire (table 1), b) the code demonstrates how an external resource (questionnaire) is linked to the activity and how the property "Prior-Knowledge" is set accordingly (table 2), c) the property "Prior-Knowledge" (local personal property in IMS-LD terms) is shown in IMS-LD syntax (table 3).

Table 1 The learner is guided to answer a prior knowledge questionnaire.

```
- <imsld:learning-activity identifier="Defining Learners' Level" isvisible="true">
  <imsld:title> What is Your Domain Prior Knowledge? Please answer the following
Questionnaire</imsld:title>
- <imsld:activity-description>
- <imsld:item identifier="item-af5 " isvisible="true" identifierref="question1_xml">
  <imsld:title>Questionnaire Resource</imsld:title>
  </imsld:item></imsld:activity-description>
- <imsld:complete-activity><imsld:user-choice />
</imsld:complete-activity></imsld:learning-activity>
```

Table 2 Calling the questionnaire and setting the property Prior-Knowledge.

```
<p>According to your answers your domain knowledge level is: <set-property
xmlns="http://www.imsglobal.org/xsd/imsld_v1p0" ref="Prior-Knowledge" property-
of="self" view="value" /> </p>
- <div class="PKnowledge_Not_Advanced">
  <p>You proceed with the normal material</p> </div>
<div class=" PKnowledge _Advanced"><font color="black">
<p>You are offered advanced material</p>
</font></div></body></html>
```

Table 3 Property Prior-Knowledge in IMS-LD syntax.

```
- <imsld:locpers-property identifier=" Prior-Knowledge">
  <imsld:title>Local Personal Property – Prior-Knowledge</imsld:title>
  <imsld:datatype datatype="string" />
  <imsld:initial-value>Select</imsld:initial-value>
  </imsld:locpers-property>
```

Then, we used Reload [40] and Recourse [41] tools to implement two basic constituents of our AP approach. We implement properties and conditions which are elements of IMS-LD Level B specification. The property that classifies the learner as “Advanced or Not” is “locprop-advanced” which is of type Boolean and has initial value false (or equally 0). Another property (i.e. “average_group_knowledge_level”) is modelling the average domain knowledge level of the group. Therefore, in the simple case of a working dyad of learners, if a learner’s domain knowledge level is 1 (i.e. he/she is advanced learner) and average_group_knowledge_level is below 1 then a scenario of Advanced-Novice activities for the learners is initiated; otherwise appropriate learning material for Advanced-Advanced group synthesis is revealed to the learners.

Table 4 Property locprop-advanced will be set by pattern rules.

```
<imsld:locpers-property identifier="locprop-advanced">
  <imsld:title>adv</imsld:title>
  <imsld:datatype datatype="boolean" />
  <imsld:initial-value>false</imsld:initial-value>
  </imsld:locpers-property>
- <imsld:locpers-property identifier=" average_group_knowledge_level ">
   <imsld:title>average_group_knowledge_level</imsld:title>
   <imsld:datatype datatype="real" />
   </imsld:locpers-property>
```

The core of the AP is presented in table 5 where the pattern rules (conditions in IMS-LD terms) are implemented. Notice that the Prior-Knowledge of the learner sets the property "locprop-advanced" to 0 or 1 and accordingly initiates a scenario that specifies the appropriate feedback to learners (see table 4). Notice that from the two rules presented in table 5 the second one implements -in a dyad of learners- in simple words the idea that: IF learner is Advanced AND average_group_knowledge_level is below 1 then Show Advanced-Novice_Study activity else Hide Advanced-Novice_Study activity. This works for a dyad of learners.

Table 5 The core of the AP implemented by two IMS-LD Level B conditions.

```
- <imsld:conditions>
....
 <imsld:title>advance_the_advanced</imsld:title> <imsld:if><imsld:is>
<imsld:property-ref ref="Prior-Knowledge" />
 <imsld:property-value>X</imsld:property-value>
 </imsld:is></imsld:if><imsld:then>
<imsld:hide>
 <imsld:class class="PKnowledge_Not_Advanced " with-control="false" /> </imsld:hide>
<imsld:show>
 <imsld:class class="PKnowledge _Advanced" with-control="false" /> </imsld:show>
- <imsld:change-property-value>
 <imsld:property-ref ref="locprop-advanced" />
 <imsld:property-value>1</imsld:property-value>
 </imsld:change-property-value>
 </imsld:then> ... <imsld:else><imsld:hide>
 <imsld:class class="PKnowledge _Advanced " with-control="false" /> </imsld:hide>
<imsld:hide>
 <imsld:class class="PKnowledge_Not_Advanced" with-control="false" />
 </imsld:hide>
<imsld:change-property-value>
 <imsld:property-ref ref="locprop-advanced" />
 <imsld:property-value>0</imsld:property-value>
 </imsld:change-property-value></imsld:else> </imsld:else>
.........
- <imsld:if>
<imsld:and>
<imsld:is>
 <imsld:property-ref ref="locprop-advanced" />
 <imsld:property-value>1</imsld:property-value>
 </imsld:is>
- <imsld:greater-than>
```

Table 5 (*continued*)

```
  <imsld:property-ref ref="locprop-advanced" />
  <imsld:property-ref ref=" average_group_knowledge_level" />
  </imsld:greater-than>
</imsld:and>
</imsld:if>
<imsld:then><imsld:show>
 <imsld:learning-activity-ref ref="Advanced-Novice _Study"/>
</imsld:show> </imsld:then>
<imsld:else><imsld:show>
 <imsld:learning-activity-ref ref="Advanced-Advanced_Study" /> </imsld:show>
</imsld:else>

...</imsld:conditions>
```

Lastly, we present the IMS-LD compliant XML code, which describes the activity triggered depending on whether the learner is “Advanced or Not”.

Table 6 Modifying the Output of the AP.

```
- <imsld:learning-activity identifier="Advanced _Study" isvisible="false">
 <imsld:title> Advanced _study </imsld:title>
 <imsld:environment-ref ref="env-11" />
- <imsld:activity-description>
- <imsld:item identifier="item-799" isvisible="true" identifierref="Advanced _study">
 <imsld:title>Resource</imsld:title>
 </imsld:item>
</imsld:activity-description>
- <imsld:complete-activity><imsld:user-choice />
</imsld:complete-activity></imsld:learning-activity>
...
- <imsld:learning-activity identifier="Novice _Study" isvisible="false">
 <imsld:title> Novice _study </imsld:title>
 <imsld:environment-ref ref="env-12" />
- <imsld:activity-description>
- <imsld:item identifier="item-800" isvisible="true" identifierref="Novice _study">
 <imsld:title>Resource</imsld:title>
 </imsld:item>
</imsld:activity-description>
- <imsld:complete-activity><imsld:user-choice />
</imsld:complete-activity></imsld:learning-activity>
```

3.1.2 Running' the Adaptation Pattern

We used the script player SLED [42] to 'run' the code of the adaptation pattern. SLED is built upon CooperCore [43]. In figure 4 the Output (i.e. the adapted user interface) of the adaptation pattern is presented, for the situation of a working dyad of learners where the first learner is classified at advanced level and the second one is classified as novice according to their answers. Thus the first learner is offered some extra material (Learning Activity in IMS-LD terms) and is prompted to study the advanced material and answer some relevant questions (figure 4). Accordingly the Novice learner is shown other material and is asked to perform a different course of actions (figure 5).

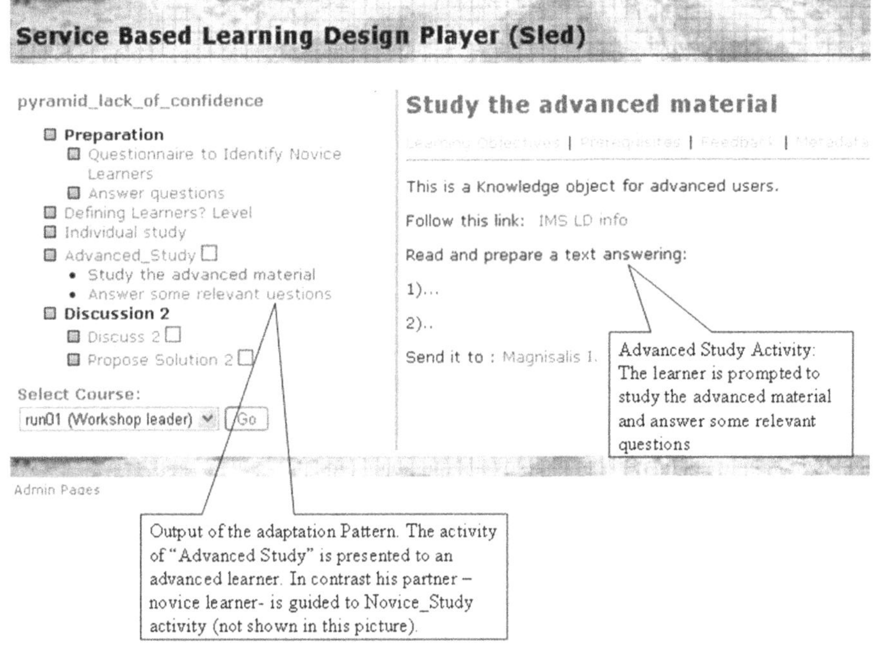

Fig. 4 SLED screenshot of the adapted user interface for the advanced learner according to the implemented adaptation pattern.

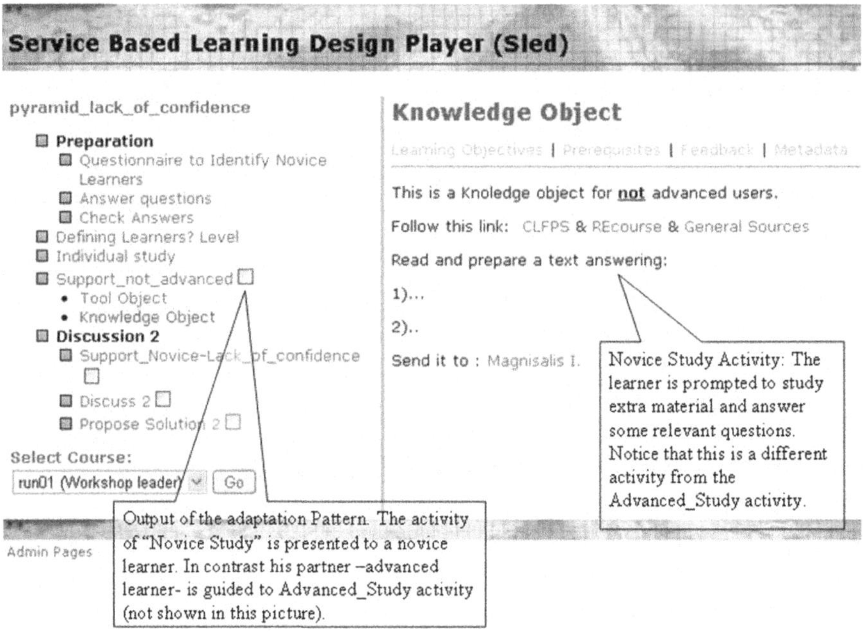

Service Based Learning Design Player (Sled)

pyramid_lack_of_confidence

- Preparation
  - Questionnaire to Identify Novice Learners
  - Answer questions
  - Check Answers
- Defining Learners? Level
- Individual study
- Support_not_advanced
  - Tool Object
  - Knowledge Object
- Discussion 2
  - Support_Novice-Lack_of_confidence
  - Discuss 2
  - Propose Solution 2

Select Course:

run01 (Workshop leader) Go

Admin Pages

Knowledge Object

Learning Objectives | Prerequisites | Feedback | Metadata

This is a Knoledge object for **not** advanced users.

Follow this link: CLFPS & REcourse & General Sources

Read and prepare a text answering:

1)...

2)..

Send it to : Magnisalis I.

Novice Study Activity: The learner is prompted to study extra material and answer some relevant questions. Notice that this is a different activity from the Advanced_Study activity.

Output of the adaptation Pattern. The activity of "Novice Study" is presented to a novice learner. In contrast his partner –advanced learner- is guided to Advanced_Study activity (not shown in this picture).

Fig. 5 SLED screenshot of the adapted user interface for the novice learner according to the implemented adaptation pattern.

3.2 Lack of Confidence

Name: Lack of Confidence

Key-idea: Support and encourage novice learners in larger groups in order to be more confident to participate.

Activation Conditions: When one (or more) novice learner participates in a large group (more than three participants and novices are minority).

What to model: (a) learners' domain knowledge, (b) group size and synthesis, (c) supportive material, (d) script alternative organizational aspects (i.e. student roles).

What to adapt: the support and encouragement offered to novice students. This may include: (a) providing specific to the task material to improve their contributions (e.g. helpful guidelines to better accomplish a task), (b) assigning specific roles to novices in order to make their contribution more clear and straightforward, (c) providing metacognitive support to novices to help them reflect and self-assess their own and others' contribution.

The pattern 'lack of confidence' expresses the simple idea that a novice learner in a specific known group needs support in the "collaborate to learn" process of a

CSCL setting (especially when this learner is the sole novice learner in a group), so that novice learners maintain their interest, interactivity and participation and the activity becomes beneficial for them [39].

According to the IRMO specification this AP is described as follows:

1. Input: the outcome of a prior knowledge questionnaire which is used as a measure of learners' expertise in both domain knowledge and communication skills,
2. Rule: IF Learner is Novice (meaning that the questionnaire outcome is below a certain level) THEN provide New supportive material AND/OR assign specific roles to novices AND/OR provide metacognitive support (e.g. messages),
3. Model: Learner's prior knowledge (Advanced/Novice) & Learning Material (Supportive) & Group size and synthesis & Roles
4. Output: provide New (supportive) material to Novice Learners AND/OR assign specific roles to novices AND/OR provide metacognitive support (e.g. messages).

Next, we design, "run" the adaptive form of the collaboration script and draw conclusions from this effort.

3.2.1 Implementing the Adaptation Pattern

Starting again from the pyramid script, we integrated in it the code of the "Lack of Confidence" AP. The key idea in this pattern is to provide support for a novice learner in a group. This is done by modeling a) the learners' domain prior knowledge (or in the same manner collaborative skills could be modeled) and b) average domain prior knowledge within group. Thus, Novice is an individual whose personal knowledge level of the domain (i.e. personal_knowledge_level property) is below the average personal knowledge level of the domain in a group of learners (i.e. average_group_knowledge_level property). Then a rule is defined to enact the adapted behaviour of the system (e.g. provide support against lack of confidence) for the novice learner.

The way we elicit and set property personal_knowledge_level is not shown as it is similar to the case of "Advance the Advanced" AP (i.e. through a questionnaire).

In the following table we present two necessary properties in order to model groups in the Pyramid script DP. These properties have to be global in IMS-LD terms. Number of learners and number of groups are necessary global properties for the script and in this example have initial values of 5 and 2 respectively. Notice that for "Lack of Confidence" AP these properties are assumed as already set. This is not true though with the next AP to be presented (i.e. "Group Heterogeneity based on Prior Domain Knowledge").

Table 7 Global Properties modelling a) number_of_learners & b) number_of_groups.

```
- <imsld:glob-property identifier=" number_of_learners ">
- <imsld:global-definition uri="">
  <imsld:title>number_of_learners</imsld:title>
  <imsld:datatype datatype="integer" />
  <imsld:initial-value>5</imsld:initial-value>
  </imsld:global-definition>
  </imsld:glob-property>

...
- <imsld:glob-property identifier=" number_of_groups ">
- <imsld:global-definition uri="">
  <imsld:title>number_of_groups</imsld:title>
  <imsld:datatype datatype="integer" />
  <imsld:initial-value>2</imsld:initial-value>
  </imsld:global-definition>
  </imsld:glob-property>
```

In table 8, a local personal property in IMS-LD terms (i.e. group_memebership) is necessary in order to denote which group exactly an individual is member of. In “Lack of Confidence” AP this is assumed as already set (according to table individual belongs to group 1). The properties that classify the learner as “Novice or Not” are personal_knowledge_level & average_group_knowledge_level which is of type Integer and Real respectively and have initial values of 10 and 4 respectively in this example case.

The core of the AP is presented in table 3 where the pattern main rule (conditions in IMS-LD terms) is implemented. Notice the comparison between the properties personal_knowledge_level & average_group_knowledge_level in order to identify the individual as Novice. Also, notice the rule implements in simple words the idea that: IF learner is Novice then Show Support_Novice-Lack_of_Confidence support activity else Support_Novice-Lack_of_Confidence support activity.

Table 8 Local Personal Properties modelling a) group_memebership & b) personal_ knowledge_level & c) average_group_knowledge_level.

```
- <imsld:locpers-property identifier=" group_memebership ">
  <imsld:title>group_memebership</imsld:title>
  <imsld:datatype datatype="integer" />
  <imsld:initial-value>1</imsld:initial-value>
  </imsld:locpers-property>
- <imsld:locpers-property identifier="personal_knowledge_level">
  <imsld:title>personal_knowledge_level</imsld:title>
  <imsld:datatype datatype="integer" />
  <imsld:initial-value>10</imsld:initial-value>
  </imsld:locpers-property>
- <imsld:locpers-property identifier=" average_group_knowledge_level ">
  <imsld:title>average_group_knowledge_level</imsld:title>
  <imsld:datatype datatype="real" />
  <imsld:initial-value>4</imsld:initial-value>
  </imsld:locpers-property>
```

Table 9 Rule of Lackof confidence AP (notice comparison between of properties personal_knowledge_level & average_group_knowledge_level).

```
- <imsld:if>
- <imsld:greater-than>
  <imsld:property-ref ref=" personal_knowledge_level " />
  <imsld:property-ref ref=" average_group_knowledge_level" />
  </imsld:greater-than>
  </imsld:if>
- <imsld:then>
- <imsld:hide>
  <imsld:support-activity-ref ref="Support_Novice-Lack_of_Confidence" />
  </imsld:hide>
  </imsld:then>
- <imsld:else>
- <imsld:show>
  <imsld:support-activity-ref ref="Support_Novice-Lack_of_Confidence" />
  </imsld:show>
  </imsld:else>
  </imsld:else>
```

Lastly, we present the IMS-LD compliant XML code, which describes the support activity triggered depending on whether the learner is "Novice or Not".

Table 10 Modifying the Output of the AP.

```
- <imsld:support-activity identifier=" Support_Novice-Lack_of_Confidence " isvisible="false">
  <imsld:title> Support_Novice-Lack_of_Confidence </imsld:title>
  <imsld:environment-ref ref="env-12" />
- <imsld:activity-description>
- <imsld:item identifier="item-798" isvisible="true" identifierref=" Support_Novice-
Lack_of_Confidence ">
  <imsld:title>Resource</imsld:title>
  </imsld:item>
</imsld:activity-description>
- <imsld:complete-activity><imsld:user-choice />
</imsld:complete-activity></imsld:support-activity>
```

3.2.2 Running the AP

We used SLED [42] to 'run' the code of the adaptation pattern. In figure 7 the Output of the adaptation pattern (i.e. the adapted user interface) is presented., This output is adapted according to his/her answers (i.e. personal_knowledge_level in figure 6) and the answers of his/her collaborators (i.e. average_group_knowledge_level) is classified as Novice or not. The learner is offered support (Support Activity in IMS-LD terms) and is prompted perform a specific course of actions (see figure 7).

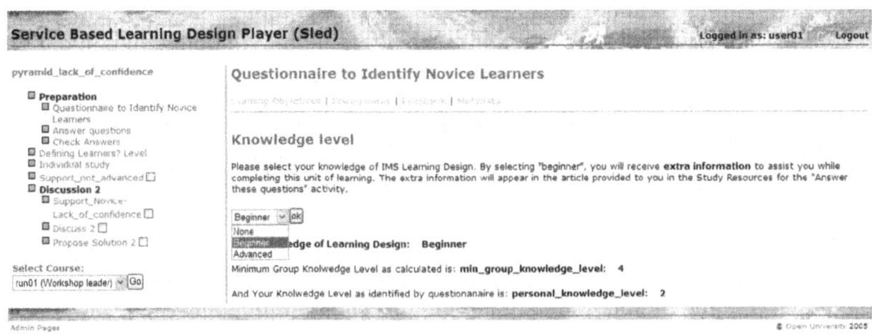

Fig. 6 Individual Answers to set personal_knowledge_level.

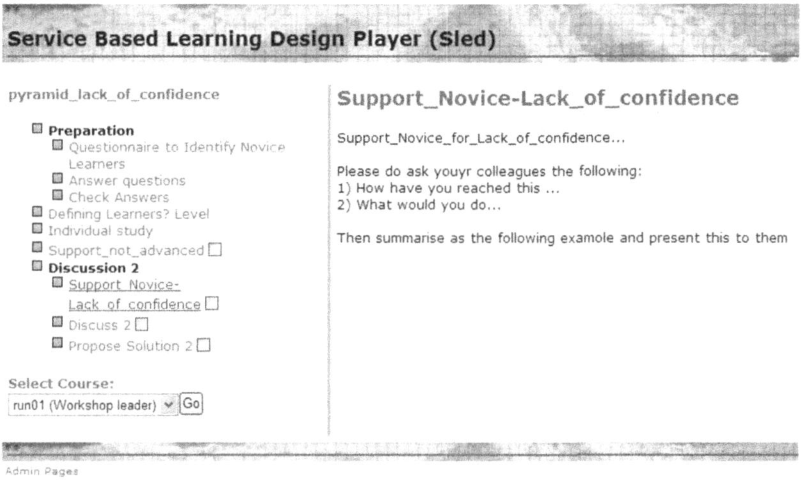

Service Based Learning Design Player (Sled)

pyramid_lack_of_confidence

- Preparation
  - Questionnaire to Identify Novice Learners
  - Answer questions
  - Check Answers
- Defining Learners? Level
- Individual study
- Support_not_advanced
- Discussion 2
  - Support Novice-Lack of confidence
  - Discuss 2
  - Propose Solution 2

Select Course:
run01 (Workshop leader) Go

Support_Novice-Lack_of_confidence

Support_Novice_for_Lack_of_confidence...

Please do ask youyr colleagues the following:
1) How have you reached this ...
2) What would you do...

Then summarise as the following examole and present this to them

Admin Pages

Fig. 7 SLED screenshot of the adapted user interface for the novice learner according to the implemented adaptation pattern.

3.3 Group Heterogeneity Based on Prior Domain Knowledge

Name: Group Heterogeneity based on Prior Domain Knowledge

Key-idea: Formation of heterogeneous groups based on partners' prior domain knowledge.

Activation conditions: When students need to work in small groups (2-3 peers) to broaden and deepen their understanding of the domain.

What to model: (a) students' prior domain knowledge, (b) group synthesis. To this end, helpful tools include: an instrument to record and analyse students' prior domain knowledge, and, in case of many students, a software module to form groups based on the principle of heterogeneity.

What to adapt: the synthesis of the group. Mildly heterogeneous groups should comprise learners whose prior domain knowledge should not be extremely different. For example, a preferable group synthesis would be novice-intermediate or intermediate-advanced but not novice-advanced

The pattern 'Group Heterogeneity based on Prior Domain Knowledge' expresses the idea that forming groups of mild heterogeneity creates favourable conditions for peer interaction. Modeling entails classifying individual learners on a specific dimension and defining "mild heterogeneity" [39].

According to the IRMO specification this AP is described as follows:

1. Input: the outcome of a prior knowledge questionnaire which is used as a measure of learners' expertise,

2. Model: Prior domain knowledge of each learner & Mean of Prior domain knowledge of all participants & number of groups & number of participants,
3. Rule: IF Group works is needed THEN provide new groups of mild heterogeneity (complex rule which entails calculations of a) number of groups, b) best distribution within them),
4. Output: Form New Groups mildly heterogeneous according to prior domain knowledge.

Next, we design, "run" the adaptive form of the collaboration script and draw conclusions from this effort. Notice that the main reason for referring to this AP is to demonstrate that there are APs so complex that can not be implemented with IMS-LD syntax and tools available up to now.

3.3.1 Implementing the Adaptation Pattern

The way we elicit and set property personal_knowledge_level is not shown as it is similar to the previous cases (i.e. through a questionnaire). This is the Input part of IRMO specification for the AP under concern.

In table 7 we have presented 2 necessary properties in order to model groups in the Pyramid script –it can be applied to any script. These properties have to be global in IMS-LD terms. Number of learners and number of groups are necessary global properties for the script. These properties are assumed not to be set until runtime, an aspect that leverages the specific AP (i.e. "Group Heterogeneity based on Prior Domain Knowledge") at the axis of design-run time towards more run-time oriented. The properties that classify the learner as more or less "expert" in the learning domain are personal_knowledge_level & average_group_knowledge_level which is of type Integer and Real numbers respectively. All these properties are the model part of IRMO specification for the AP under concern and have already been codified and shown in the previous AP.

In table 8, a local personal property in IMS-LD terms (i.e. group_memebership) was mentioned as necessary in order to denote which group exactly an individual is member of. In "Group Heterogeneity based on Prior Domain Knowledge" AP this property is also required to be set during execution of the script. This is the Output part of IRMO specification for the AP under concern.

The great difference of this AP with the previous ones is the complexity of the Rules to be modeled and implemented. To briefly describe this let's assume a specific case where:

- initial group of learners where their scores in personal_knowledge_level are 7, 2, 4, 9, and 6
- number_of_learners is 5
- number of wished groups is 2

In this case the easy part is the calculation of average_group_knowledge_level which equals 5.6 and that the groups should be 2 with 2 and 3 members respectively. But, it is not trivial at all to calculate group_memebership, that is who belongs to which group, having in mind that we cater for mild heterogeneous groups. The solution – which is Group1: Individuals with personal_knowledge_level 9, 7,

4 and Group2: Individuals with personal_knowledge_level 6, 2- is out of the scope of this work. Nevertheless, this study has demonstrated that there are cases where more advanced programming structures (than those offered by IMS-LD) are needed for complex algorithms to be implemented. The Rule part of IRMO specification for this specific AP could not be implemented with IMS-LD.

3.3.2 Running the AP

We could not run 'Group Heterogeneity based on Prior Domain Knowledge', simply because the Rule part of IRMO specification for this AP could not be implemented in IMS-LD. Group formation is a broad research issue in itself and is mostly accomplished by developing independent systems focused solely on forming the desired groups for a CSCL activity (Sancho et al. in [12] present NUCLEO system which facilitates adaptive group formation in Role Game Based Scenarios, Alfonseca et al. in [44] Paredes and Rodriguez in [45] investigate the role of learning styles in forming groups for fruitful collaboration with successful results, Tourtoglou and Virvou in [46] form groups based on specific user models aiming the learning domain of UML while Pollalis and Mavrommatis in [47] and Christodoulopoulos and Papanikolaou in [48] all use and implement specific tools and strategies for forming groups in CSCL settings). We reckon that such systems if connected to an IMS-LD compliant script editor can facilitate implementation of more complex APs. Such issues are discussed next.

4 Discussion and Future Work

In this work we have presented our latest efforts toward linking the "adaptation pattern" (AP) approach with the IMS-LD modeling language in order to build powerful and interoperable tools to support the design of flexible collaborative learning environments.

We have argued that the AP approach employs powerful pedagogical ideas and support teachers in understanding what could be successfully adapted in situations of group learning. This is in contrast to simply providing teachers with the capability of altering various parameters of the CSCL system (a technologically driven design which usually does not help teachers to grasp the pedagogical value of their possible actions).

Also, an AP is meant to provide a supportive and guiding framework to teachers so that they can easily apply the available adaptations. However, when teachers reach a higher level of expertise they could always intervene to the AP code and create their own versions of existing or totally new adaptation patterns.

Finally, we anticipate that the pattern approach will enhance the transferability of adaptation ideas and practices to various technological tools. An AP developed in the form of a software component could be, in principle, integrated in any LD compliant editor.

Moreover, we have dealt with some important technical issues during our efforts to tackle the hindrances of implementing IMS-LD compliant APs. A concise compilation of these issues (along with possible solutions, whether actually applied or simply proposed) follows:

I1: How can we import to an IMS-LD running script, properties like number of groups or number of learners?

This has been achieved through a question and a global variable through which we can input properties not set at design-time. This is a solution applied in the presented case studies.

I2: How can we calculate properties like average knowledge of all participants?

In IMS-LD there are no constructs to calculate (e.g. for...each) complex outcomes, especially when these are dependent on run-time known values. An applicable solution to this issue would be through an external algorithm -programmed in a common language e.g. Java- that can calculate expressions and then insert the outcome to a property of IMS-LD. This is a proposal that is planned to be tested. Another possible direction –not proposed by this study because it makes IMS-LD more complex- is to extend IMS-LD specification with specific constructs.

I3: IMS-LD runtime engine provides hitherto no interface (i.e. in the form of an API) to allow for dynamic role creation. This can only be performed manually as shown in figure 8. That is the reason that one can not assign dynamically (i.e. at run-time) a role to an individual (recall assigning specific roles to novices in order to make their contribution more clear and straightforward in "Lack of Confidence" AP). Towards finding an applicable solution to this issue, one can find that in literature there are studies dealing with Group instantiation during run-time like Perez-Sanagustin et al. depict in [27]. Others like Hernandez et al. in [49] tackle this issue with role changing. From these works it can be concluded that IMS-LD needs a run-time model and mechanisms to allow for adaptations on the DP at run-time. One can talk for even introducing a role at run-time that was not implemented at design-time. Until now there is no-way to change group synthesis at run-time (for example, in case that a student leaves at a later stage of a script and regrouping is needed).

Two issues emerged from the effort to unsuccessfully implement the 'Group Heterogeneity based on Prior Domain Knowledge' pattern:

I4: IMS-LD has no programming constructs to facilitate the modeling of complex rules. For instance, IMS-LD has no "for", "while" or similar constructs met in programming languages in order to implement loops and complex algorithms. This issue can be dealt with through external autonomous and loosely coupled software components one can facilitate exchange of wished information and implementation of complex rules. This is a proposal that is of high implementation priority in our study.

Fig. 8 SLED screenshot of the administration interface where user01 is manually set to Role of a Student in a specific run –called "run_test" of the AP "Lack_of_Confidence"

I5: IMS-LD has no obvious mechanism hitherto (at least in the tools we used) of communicating with external systems. A running LD cannot get a value, e.g. from a forum and set a property accordingly. Moreover, an external system cannot see (i.e. get and set) a published IMS-LD property. A possible applicable solution would be to introduce APIs in the form of web-services. This is a promising mechanism of publishing IMS-LD design-time properties. Then, an external system can get and set the wished property. The same applies the way around, that is systems should expose their capabilities in order for an IMS-LD discover what is offered and 'call' it. Again this is a research proposal and a work in progress.

We are currently developing an early prototype of an AP software component that could be used as an add-on (or plug-in) to an LD Design Tool. Such candidate tools to work upon are Reload LD Editor and/or Re-Course Editor. Both are open source tools and we have already used them in developing our showcase. However, Recourse has proved to be more user-friendly for defining properties and especially for handling IMS-LD Level B conditions. Although we found out some bugs in the tool (e.g. when defining an extra Learning Activity we could not give our own reference to it), we were able with a simple Notepad application to change the corresponding XML code. This, of course, proves that such tools are still far from being easily used by everyday teachers, who wish to design their courses.

In our future research we plan to implement more APs and experiment with situations ranging towards more complex and "groupalised" learning that demands run-time (or "on the fly") adaptations. To achieve this, the following roadmap of research is planned:

- Automatize the process of producing the IMS-LD code for implementing the adaptation pattern. This means that when users operate a GUI to link an adaptation pattern to their design, the tool should automatically create the necessary code. More specifically, when users operate a tool like Collage, system presents a GUI. A possible interface is depicted in figure 9. Thus, we aim to link the adaptation pattern approach to the capacity of specific tools. We plan to introduce adaptation patterns as components/services/tools in the form of a toolbox. Interfaces have to be built in accordance to IRMO specification. Also, we plan to engage teachers in user studies providing feedback on the usefulness and usability of these tools

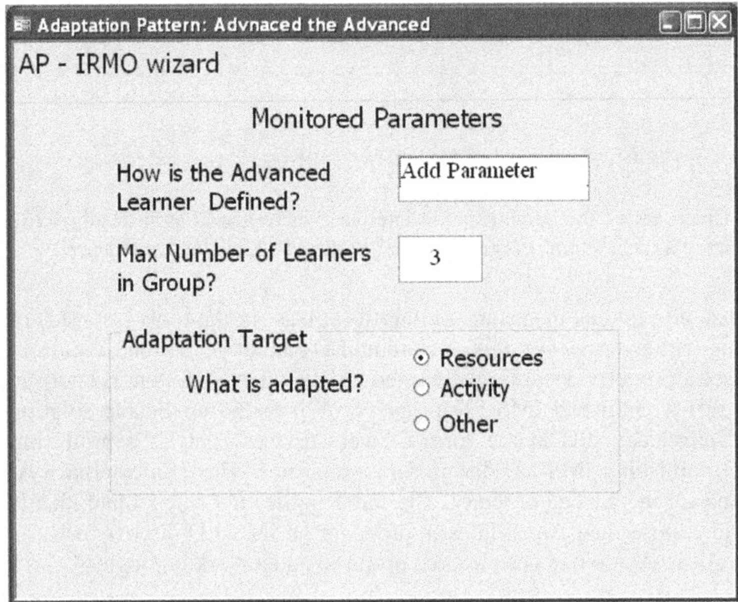

Fig. 9 A possible interface of a s/w component facilitating AP application.

- Design AP software components so that they facilitate the distinction between intrinsic and extrinsic script features [7]. This, we expect, to help manage all adaptations – extrinsic features- only through APs, with the core part of the design remaining intact. The advantages stemming from it are: a) users are protected against modifying a parameter that might make the collaborative script collapse, b) if, however, such a situation occurs then the system can always resort to the intact core design (rejecting introduced variations by APs).
- Connect-extend IMS-LD to external systems. These systems are expected not only to behave as services to be called by IMS-LD but also as systems setting IMS-LD properties during run-time. Studies exemplifying such directions are found in literature (Moreno-Ger et al. attempt to adapt games in [50], Sharples

et al. in [51] and Wilson et al. in [52] extend IMS-LD with software components called widgets, Valentin Luis et al. in [53] present technical architectures to extend IMS-LD capabilities and Miao et al. in [54] describe ways of representation of Coordination Mechanisms in IMS-LD).

6 Conclusion

In this study we have exemplified that it is possible to implement the core specification of an Adaptation Pattern (Input, Rules, Model and Output) on the basis of using tools and technologies compliant to the IMS-LD language. Thus, we have proven, at least in theory, that it is possible to have services (possibly as add-ons to existing tools) that support a teacher to apply an adaptation pattern during the design of a collaborative task scenario. As important next steps we argued that it is necessary to develop and experiment with more complex APs and automatize the process of producing IMS-LD code implementing the adaptive capabilities of the patterns.

References

1. Gweon, G., Rosé, C.P., Carey, R., Zaiss, Z.S.: Providing Support for Adaptive Scripting in an On-Line Collaborative Learning Environment. In: Proceedings of the SIGCHI Conference on Human Factors in Computing Systems, Montreal, Quebec, Canada, pp. 251–260 (2006)
2. Harrer, A., Malzahn, N., Wichmann, A.: The remote control approach - An architecture for adaptive scripting across collaborative learning environments. JUCS 14(1), 148–173 (2008)
3. Karakostas, A., Demetriadis, S.: Systems for Adaptive Collaboration Scripting: Architecture and Design. In: Adaptive Collaboration Support Workshop in 5th International Conference on Adaptive Hypermedia and Adaptive Web-Based Systems, pp. 7–12 (2008), http://www.ah2008.org/index.php?section=62 (accessed January 5, 2011)
4. Hewitt, J.: Toward an understanding of how threads die in asynchronous computer conferences. JLS 7(4), 567–589 (2005)
5. Kobbe, L., Weinberger, A., Dillenbourg, P., Harrer, A., Hämäläinen, R., Häkkinen, P., Fischer, F.: Specifying computer-supported collaboration scripts. ijCSCL 2(2), 211–224 (2007)
6. Bote-Lorenzo, M.L., Gomez-Sanchez, E., Vega-Gorgojo, G., Dimitriadis, Y.A., Asensio-Perez, J.I., Jorrin-Abellan, I.M.: Gridcole: A tailorable grid service based system that supports scripted collaborative learning. Computers and Education 51(1), 155–172 (2008)
7. Dillenbourg, P., Tchounikine, P.: Flexibility in macro-scripts for computer-supported collaborative learning. JCAL 23(1), 1–13 (2007)
8. Demetriadis, S., Karakostas, A.: Adaptive collaboration scripting: A conceptual framework and a design case study. In: Xhafa, F., Barolli, L. (eds.) Proceedings of the CISIS 2008: 2nd International Conference on Complex, Intelligent and Software Intensive Systems, pp. 487–492. IEEE Computer Society, Los Alamitos (2008)

9. Demetriadis, S., Magnisalis, I., Karakostas, A.: Adaptation Patterns in Systems for Collaborative Learning and the Role of the Learning Design Specification. In: Scripted vs. Free CS Collaboration: Alternatives and Paths for Adaptable and Flexible CS Scripted Collaboration Workshop in CSCL 2009, Rhodes, pp. 43–47 (2009), `http://mlab.csd.auth.gr/cscl2009/sfc-workshop.htm#proceedings` (accessed January 5, 2011)
10. Brusilovsky, P., Peylo, C.: Adaptive and intelligent Web-based educational systems. IJAIED, Special Issue on Adaptive and Intelligent Web-based Educational Systems 13(2-4), 159–172 (2003)
11. Tsovaltzi, D., Rummel, N., Pinkwart, N., Scheuer, O., Harrer, A., Braun, I., McLaren, B.M.: CoChemEx: Supporting Conceptual Chemistry Learning via Computer-Mediated Collaboration Scripts. In: Dillenbourg, P., Specht, M. (eds.) EC-TEL 2008. LNCS, vol. 5192, pp. 437–448. Springer, Heidelberg (2008)
12. Sancho, P., Fuentes-fernández, R., Fernández-manjón, B.: NUCLEO: Adaptive Computer Supported Collaborative Learning in a Role Game Based Scenario. In: Eighth IEEE International Conference on Advanced Learning Technologies, ICALT 2008, pp. 671–675 (2008)
13. Vygotsky, L.S.: Mind in society: The development of higher psychological processes. Harvard University Press, Cambridge (1978)
14. Dillenbourg, P.: What do you mean by collaborative leraning? In: Dillenbourg, P. (ed.) Collaborative-learning: Cognitive and Computational Approaches, pp. 1–19 (1999)
15. Laurillard, D.: The pedagogical challenges to collaborative technologies. Computer-Supported Collaborative Learning 4, 5–20 (2009)
16. Caballé, S., Daradoumis, T., Xhafa, F.: Efficient Embedding of Information and Knowledge into CSCL Applications. In: Hui, K.-c., Pan, Z., Chung, R.C.-k., Wang, C.C.L., Jin, X., Göbel, S., Li, E.C.-L. (eds.) EDUTAINMENT 2007. LNCS, vol. 4469, pp. 548–559. Springer, Heidelberg (2007)
17. Nardi, B.A.: Context and Consciousness: activity theory and human-computer interaction. The MIT Press, Cambridge (1996)
18. Ronen, M., Kohen-Vacs, D.: Designing and Applying Adaptation Patterns Embedded in the Script. In: International Conference on Intelligent Networking and Collaborative Systems, INCOS 2009, pp. 306–310 (2009)
19. Furugori, N., Sato, H., Ogata, H., Ochi, Y., Yano, Y.: COALE: Collaborative and Adaptive Learning Environment. In: Proceedings of CSCL 2002, pp. 493–494 (2002)
20. Walker, E., Rummel, N., Koedinger, K.R.: Beyond explicit feedback: new directions in adaptive collaborative learning support. In: Proceedings of the 9th International Conference on CSCL 2009, pp. 552–556 (2009)
21. Miao, Y., Hoppe, U.: Adapting Process-Oriented Learning Design to Group cjaracteristics. In: Looi, C., McCalla, G., Bredeweg, B., Breuker, J. (eds.) Proceedings of Artificial Intelligence in Education, pp. 475–482. IOS Press, Amsterdam (2005)
22. Walker, E., Rummel, N., Koedinger, K.R.: Modeling Helping Behavior in an Intelligent Tutor for Peer Tutoring. In: Proceedings of AIED, pp. 341–348 (2009)
23. Baghaei, N., Mitrovic, T., Irwin, W.: Supporting Collaborative Learning and Problem Solving in a Constraint-based CSCL Environment for UML Class Diagrams. International Journal of Computer-Supported Collaborative Learning 2(2-3), 159–190 (2007)
24. Karakostas, A., Demetriadis, S.: Adaptation Patterns as a Conceptual Tool for Designing the Adaptive Operation of CSCL Systems. Educational Technology Research & Development 23(1), 1042–1629 (2010)

25. Hernández-Leo, D., Asensio-Perez, J.I., Dimitriadis, Y.: Computational Representation of Collaborative Learning Flow Patterns using IMS Learning Design. JETS 8(4), 75–89 (2005a), http://www.ifets.info/issues.php?id=29 (accessed January 5, 2011)
26. Hernández-Leo, D., Villasclaras-Fernández, E.D., Jorrín-Abellán, I.M., Asensio-Pérez, J.I., Dimitriadis, Y., Ruiz-Requies, I., Rubia-Avi, B.: Collage, a Collaborative Learning Design Editor Based on Patterns Special Issue on Learning Design. JETS 9(1), 58–71 (2006), http://www.ifets.info/issues.php?id=30 (accessed January 5, 2011)
27. Perez-Sanagustin, M., Burgos, J., Hernandez-Leo, D., Blat, J.: Considering the Intrinsic Constraints for Groups Management of TAPPS and Jigsaw CLFPs. In: International Conference on Intelligent Networking and Collaborative Systems, INCOS 2009, pp. 317–322 (2009)
28. Hinze, U., Bischoff, M., Blakowski, G.: Jigsaw Method in the Context of CSCL. In: Proc. of World Conference in Educational Multimedia, Hypermedia and Telecommunications, pp. 789–794. AACE, Chesapeake (2002)
29. Collage. Collaborative learning design editor – Collage (2009), http://ulises.tel.uva.es/collage/ (accessed January 5, 2011)
30. Halm, J., Olivier, B., Farooq, U., Hoadley, C.: Collaboration in Learning Design Using Peer-to-peer technologies. In: Koper, R., Tattersall, C. (eds.) Learning Design: A Handbook on Modelling and Delivering Networked Education and Training, pp. 203–213. Springer, Heidelberg (2005)
31. Koper, E.J.R., Olivier, B.: Representing the Learning Design of Units of Learning. JETS 7(3), 97–111 (2004)
32. LAMS. Learning Activity management System (2010), http://www.lamsinternational.com/ (accessed January 5, 2011)
33. IMS LD. IMS Global Learning Consortium: Learning Design Specification (2003), http://www.imsglobal.org/specifications.html (accessed January 5, 2011)
34. Burgos, D., Tattersall, C., Koper, E.J.R.: Representing adaptive and adaptable Units of Learning. How to model personalized eLearning in IMS Learning Design. In: Fernández Manjon, B., Sanchez Perez, J.M., Gómez Pulido, J.A., Vega Rodriguez, M.A., Bravo, J. (eds.) Computers and Education: E-learning - from Theory to Practice, Kluwer, Germany (2006)
35. Richards, G.: Designing Educational Games. In: Koper, R., Tattersall, C. (eds.) Learning Design: A Handbook on Modelling and Delivering Networked Education and Training, pp. 227–237. Springer, Heidelberg (2005)
36. Paramythis, A.: Adaptive Support for Collaborative Learning with IMS Learning Design: Are We There Yet? In: Proceedings of the Adaptive Collaboration Support Workshop, Held in Conjunction with the Adaptive Hypermedia 2008 Conference, Hannover, Germany, July 29, pp. 17–29 (2008)
37. Towle, B., Halm, M.: Designing Adaptive Learning Environments with Learning Design. In: Koper, R., Tattersall, C. (eds.) Learning Design: A Handbook on Modelling and Delivering Networked Education and Training, pp. 216–226. Springer, Heidelberg (2005)
38. Durand, G., Downes, S.: Toward Simple Learning Design 2.0. Paper presented at the 4th International Conference on Computer Science & Education (ICCSE 2009), Nanning, Guangxi, China, July 25-28 (2009), doi:10.1109/ICCSE.2009.5228214

39. Karakostas, A., Demetriadis, S.: Adaptation patterns in systems for scripted collaboration. In: O'Malley, C., Suthers, D., Reimann, P., Dimitracopoulou, A. (eds.) Computer Supported Collaborative Learning Practices: CSCL 2009 Conference Proceedings, pp. 477–481 (2009)
40. Reload. Reload Learning Design Editor (2005), http://www.reload.ac.uk/ldeditor.html (accessed January 5, 2005)
41. Recourse. Recourse Learning Design Editor (2009), http://www.tencompetence.org/ldauthor/ (accessed January 5, 2011)
42. SLeD. Service Based Learning Design Player (2005), http://sled.open.ac.uk/ (accessed January 5, 2011)
43. Coopercore. The IMS LD Engine (2008), http://coppercore.sourceforge.net (accessed January 5, 2011)
44. Alfonseca, E., Carro, R.M., Martín, E., Ortigosa, A., Paredes, P.: The impact of learning styles on student grouping for collaborative learning: a case study. User Modeling and User-Adapted Interaction 16(3-4), 377–401 (2006)
45. Paredes, P., Rodriguez, P.: The application of learning styles in both individual and collaborative learning. In: Proceedings of the Sixth International IEEE Conference on Advanced Learning Technologies, pp. 1141–1142 (2006)
46. Tourtoglou, K., Virvou, M.: User Modelling in a Collaborative Learning Environment for UML. In: Fifth International Conference on Information Technology: New Generations, ITNG 2008, pp. 1257–1258 (2008)
47. Pollalis, Y.A., Mavrommatis, G.: Using similarity measures for collaborating groups formation: A model for distance learning environments. EJOR 193, 626–636 (2009)
48. Christodoulopoulos, C.E., Papanikolaou, K.A.: A Group Formation Tool in a E-Learning Context. In: Proceedings of the 19th IEEE ICTAI 2007, pp. 117–123 (2007)
49. Hernandez-Gonzalo, J.A., Villasclaras-Fernandez, E.D., Hernandez-Leo, D., Asensio-Perez, J.I., Dimitriadis, Y.: InstanceCollage: A Graphical Tool for the Particularization of Role/Group Structures in Pattern-Based IMS-LD Collaborative Scripts. In: Eighth IEEE International Conference on Advanced Learning Technologies ICALT 2008, pp. 506–510 (2008)
50. Moreno-Ger, P., Sancho, P., Martínez-Ortiz, I., Sierra, J.L., Fernández-Manjón, B.: Adaptive Units of Learning and Educational Video Games. JIME 3, 252–268 (2007)
51. Sharples, P., Griffiths, D., Scott, W.: Using Widgets to Provide Portable Services for IMS Learning Design. In: Koper, R., Stefanov, K., Dicheva, D. (eds.) Proceedings of the 5th International TENCompetence Open Workshop "Stimulating Personal Development and Knowledge Sharing", pp. 57–60. TENCompetence Workshop, Sofia (2008)
52. Wilson, S., Sharples, P., Griffiths, D.: Extending IMS Learning Design services using Widgets: Initial findings and proposed architecture. In: Proceedings of the 3rd TEN Competence Open Workshop on Current Research on IMS Learning Design and Lifelong Competence Development Infrastructures (2007), http://dspace.ou.nl/handle/1820/963 (accessed January 5, 2011)
53. de la Fuente Valentin, L., Miao, Y., Pardo, A., Delgado Kloos, C.: A supporting architecture for generic service integration in IMS learning design. In: Dillenbourg, P., Specht, M. (eds.) EC-TEL 2008. LNCS, vol. 5192, pp. 467–473. Springer, Heidelberg (2008), doi:10.1007/978-3-540-87605-2_52
54. Miao, Y., Burgos, D., Griffiths, D., Koper, R.: Representation of Coordination Mechanisms in IMS-LD. In: Lockyer, L., Bennet, S., Agostinho, S., Harper, B. (eds.) Handbook of Research on Learning Design and Learning Objects: Issues, Applications and Technologies, Idea Group Inc., Wollongong (2008), http://hdl.handle.net/1820/930 (accessed January 5, 2011)

Author Index